Lecture Notes in Computer Science 12806

More information about this subseries at http://www.springer.com/series/7407

Christos Kaklamanis · Asaf Levin (Eds.)

Approximation and Online Algorithms

18th International Workshop, WAOA 2020
Virtual Event, September 9–10, 2020
Revised Selected Papers

Editors
Christos Kaklamanis
Computer Technology Institute
University of Patras
Patras, Greece

Asaf Levin
Faculty of Industrial Engineering
and Management
Technion - Israel Institute of Technology
Haifa, Israel

ISSN 0302-9743 ISSN 1611-3349 (electronic)
Lecture Notes in Computer Science
ISBN 978-3-030-80878-5 ISBN 978-3-030-80879-2 (eBook)
https://doi.org/10.1007/978-3-030-80879-2

LNCS Sublibrary: SL1 – Theoretical Computer Science and General Issues

This Springer imprint is published by the registered company Springer Nature Switzerland AG
The registered company address is: Gewerbestrasse 11, 6330 Cham, Switzerland

Preface

The 18th Workshop on Approximation and Online Algorithms (WAOA 2020) focused on the design and analysis of algorithms for online and computationally hard problems. Both kinds of problems have a large number of applications in a variety of fields. Due to the COVID-19 pandemic WAOA 2020 took place virtually in Pisa, Italy, during September 9–10 2020, and was a success: it featured many interesting presentations and provided opportunity for stimulating interactions. WAOA 2020 was part of the ALGO 2020 event that also hosted ALGOSENSORS, ATMOS, ESA, and WABI.

Topics of interest for WAOA 2020 were graph algorithms, inapproximability results, network design, packing and covering, paradigms for the design and analysis of approximation and online algorithms, parameterized complexity, scheduling problems, algorithmic game theory, algorithmic trading, coloring and partitioning, competitive analysis, computational advertising, computational finance, cuts and connectivity, geometric problems, mechanism design, resource augmentation, and real-world applications.

In response to the call for papers we received 40 submissions. Two submissions were rejected as out of scope right away, and each of the remaining submissions was reviewed by at least three referees, and many of the submissions were reviewed by more than three referees. The submissions were mainly judged on originality, technical quality, and relevance to the topics of the conference. Based on the reviews, the Program Committee selected 15 papers. This volume contains final revised versions of those papers as well as an invited contribution by our invited speaker Cliff Stein.

The EasyChair conference system was used to manage the electronic submissions, the review process, and the electronic Program Committee discussions. It made our task much easier.

We would like to thank all the authors who submitted papers to WAOA 2020 and all attendees of WAOA 2020, including the presenters of the accepted papers. A special thank you goes to the plenary invited speaker Cliff Stein for accepting our invitation despite the uncertainty involved due to the pandemic and giving a very nice talk. We would also like to thank the Program Committee members and the external reviewers for their diligent work in evaluating the submissions and their contributions to the electronic discussions. Furthermore, we are grateful to all the local organizers of ALGO 2020, especially the local chair of the organizing committee Roberto Grossi.

May 2021

Christos Kaklamanis
Asaf Levin

Organization

Program Committee

Antonios Antoniadis	University of Cologne, Germany
Christoph Dürr	Sorbonne University, France
Lene Monrad Favrholdt	University of Southern Denmark, Denmark
Archontia Giannopoulou	National and Kapodistrian University of Athens, Greece
Christos Kaklamanis (Co-chair)	University of Patras and CTI "Diophantus", Greece
Panagiotis Kanellopoulos	University of Essex, UK
Tamás Király	Eötvös Loránd University, Hungary
Danny Krizanc	Wesleyan University, USA
Asaf Levin (Co-chair)	Technion, Israel
Bodo Manthey	University of Twente, the Netherlands
George Mertzios	Durham University, UK
Zeev Nutov	Open University of Israel, Israel
Kirk Pruhs	University of Pittsburgh, USA
Lars Rohwedder	EPFL, Switzerland
Laura Sanità	University of Waterloo, Canada, and Eindhoven University of Technology, the Netherlands
Piotr Sankowski	University of Warsaw, Poland
Kevin Schewior	University of Cologne, Germany
Baruch Schieber	New Jersey Institute of Technology, USA
Melanie Schmidt	University of Cologne, Germany
José Verschae	Pontificia Universidad Católica, Chile
Pavel Veselý	University of Warwick, UK

Additional Referees

Paritosh Garg
Sebastian Berndt
Adam Kurpisz
Kim-Manuel Klein
Nick Harvey
Hendrik Molter
István Miklós
Moritz Buchem
Ruben Hoeksma
Franziska Eberle
Hans-Joachim Boeckenhauer
Moran Feldman
Dimitris Zoros
Andreas Wiese
Malin Rau
Victor Verdugo
Ulrike Schmidt-Kraepelin
Andreas Emil Feldmann
Euiwoong Lee
Rob van Stee
Mordechai Shalom
Dachuan Xu

Argyrios Deligkas
Parinya Chalermsook
R. Ravi
Jannik Matuschke
Gyorgy Dosa
Malte Renken
Daniel R. Schmidt
Etienne Bamas
Andre Nichterlein
Erik Jan van Leeuwen
Christopher Harshaw
An Zhang
Bhargav Samineni
Leon Ladewig
Andreas Abels
Lukasz Jez
Oleg Pikhurko
Sasanka Roy
Spyridon Maniatis
Pranabendu Misra

Parallel Approximate Undirected Shortest Paths Via Low Hop Emulators (Invited Talk)

Clifford Stein
Columbia University, New York, NY 10027, USA
cliff@ieor.columbia.edu

Abstract. Although sequential algorithms with (nearly) optimal running time for finding shortest paths in undirected graphs with non-negative edge weights have been known for several decades, near-optimal parallel algorithms have turned out to be a much tougher challenge.

In this talk, we present a (1+epsilon)-approximate parallel algorithm for computing shortest paths in undirected graphs, achieving polylog depth and near-linear work. All prior all prior (1+epsilon)-algorithms with polylog depth perform at least superlinear work. Improving this long-standing upper bound obtained by Cohen (STOC'94) has been open for 25 years. Our algorithm uses several new tools. Prior work uses hopsets to introduce shortcuts in the graph. We introduce a new notion that we call low hop emulators. We also introduce compressible preconditioners, which we use in conjunction with Serman's framework (SODA '17) for the uncapacitated minimum cost flow problem.

Contents

LP-Based Algorithms for Multistage Minimization Problems 1
Evripidis Bampis, Bruno Escoffier, and Alexander Kononov

A Faster FPTAS for Knapsack Problem with Cardinality Constraint 16
Wenxin Li, Joohyun Lee, and Ness Shroff

Distributed Algorithms for Matching in Hypergraphs. 30
Oussama Hanguir and Clifford Stein

Online Coloring and a New Type of Adversary for Online Graph Problems 47
Yaqiao Li, Vishnu V. Narayan, and Denis Pankratov

Maximum Coverage with Cluster Constraints: An LP-Based Approximation Technique 63
Guido Schäfer and Bernard G. Zweers

A Constant-Factor Approximation Algorithm for Vertex Guarding a WV-Polygon 81
Stav Ashur, Omrit Filtser, and Matthew J. Katz

Lasserre Integrality Gaps for Graph Spanners and Related Problems 97
Michael Dinitz, Yasamin Nazari, and Zeyu Zhang

To Close Is Easier Than To Open: Dual Parameterization To k-Median 113
Jarosław Byrka, Szymon Dudycz, Pasin Manurangsi, Jan Marcinkowski, and Michał Włodarczyk

Explorable Uncertainty in Scheduling with Non-uniform Testing Times. 127
Susanne Albers and Alexander Eckl

Memoryless Algorithms for the Generalized k-server Problem on Uniform Metrics. 143
Dimitris Christou, Dimitris Fotakis, and Grigorios Koumoutsos

Tight Bounds on Subexponential Time Approximation of Set Cover and Related Problems 159
Magnús M. Halldórsson, Guy Kortsarz, and Marek Cygan

Concave Connection Cost Facility Location and the Star Inventory Routing Problem. 174
Jarosław Byrka and Mateusz Lewandowski

An Improved Approximation Algorithm for the Uniform Cost-Distance Steiner Tree Problem 189
Ardalan Khazraei and Stephan Held

A Constant-Factor Approximation Algorithm for Red-Blue Set Cover with Unit Disks 204
Raghunath Reddy Madireddy and Apurva Mudgal

2-Node-Connectivity Network Design 220
Zeev Nutov

Author Index 237

LP-Based Algorithms for Multistage Minimization Problems

Evripidis Bampis[1], Bruno Escoffier[1,2(✉)], and Alexander Kononov[3,4]

[1] CNRS, LIP6 UMR, Sorbonne Université, 7606, 4 Place Jussieu, 75005 Paris, France
{evripidis.bampis,bruno.escoffier}@lip6.fr
[2] Institut Universitaire de France, Paris, France
[3] Sobolev Institute of Mathematics, Novosibirsk, Russia
[4] Novosibirsk State University, Novosibirsk, Russia
alvenko@math.nsc.ru

Abstract. We consider a *multistage* framework introduced recently (Eisenstat et al., Gupta et al., both in ICALP 2014), where given a time horizon $t = 1, 2, \dots, T$, the input is a sequence of instances of a (static) combinatorial optimization problem $I_1, I_2, \dots, I_T$, (one for each time step), and the goal is to find a sequence of solutions $S_1, S_2, \dots, S_T$ (one for each time step) reaching a tradeoff between the quality of the solutions in each time step and the stability/similarity of the solutions in consecutive time steps. For several polynomial-time solvable problems, such as Minimum Cost Perfect Matching, the multistage variant becomes hard to approximate (even for two time steps for Minimum Cost Perfect Matching). In this paper, we study multistage variants of some important discrete minimization problems (including Minimum Cut, Vertex Cover, Prize-Collecting Steiner Tree, Prize-Collecting Traveling Salesman). We focus on the natural question of whether linear-programming-based methods may help in developing good approximation algorithms in this framework.

We first introduce a new two-threshold rounding scheme, tailored for multistage problems. Using this rounding scheme we propose a 3.53-approximation algorithm for a multistage variant of the Prize-Collecting Steiner Tree problem, and a 3.034-approximation algorithm for a multi-stage variant of the Prize-Collecting Metric Traveling Salesman problem. Then, we show that Min Cut remains polytime solvable in its multi-stage variant, and Vertex Cover remains 2-approximable, as particular cases of a more general statement which easily follows from the work of (Hochbaum, EJOR 2002) on monotone and IP2 problems.

Keywords: Multistage optimization · Approximation algorithms · LP-rounding

1 Introduction

In many applications, data are evolving with time and the systems have to be adapted in order to take into account this evolution by providing near optimal

C. Kaklamanis and A. Levin (Eds.): WAOA 2020, LNCS 12806, pp. 1–15, 2021.
https://doi.org/10.1007/978-3-030-80879-2_1

solutions over the time. However, moving from a solution to a new one may induce non-negligible transition costs. This cost may represent e.g. the cost of turning on/off the servers in a data center [1], the cost of changing the quality level in video streaming [21], or the cost for turning on/off nuclear plants in electricity production [25]. Various models and algorithms have been proposed for modifying (re-optimizing) the current solution by making as few changes as possible (see [17,22,23] and the references therein). In this paper, we follow a recent trend, popularized by the works of Eisenstat et al. [13] and Gupta et al. [18], known as the *multistage model*: Given a time horizon $t = 1, 2, \ldots, T$ and a sequence of instances $I_1, I_2, \ldots, I_T$, (one for each time step), the goal is to find a sequence of solutions $S_1, S_2, \ldots, S_T$ (one for each time step) optimizing the quality of the solution in each time step and the stability (transition cost or profit) between solutions in consecutive time steps. Surprisingly, even in the *offline case*, where the sequence of instances is known in advance, some classic combinatorial optimization problems become much harder in the multistage model [2,5,13,18]. For instance, the Minimum Cost Perfect Matching problem, which is polynomially-time solvable in the one-step case, becomes hard to approximate even for bipartite graphs and for only two time steps [5]. In a more recent work, Fluschnik et al. [14] study multistage optimization problems from a parameterized complexity point of view, focusing on the multistage Vertex Cover problem. The motivation to consider the offline case is based on the fact that in many applications, a good forecast on the data is available. This is the case for instance in electricity production planning, where on a small time window (a few days) a good estimation of the demand is known.

In this article, we focus on the *complexity* and the *approximability* of various multistage minimization problems including Min Cut, Vertex Cover, Prize-Collecting Steiner Tree and Prize-Collecting Metric Traveling Salesman. The central question that we address in this work is to determine to what extent linear-programming based methods can be used in the multistage framework. Arguably one of the main techniques for the design of approximation algorithms for usual (static) problems, LP-based methods have already been fruitfully applied in the multistage setting for facility location problems in [2,13,15].

Our Contribution. We introduce a new rounding scheme, called *two-threshold rounding scheme*, designed for multistage problems in the sense that it is able to take into account both the transition and the individual costs of the solutions when rounding. The main result of this article is the use of this new rounding scheme for a multistage variant of two well known prize-collecting problems: the multistage Prize-Collecting Steiner Tree problem, for which we obtain a 3.53-approximation algorithm, and the multistage Prize-Collecting Metric Traveling Salesman problem, for which we obtain a 3.034-approximation algorithm.

We then note that the multistage variants of monotone IP2 minimization problems (as defined in [20], see Sect. 3) such as Min Cut, or (non-monotone) IP2 minimization problems [20] such as Vertex Cover, remain (monotone) IP2, respectively. As a consequence, the multistage variant of min cut (as well as the ones of monotone problems) can be solved in polynomial time, while the multistage variant of vertex cover has the half-integrality property and consequently can be solved by a 2-approximation algorithm. Though obtained using simple arguments, we quote these results as they contrast with the several hardness results for multistage versions of classical problems such as Spanning Tree or Minimum Cost Perfect Matching. Indeed, Min Cut is, to the best of our knowledge, the first problem that is shown to remain polytime solvable in the multistage setting. Also, multistage Vertex Cover has the same approximation guarantee (2-approximation) as the static version (the ratio being tight under UGC). The results are summarized in Table 1 (last column).

Table 1. Comparison of approximation ratios for static and multistage versions.

Problem	Static ($T = 1$)	Multistage
Min cut	Polytime	Polytime
Min vertex cover	2	2
Prize-collecting Steiner tree	1.991 [3]	3.53
Prize-collecting metric traveling Salesman	1.990 [3]	3.03

Note that while the approximation lower bounds for the static case obviously are also valid for the multistage version, finding specific (larger) lower bounds for multistage Prize-Collecting problems is an interesting direction that we leave as an open question.

Note: during the time between submission and this final version, we got some improved results, namely ratios 2.54 for Prize-Collecting Steiner Tree and 2.06 for Prize-Collecting Metric Traveling Salesman.

Related Work

Multistage Framework. The multistage model considered in this paper has been introduced in Eisenstat et al. [13] and Gupta et al. [18]. Eisenstat et al. [13] studied the multistage version of facility location problems for which they proposed logarithmic approximation algorithms. An et al. [2] obtained constant factor approximation algorithms for some related problems. Interestingly, these results are obtained using LP-based techniques, and in particular (randomized) roundings. These roundings are tailored for facility location problems, and thus quite different from the approach followed in this article.

Gupta et al. [18] studied the multistage Maintenance Matroid problem (which includes multistage Spanning tree as a subcase) for both the offline and the online settings, providing a logarithmic approximation algorithm for this problem, and also introduced the multistage Minimum Cost Perfect Matching problem for which they proved that it is hard to approximate even for a constant number of stages in the offline case. Bampis et al. [5] improved this negative result by showing that the problem is hard to approximate even for bipartite graphs and for the case of two time steps. Olver et al. [24] studied a multistage version of the Minimum Linear Arrangement problem, which is related to a variant of the List Update problem [27], both in the online and offline version. Very recently, Fluschnik et al. [14] studied the multistage Vertex Cover problem under a parameterized complexity perspective, and Fotakis et al. considered in [15] a facility reallocation problem in the multistage setting. Some multistage maximization problems have been studied as well, such as the multistage Max-Min Fair Allocation problem [6], and a multistage variant of the Knapsack problem [8].

More recently, in [7], general techniques for a large family of online multistage problems, called Subset Maximization problems, have been developed leading to a characterization of the variants that admit a constant-competitive online algorithm. Finally, let us mention the works by Buchbinder et al. [11] and Buchbinder, Chen and Naor [10] who also considered a multistage model and studied the relation between the online learning and competitive analysis frameworks, mostly for fractional optimization problems.

Static Framework. In [20] (see also Sect. 3), Hochbaum described an easily recognizable class of integer programming problems solvable in polynomial time. She also considered a larger class for which she proposed a polynomial time algorithm that finds (fractional) optimal solutions that are half integral. These solutions can be used in order to devise 2-approximate solutions.

For the Prize-Collecting Metric Traveling Salesman problem that has been introduced in [4], Bienstock et al. [9] proposed a 2.5-approximation algorithm. They have also proposed a 3-approximation algorithm for the Prize-Collecting Steiner Tree problem. Both algorithms use a rounding procedure of appropriate linear programming relaxations. Goemans and Williamson [16] improved these results providing a 2-approximation for both problems, based on the primal-dual scheme. Then, Archer et al. [3] devised a $(2-\epsilon)$-approximation algorithm, for some $\epsilon > 0$, for both problems. All these results hold for both the *rooted* and *unrooted* versions of the problems. This is true since there are approximation-preserving reductions from the rooted to the unrooted version of the problems, and vice versa [3].

2 Rounding Scheme for Some Prize-Collecting Problems

We consider in this section Prize-Collecting Steiner Tree and Prize-Collecting Metric Traveling Salesman. In both problems, some clients, represented as (some) vertices of a graph, address a request. We have to select some edges in the graph, to serve *some* of the requests. Each selected edge has some cost. The non-served requests come with a penalty. The goal is to find a solution minimizing the sum of the service cost (sum of costs of selected edges) and the penalties.

In the multistage version we consider here, cost of edges and penalties evolve with time, in a time horizon $1, \dots, T$. Depending on these costs, we may decide to serve different requests at different time steps. However, changing this decision to serve client/vertex v between t and $t+1$ comes with a *transition cost* w_v.

Then, the goal is to find a sequence of solutions (one for each time step) which minimizes an objective function which is the sum of:

- The sum over t of the global cost (service cost plus penalties) of each individual solution.
- The transition cost, which can be written as $\sum_t \sum_v w_v |s_v^t - s_v^{t-1}|$, where $s_v^t = 1$ if the request of v is satisfied at time t, and 0 otherwise.

We propose a new rounding method, called 2-threshold rounding scheme, which allows to take into account both individual costs of solutions and transition costs in the multistage setting. We use this 2-threshold rounding scheme in Sects. 2.2 and 2.3 and obtain a 3.53-approximation algorithm for Prize-Collecting Steiner Tree and a 3.034-approximation algorithm for the Prize-Collecting Metric Traveling Salesman.

We note that these results do not transfer to multistage versions of Steiner trees and TSP, as we focus on transition costs based on the set of served clients/requests. For instance, a multistage Steiner tree problem with transition costs based on *edge* modifications is hard to approximate, as it is already the case for the corresponding multistage version of the spanning tree problem [18].

2.1 Rounding Scheme

A classical rounding scheme for a static problem expressed as an ILP is as follows: starting from an optimal solution $\hat{x} = (\hat{x}_1, \dots, \hat{x}_n)$ of a continuous relaxation, fix a threshold $\alpha \in (0, 1)$, and fix $x_i = 1$ if $\hat{x}_i \geq \alpha$, and $x_i = 0$ otherwise. However, this rounding is not suitable for a multistage problem as it may induce very large transition cost: if $\hat{x}_i^t = \alpha$ and $\hat{x}_i^{t+1} = \alpha - \epsilon$, then $|x_i^{t+1} - x^t| = 1$ while $|\hat{x}_i^{t+1} - \hat{x}^t| = \epsilon$.

To overcome this, we introduce the following 2-threshold rounding scheme which, as we will see, allows to bound both transition costs and individual costs of solutions when rounding.

Definition 1. *(2-threshold rounding scheme) Let $x = (x^1, \dots, x^T)$ a sequence of T values in $[0, 1]$, and α, β two parameters with $1 \geq \alpha > \beta \geq 0$. Then RS$(x, \alpha, \beta)$ is the sequence $y = (y^1, \dots, y^T)$ of T values in $\{0, 1\}$ defined as:*

- *If $x^t \geq \alpha$ then $y^t = 1$.*
- *If $x^t \leq \beta$ then $y^t = 0$.*
- *Consider a maximal interval $[t_1, t_2]$ with $\beta < x^t < \alpha$ for all $t \in [t_1, t_2]$. Then: if simultaneously (1) $t_1 = 1$ or $x^{t_1 - 1} \geq \alpha$, and (2) $t_2 = T$ or $x^{t_2 + 1} \geq \alpha$, then fix $y^t = 1$ for all $t \in [t_1, t_2]$. Otherwise ((1) or (2) is not verified) then fix $y^t = 0$ for all $t \in [t_1, t_2]$.*

2.2 Prize-Collecting Steiner Tree

In the (static) Prize-Collecting Steiner Tree problem we have: a graph $G = (V, E)$, a root $r \in V$, each edge has a cost $c(e) \geq 0$, each vertex v has a penalty

$\pi(v) \geq 0$. The goal is to connect *some* vertices to the root using edges, while minimizing the cost of edges plus the penalties for vertices *not* connected to r.

Formally, we can define a solution as a subset $F \subseteq E$ of edges. By denoting $C(F)$ the set of vertices connected to the root by this set of edges, the value $z(F)$ associated to F is $z(F) = \sum_{e \in F} c(e) + \sum_{v \notin C(F)} \pi(v)$. Given that costs are nonnegative, there always exists an optimal solution which forms a tree (with the root inside). Note that vertices with penalty 0 correspond to Steiner (i.e., optional) vertices in the classical Steiner tree problem.

We formulate the (static) problem with the following ILP, where $x_e = 1$ iff e is taken in the solution, and $s_v = 1$ iff v is connected to the root. We denote by $\mathcal{P}_v$ the set of sets of vertices S with $v \in S$ and $r \notin S$, and by $\delta(S)$ the set of edges with one endpoint in S and one endpoint outside S.

$$\begin{cases} \min \sum_{e \in E} c(e)x_e + \sum_{v \in V} \pi(v)(1 - s_v) \\ s.t. \left| \begin{array}{ll} \sum_{e \in \delta(S)} x_e & \geq s_v \ \forall v \in V, \forall S \in \mathcal{P}_v \\ s_r = 1 & \\ x_e, s_v \in \{0, 1\} & \end{array} \right. \end{cases}$$

The constraint states that whenever $s_v = 1$, for any $S \in \mathcal{P}_v$ at least one edge goes outside S: by definition of $\mathcal{P}_v$ this ensures that v is indeed connected to the root.

As explained earlier, we consider a multistage version where we have a transition cost w_v induced by modifying our decision about vertex v between time steps t and $t+1$. We formulate the problem as an ILP as follows: $x_e^t = 1$ if edge e is taken at time t, $s_v^t = 1$ if v is connected to r at time t, $z_v^t = 1$ if $s_v^t \neq s_v^{t+1}$ (transition cost).

$$\begin{cases} \min \sum_{t=1}^{T} \sum_{e \in E} c^t(e)x_e^t + \sum_{t=1}^{T} \sum_{v \in V} \pi^t(v)(1 - s_v^t) + \sum_{t=1}^{T-1} \sum_{v \in V} w_v z_v^t \\ s.t. \left| \begin{array}{lll} \sum_{e \in \delta(S)} x_e^t & \geq s_v^t & \forall t \in \{1, \dots, T\}, \forall v \in V, \forall S \in \mathcal{P}_v \\ s_r^t & = 1 & \forall t \in \{1, \dots, T\} \\ z_v^t & \geq s_v^t - s_v^{t+1} & \forall t \in \{1, \dots, T-1\}, \forall v \in V \\ z_v^t & \geq s_v^{t+1} - s_v^t & \forall t \in \{1, \dots, T-1\}, \forall v \in V \\ x_e^t, s_v^t, z_v^t \in \{0, 1\} & & \end{array} \right. \end{cases}$$

Let us consider the algorithm RS-MPCST (Rouding Scheme for Multistage Prize-Collecting Steiner Tree) which works as follows.

1. Find an optimal solution $(\hat{x}, \hat{s}, \hat{z})$ to the relaxation of the ILP where variables are in $[0, 1]$.
2. Apply the rounding scheme RS to each sequence of variables $\hat{s}_u = (s_u^1, \dots, s_u^T)$ with parameters α and β (to be specified), let $\tilde{s}_u = (\tilde{s}_u^1, \dots, \tilde{s}_u^T)$ be the corresponding (integer) values. At this point, we have decided for each time t which vertices we want to connect to the root r (those for which $\tilde{s}_u^t = 1$).
3. For each $t = 1, \dots, T$, apply the algorithm ST-Algo of [16] for the Steiner Tree problem on the instance whose costs on edges are the ones at time t, and the set of terminals is $\{v : \tilde{s}_v^t = 1\}$ (those we have to connect). This gives the set of edges chosen at time t.

The algorithm ST-Algo has the following property [16]: it computes a (intregal) solution the value of which is at most 2 times the value of a (feasible) solution of a dual formulation of the problem. More precisely, consider the following ILP formulation of an instance of Steiner Tree, where T is the set of terminals, and $\mathcal{P}$ is the set of sets S such that $S \cap T \neq \emptyset$ and $r \notin S$.

$$(ILP - ST) \begin{cases} \min \sum_{e \in E} c(e) x_e \\ s.t. \left| \begin{array}{l} \sum_{e \in \delta(S)} x_e \geq 1 \; \forall S \in \mathcal{P} \\ x_e \in \{0,1\} \end{array} \right. \end{cases}$$

Associating a variable y_S to each $S \in \mathcal{P}$, the following LP $(D - ST)$ is the dual of the relaxation (LP-ST) of (ILP-ST) where we relax $x_e \in \{0,1\}$ as $x_e \geq 0$.

$$(D - ST) \begin{cases} \max \sum_{S \in \mathcal{P}} y_S \\ s.t. \left| \begin{array}{l} \sum_{S \in \mathcal{P}: e \in \delta(S)} y_S \leq c(e) \; \forall e \in E \\ y_S \geq 0 \end{array} \right. \end{cases}$$

Then ST-Algo computes a feasible solution y_S of the dual and an integral solution $\hat{x}$ of the primal such that $\sum_e c(e)\hat{x}_e \leq 2 \sum_S y_S$.

Theorem 1. *With* $\alpha = 3/4$ *and* $\beta = 1/2$, *RS-MPCST is a* 4*-approximation algorithm.*

Proof. We note that the LP can be solved in polynomial time by the Ellipsoid method. The separation problem for the constraints $\sum_{e \in \delta(S)} x_e^t \geq s_v^t$ is reduced to the Min Cut problem and can be solved in polynomial time.

By solving a Steiner Tree problem in step 3, we connect at each time step the vertices we chose in step 2 to connect to r, so the solution is clearly feasible.

Let $(\tilde{x}, \tilde{s})$ be the computed solution ($\tilde{z}$ being immediately deduced from $\tilde{s}$). First, we note that in the rounding scheme, if $\hat{s}_i^t \geq \alpha$ then $\tilde{s}_i^t = 1$, so for any i, t we have $(1 - \tilde{s}_i^t) \leq \frac{1 - \hat{s}_i^t}{1 - \alpha}$, and we can bound the overall penalty of non-connected vertices:

$$\sum_{t=1}^{T} \sum_{v \in V} \pi^t(v)(1 - \tilde{s}_v^t) \leq \frac{\sum_{t=1}^{T} \sum_{v \in V} \pi^t(v)(1 - \hat{s}_v^t)}{1 - \alpha} \tag{1}$$

Now, a variable $\tilde{s}_u^t$ jumps (once) from 0 to 1 only if $\hat{s}_u^t \leq \beta$, $\hat{s}_u^{t'} \geq \alpha$ for some $t' > t$ and $\beta < \hat{s}_u^{t''} < \alpha$ for all $t < t'' < t'$ (or $t' = t + 1$). But then, the global transition cost of $\hat{s}_u^t$ on this period between t and t' is at least $w_u(\hat{s}_u^{t'} - \hat{s}_u^t) \geq w_u(\alpha - \beta)$, while the transition cost of $\tilde{s}_u^t$ is w_u. The same argument holds for the jumps of $\tilde{s}_u^t$ from 1 to 0. Then we can bound the loss on the transition costs:

$$\sum_{t=1}^{T-1} \sum_{v \in V} w_v \tilde{z}_v^t \leq \frac{\sum_{t=1}^{T-1} \sum_{v \in V} w_v \hat{z}_v^t}{\alpha - \beta} \tag{2}$$

Now we have to deal with the cost of taking edges. We consider some time step t. Since $(\hat{x}, \hat{s}, \hat{z})$ is feasible, for all $v \in V$, all $S \in \mathcal{P}_v$, $\sum_{e \in \delta(S)} \hat{x}_e^t \geq \hat{s}_v^t$.

Since $\tilde{s}_v^t = 0$ if $\hat{s}_v^t \leq \beta$, we have $\tilde{s}_v^t \leq \frac{\hat{s}_v^t}{\beta}$ for all t, v. Then $\sum_{e\in\delta(S)} \beta^{-1}\hat{x}_e^t \geq \beta^{-1}\hat{s}_v^t \geq \tilde{s}_v^t$.

Considering the formulation (LP-ST) of the Steiner Tree problem where the set of terminals is $T = \{v : \tilde{s}_v^t = 1\}$, we have that $\sum_{e\in\delta(S)} \beta^{-1}\hat{x}_e^t \geq 1$ for all S containing at least one terminal (one v with $\tilde{s}_v^t = 1$), and not containing r. In other words, $\beta^{-1}\hat{x}_e^t$ is a (continuous) feasible solution for (LP-ST) (with $T = \{v : \tilde{s}_v^t\} = 1)$. By duality, its value is at least the value of any dual feasible solution y_S:

$$\sum_{e\in E} c^t(e)\beta^{-1}\hat{x}_e^t \geq \sum_S y_S$$

Since ST-Algo computes a solution $\tilde{x}$ of cost at most $2\sum_S y_S$, we get, for all t:

$$\sum_{e\in E} c^t(e)\tilde{x}_e^t \leq \frac{2}{\beta}\sum_{e\in E} c^t(e)\hat{x}_e^t \tag{3}$$

In all, the computed solution has ratio at most $\frac{1}{1-\alpha}$ for the cost of not connecting vertices, ratio at most $\frac{1}{\alpha-\beta}$ for the transition costs, and ratio at most $\frac{2}{\beta}$ for the cost of edges. With $\beta = \frac{1}{2}$ and $\alpha = \frac{3}{4}$, we get ratio at most 4 in each case.

Improvement via (de)randomization. Following an approach used for the classical Prize-Collecting Steiner Tree problem (see [28]) we give here a randomized algorithm for the multistage Prize-collecting Steiner Tree problem leading to a better (expected) ratio. We then show how it can be derandomized.

The randomized algorithm is the same as RS-MPCST except that we choose α at random uniformly from the range $[\gamma, 1]$, where $\gamma = e^{-\frac{1}{3}} > 0$, and we set $\beta = \frac{2}{3}\alpha$. The probability density function for a uniform random variable over $[\gamma, 1]$ is the constant $1/(1-\gamma)$. We have

$$\begin{aligned}
\mathbb{E}[\sum_{t=1}^{T}\sum_{v\in V}\pi^t(v)(1-\tilde{s}_v^t)] &= \mathbb{E}[\sum_{t=1}^{T}(\sum_{v|\hat{s}_v^t<\gamma}\pi^t(v)(1-\tilde{s}_v^t) + \sum_{v|\hat{s}_v^t\geq\gamma}\pi^t(v)(1-\tilde{s}_v^t))] \\
&\leq \sum_{t=1}^{T}\sum_{v|\hat{s}_v^t<\gamma}\pi^t(v) + \sum_{t=1}^{T}\sum_{v|\hat{s}_v^t\geq\gamma}\pi^t(v)Pr(\tilde{s}_v^t = 0) \\
&\leq \sum_{t=1}^{T}\left(\sum_{v|\hat{s}_v^t<\gamma}\frac{\pi^t(v)(1-\hat{s}_v^t)}{(1-\hat{s}_v^t)} + \sum_{v|\hat{s}_v^t\geq\gamma}\frac{\pi^t(v)(1-\hat{s}_v^t)}{1-\gamma}\right) \leq \frac{1}{1-\gamma}\sum_{t=1}^{T}\sum_{v\in V}\pi^t(v)(1-\hat{s}_v^t)
\end{aligned} \tag{4}$$

From (2) and linearity of expectations, we obtain:

$$\mathbb{E}[\sum_{t=1}^{T-1}\sum_{v\in V} w_v\tilde{z}_v^t] \leq \sum_{t=1}^{T-1}\sum_{v\in V} w_v\hat{z}_v^t\mathbb{E}[\frac{1}{\alpha-\beta}] = \sum_{t=1}^{T-1}\sum_{v\in V} w_v\hat{z}_v^t\mathbb{E}[\frac{3}{\alpha}]$$

$$= \left(\frac{3}{1-\gamma}\ln\frac{1}{\gamma}\right)\sum_{t=1}^{T-1}\sum_{v\in V} w_v \hat{z}_v^t = \frac{1}{1-\gamma}\sum_{v\in V} w_v \hat{z}_v^t \quad (5)$$

Finally, from (3) we have

$$\mathbb{E}[\sum_{e\in E} c^t(e)\overline{x}_e^t] \leq \mathbb{E}[\frac{2}{\beta}\sum_{e\in E} c^t(e)\hat{x}_e^t] = \sum_{e\in E} c^t(e)\hat{x}_e^t\mathbb{E}[\frac{3}{\alpha}] = \frac{1}{1-\gamma}\sum_{e\in E} c^t(e)\hat{x}_e^t$$

Thus, the expected cost of the solution is at most $\frac{1}{1-e^{-\frac{1}{3}}}OPT \leq 3.53OPT$.

Now, let us deal with derandomization. The random selection of α determines the splitting of the values $\hat{s}_u^t$ into three sets $\mathcal{V}^+(\alpha) = \{\hat{s}_u^t | \hat{s}_u^t \geq \alpha\}$, $\mathcal{V}^-(\alpha) = \{\hat{s}_u^t | \hat{s}_u^t \leq \beta\}$, and $\mathcal{V}(\alpha) = \{\hat{s}_u^t | \beta < \hat{s}_u^t < \alpha\}$. Since $\beta = \frac{2}{3}\alpha$, any possible value of α corresponds to a splitting of the set of values into three sets. It is clear that there are at most $(|V|T)^2$ such partitions (α or β taking one of the at most $|V|t$ values $\hat{s}_u^t$). Let us call `I-RS-MPCST` (for improved `RS-MPCST`) the corresponding derandomized algorithm. Then we get:

Theorem 2. *I-RS-MPCST is a (deterministic) 3.53-approximation algorithm.*

2.3 Prize-Collecting Metric Traveling Salesman

In this section we consider the Prize-Collecting Metric Traveling Salesman problem. We have a complete graph $G = (V, E)$, a depot $r \in V$, each edge has a cost $c(e)$, each vertex v has a penalty $\pi(v)$. We assume that the vertex r must be in the tour, i.e. the tour starts and ends at vertex r. The edge costs are assumed to satisfy the triangle inequality. In the Price Collecting Metric Traveling Salesman problem, it is required to find a tour that visits a subset of the vertices such that the length of the tour plus the sum of penalties of all vertices not in the tour is as small as possible. We consider a multistage version of the problem in which the costs of edges and the penalties may change over time. Additionally, we have a transition cost induced by modifying our decision about vertex v between time t and $t+1$. Namely, we pay a cost w_v if we visit v at time step t but not at time step $t+1$, or vice-versa.

We adapt the ILP for the Prize-Collecting Travelling Salesman problem introduced in [9] to the multistage setting. Let $x_e^t = 1$ if edge e is in the tour at time t and zero otherwise, $s_v^t = 1$ if v is in the tour at time t and zero otherwise, $z_v^t = 1$ if $s_v^t \neq s_v^{t+1}$ (transition cost) and 0 otherwise.

$$ILP\text{-}MPCTSP \begin{cases} \min \sum_{t=1}^{T}\sum_{e\in E} c^t(e)x_e^t + \sum_{t=1}^{T}\sum_{v\neq r}\pi^t(v)(1-s_v^t) \\ \qquad + \sum_{t=1}^{T-1}\sum_{v\neq r} w_v z_v^t \\ s.t. \left| \begin{array}{lll} \sum_{e\in\delta(\{v\})} x_e^t = 2s_v^t & & \forall t\in\{1,\ldots,T\}, \forall v\in V \\ \sum_{e\in\delta(S)} x_e^t & \geq 2s_v^t & \forall t\in\{1,\ldots,T\}, \forall v\in V, \forall S\subset V \\ & & \text{such that } |S\cap\{r,v\}| = 1 \\ z_v^t & \geq s_v^t - s_v^{t+1} & \forall t\in\{1,\ldots,T-1\}, \forall v\in V \\ z_v^t & \geq s_v^{t+1} - s_v^t & \forall t\in\{1,\ldots,T-1\}, \forall v\in V \\ x_e^t, s_v^t, z_v^t & \in\{0,1\} & \end{array}\right. \end{cases}$$

Let us consider the algorithm RS-MPCTSP (Rounding Scheme for multistage Prize-Collecting TSP).

1. Find an optimal solution $(\hat{x}, \hat{s}, \hat{z})$ to the relaxation of ILP-MPCTSP where variables are in $[0, 1]$.
2. Apply the rounding scheme RS to each sequence of variables $\hat{s}_u = (\hat{s}_u^1, \dots, \hat{s}_u^T)$ with parameters α and β (to be specified), let $\tilde{s}_u = (\tilde{s}_u^1, \dots, \tilde{s}_u^T)$ be the corresponding (integer) values.
3. Let $R_t = \{v : \tilde{s}_v^t = 1\}$. For each $t = 1, \dots, T$, construct a Traveling Salesman tour through all vertices in R_t using Christofides' algorithm [12].

Theorem 3. *With parameters $\alpha = 5/7$ and $\beta = 3/7$, RS-MPCTSP is a 3.5-approximation algorithm.*

Proof. Let $L^*(R_t)$ be the length of an optimal tour through the subset of vertices R_t and $L(R_t)$ be the length of the tour through the subset of vertices R_t produced by Christofides' algorithm. The following linear program was introduced by Bienstock et al. [9].

$$LP - BGSW(t) \begin{cases} \min \sum_{e \in E} c^t(e) x_e^t \\ s.t. \left| \begin{array}{ll} \sum_{e \in \delta(S)} x_e^t & \geq 2 \ \forall S \subset V : R_t \cap S \neq \emptyset,\ R_t \cap (V \setminus S) \neq \emptyset \\ x_e^t & \geq 0 \ \forall e \in E \end{array} \right. \end{cases}$$

Let $\bar{x}^t$ be an optimal solution of $LP - BGSW(t)$ and $Z(\bar{x}^t)$ be the cost of this solution. Bienstock et al. proved that $Z(\bar{x}^t)$ coincides with the well-known lower bound obtained by Held and Karp [19]. It follows that $Z(\bar{x}^t) \leq L^*(R_t)$ and $L(R_t) \leq \frac{3}{2} Z(\bar{x}^t)$. The second inequality follows from a similar result obtained for the Held and Karp lower bound obtained in [26,29]. Thus, we have

$$L(R_t) \leq \frac{3}{2} \sum_{e \in E} c^t(e) \bar{x}_e^t$$

Define new vectors $\tilde{x}_e^t = \frac{\hat{x}_e^t}{\beta}$ for all $e \in E$ and $t = 1, \dots, T$. Consider some time step t. We now show that $\tilde{x}_e^t$ is feasible for $LP - BGSW(t)$. To prove that the first constraints are satisfied, we consider any $S \subset V$ such that $v \in R_t \cap S$ and $r \in R_t \setminus S$. In this case, we have

$$\sum_{e \in \delta(S)} \tilde{x}_e^t = \frac{1}{\beta} \sum_{e \in \delta(S)} \hat{x}_e^t \geq \frac{2\hat{s}_v^t}{\beta} \geq 2$$

The first inequality follows from the feasibility of $(\hat{x}, \hat{s}, \hat{z})$. According to the rounding scheme RS $\hat{s}_v^t \geq \beta$ for all $v \in R_t$ and the last inequality also holds. From optimality of $\bar{x}$ we obtain $\sum_{e \in E} c^t(e) \bar{x}_e^t \leq \sum_{e \in E} c^t(e) \tilde{x}_e^t$. Hence:

$$L(R_t) \leq \frac{3}{2} \sum_{e \in E} c^t(e) \bar{x}_e^t \leq \frac{3}{2} \sum_{e \in E} c^t(e) \tilde{x}_e^t \leq \frac{3}{2\beta} \sum_{e \in E} c^t(e) \hat{x}_e^t \tag{6}$$

Repeating the reasoning of the previous section we obtain that the computed solution has a ratio of at most $\frac{1}{1-\alpha}$ for the cost of the set of the unserved vertices, and a ratio of at most $\frac{1}{\alpha-\beta}$ for the transition costs. Finally, by choosing $\beta = \frac{3}{7}$ and $\alpha = \frac{5}{7}$, we get a ratio of at most 3.5 in each case.

Improvement via (de)randomization. We use the same technique as the one of multistage Prize-Collecting Steiner Tree in order to improve the ratio. First we consider a randomized algorithm where we choose α uniformly from the range $[\gamma, 1]$, where $\gamma = e^{-\frac{2}{5}} > 0$ and set $\beta = \frac{3}{5}\alpha$. The probability density function for a uniform random variable over $[\gamma, 1]$ is the constant $1/(1-\gamma)$. As previously, we have:

$$\mathbb{E}[\sum_{t=1}^{T}\sum_{v\in V}\pi^t(v)(1-\overline{s}_v^t)] \leq \frac{1}{1-\gamma}\sum_{t=1}^{T}\sum_{v\in V}\pi^t(v)(1-\hat{s}_v^t) \tag{7}$$

From (2) and linearity of expectations, we obtain:

$$\mathbb{E}[\sum_{t=1}^{T-1}\sum_{v\in V}w_v\overline{z}_v^t] \leq \sum_{t=1}^{T-1}\sum_{v\in V}w_v\hat{z}_v^t\mathbb{E}[\frac{1}{\alpha-\beta}] = \sum_{t=1}^{T-1}\sum_{v\in V}w_v\hat{z}_v^t\mathbb{E}[\frac{5}{2\alpha}]$$

$$= \left(\frac{5}{2(1-\gamma)}\ln\frac{1}{\gamma}\right)\sum_{t=1}^{T-1}\sum_{v\in V}w_v\hat{z}_v^t = \frac{1}{1-\gamma}\sum_{v\in V}w_v\hat{z}_v^t \tag{8}$$

Finally from (6), we have:

$$\mathbb{E}[\sum_{e\in E}c^t(e)\overline{x}_e^t] \leq \mathbb{E}[\frac{3}{2\beta}\sum_{e\in E}c^t(e)\hat{x}_e^t] = \sum_{e\in E}c^t(e)\hat{x}_e^t\mathbb{E}[\frac{5}{2\alpha}] = \frac{1}{1-\gamma}\sum_{e\in E}c^t(e)\hat{x}_e^t$$

Then, the expected cost of the obtained solution is no greater than $\frac{1}{1-e^{-\frac{2}{5}}}OPT \leq 3.034OPT$.

The derandomization can be done in a very similar way as for Prize-Collecting Steiner Tree, leading to the algorithm `I-RS-MPCTSP` and the following result.

Theorem 4. *`I-RS-MPCTSP` is a (deterministic) 3.034-approximation algorithm.*

3 Min Cut, Vertex Cover and IP2 Linear Problems

Multistage Min Cut. In the Min Cut problem, we are given an undirected graph $G = (V, E)$, two distinguished vertices s and p of V, and a function $c : E \to \mathbb{N}$ associating a weight to each edge of the graph. A cut (A, B) of G is a partition of V into A and $B = V \setminus A$ such that $s \in A$ and $p \in B$. The cost of the cut (A, B) is $c(A, B) = \sum_{u\in A}\sum_{v\in B}c(\{u, v\})$. A minimum cut in a graph is a cut whose cost is minimum over all cuts of the graph. In the multistage version, we are given a sequence of T cost functions c^t, $t = 1, \ldots, T$, and the goal is to find a sequence (A_t, B_t) of cuts. There is a transition cost w_i to move vertex v_i from one side of the partition at time $t-1$ to the other one at time t. Thus the global transition cost can be written as $\sum_{t=1}^{T-1}\sum_i w_i|x_i^{t+1} - x_i^t|$, where $x_i^t = 1$ if v_i is in A_t, and 0 if it is in B_t.

Note that by putting some edge cost to 0 this model captures also the case where the edge set changes over time.

Multistage Vertex Cover. In the (weighted) vertex cover problem, we are given a vertex-weighted graph $G = (V, E)$, where $c_i \geq 0$ is the weight of vertex v_i, and the goal is to find a subset of vertices spanning all the edges with minimal cost. We consider the multistage version where the cost of vertices evolve with time: we are given a sequence of T cost functions c_i^t, $t = 1, \ldots, T$. The transition cost is here $\sum_{t=1}^{T-1}\sum_i w_i|x_i^{t+1} - x_i^t|$, where $x_i^t = 1$ if v_i is taken in the cover at time t, and 0 otherwise.

(Monotone) IP2 Problems. It is well known that (static) Min Cut can be solved in polynomial time, and that extremal solutions of the continuous relaxation of the ILP formulation of vertex cover is semi-integral, i.e. every variable has value in $\{0, 1/2, 1\}$. Hochbaum [20] generalized these results to classes of problems, called (monotone) IP2 problems. A *linear*[1] *IP2 problem* is a minimization problem on integer variables $x_1, \ldots, x_{n_1}$ and $y_1, \ldots, y_{n_2}$, such that:

- The objective is to minimize $\sum_{i=1}^{n_1} c_i x_i + \sum_{k=1}^{n_2} w_k y_k$, with $c_i \geq 0$ and $w_k \geq 0$.
- $l_i \leq x_i \leq u_i$, $i = 1, \ldots, n_1$, and $d_k \leq y_k \leq e_k$, $k = 1, \ldots, n_2$.
- Each other constraint is either $ax_i + bx_j \leq c + y_k$ (type I), or $ax_i + bx_j \leq c$ (type II). There is no restriction on the signs of a and b.
- Each variable y_j appears in (at most) one constraint of type I.

If, moreover, in each constraint of type I $a \geq 0$ and $b \leq 0$ then the problem is called *(linear IP2) monotone.*

Let $U = \max\{|u_i - l_i|\}$, $n = n_1 + n_2$ be the number of variables and m be the number of constraints. The following result is shown in [20] (Theorem 1.1).

Theorem 5. *[20] A monotone linear IP2 problem is solvable in time poly(U, n, m). The continuous relaxation of an IP2 problem has an half-integral optimal solution, which is computable in time poly(U, n, m).*

Let us now go back to the multistage setting. Consider a minimization problem where we want to minimize $\sum_{i=1}^{n} c_i x_i$ under some set of constraints defined as $Ax \geq b$ and $x_i \in \{0, 1\}$. In the multistage model, we are given a sequence of T instances of the problem, i.e., T cost functions c_i^t, $t = 1, \ldots, T$, and T matrices A_t and b_t, $t = 1, \ldots, T$. The goal is to find a sequence of feasible solutions $x^t, t = 1, \ldots, T$ (i.e., $A_t x^t \geq b_t$ and $x^t \in \{0, 1\}^n$), so as to minimize an objective function which is the sum of (1) the costs of individual solutions, i.e., $\sum_t \sum_i c_i^t x_i^t$ and (2) a transition cost $\sum_{t=1}^{T-1}\sum_i w_i|x_i^{t+1} - x_i^t|$, where $w_i \geq 0$ is a transition cost associated to x_i.

[1] Note that the results in [20] consider also the nonlinear case.

Note that this models allows to put transition costs only on some specific variables (by putting $w_i = 0$ to the other ones), such as variables related to vertices/clients in prize-collecting problems or vertices in min cut.

We now show that the multistage setting keeps the structure of (monotone) IP2 problems, as stated in the following Lemma.

Lemma 1. *If an ILP is IP2 (resp. monotone IP2), then its multistage version is IP2 (resp. monotone IP2).*

Proof. The multistage problem can be modeled as an ILP as follows.

$$\begin{cases} \min \sum_{t=1}^{T} \sum_{i=1}^{n} c_i^t x_i^t + \sum_{t=1}^{T-1} \sum_{i=1}^{n} w_i (z_i^t + z_i'^t) \\ s.t. \left| \begin{array}{lll} A_t x^t & \geq b_t & \forall t \in \{1, \ldots, T\} \\ z_i^t & \geq x_i^t - x_i^{t+1} & \forall t \in \{1, \ldots, T-1\}, \forall i \in \{1, \ldots, n\} \\ z_i'^t & \geq x_i^{t+1} - x_i^t & \forall t \in \{1, \ldots, T-1\}, \forall i \in \{1, \ldots, n\} \\ x_i^t, z_i^t, z_i'^t & \in \{0, 1\} & \end{array} \right. \end{cases}$$

We have T 'independent' sets of constraints $A_t x^t \geq b_t$ (that by assumption fulfills the conditions of (monotone) IP2 membership). There are $2n$ variables z_i^t and $z_i'^t$. Each one appears in exactly one constraint (besides the fact that it is binary), and this constraint is of type I, with $a \geq 0$ and $b \leq 0$. □

As an immediate corollary, we can state the following result, that easily generalizes to other (monotone) IP2 problems.

Theorem 6. *Multistage Min Cut is solvable in polynomial time. Multistage Vertex Cover is 2-approximable in polynomial time.*

Proof. As Min Cut is monotone IP2, by Lemma 1 multistage Min Cut is also monotone IP2. The range U of values of variables is 1, so multistage Min Cut is polynomially solvable.

As Vertex Cover is IP2, so is multistage Vertex Cover (Lemma 1). We can compute in polynomial time (here again, $U = 1$) a half integral solution, and round it the usual way: put a variable to 1 if it is 1/2 or 1. Note that the constraints involving variables z_t^i and $z_t'^i$ are satisfied. The rounded solution is feasible, and (trivially) 2-approximate. □

Acknowledgement. For the first two authors, this work was partially funded by the grant ANR-19-CE48-0016 from the French National Research Agency (ANR). The research of the third author was partially supported by RFBR grant 20-07-00458.

References

1. Albers, S.: On energy conservation in data centers. In: Proceedings of SPAA, pp. 35–44 (2017)
2. An, H., Norouzi-Fard, A., Svensson, O.: Dynamic facility location via exponential clocks. ACM Trans. Algorithms **13**(2), 21:1–21:20 (2017)

3. Archer, A., Bateni, M., Hajiaghayi, M., Karloff, H.J.: Improved approximation algorithms for prize-collecting Steiner tree and TSP. SIAM J. Comput. **40**(2), 309–332 (2011)
4. Balas, E.: The prize collecting traveling salesman problem. Networks **19**(6), 621–636 (1989)
5. Bampis, E., Escoffier, B., Lampis, M., Paschos, V.T.: Multistage matchings. In: Proceedings of the SWAT, pp. 7:1–7:13 (2018)
6. Bampis, E., Escoffier, B., Mladenovic, S.: Fair resource allocation over time. In: Proceedings of the AAMAS, pp. 766–773 (2018)
7. Bampis, E., Escoffier, B., Schewior, K., Teiller, A.: Online multistage subset maximization problems. In: Proceedings of the ESA, pp. 11:1–11:14 (2019)
8. Bampis, E., Escoffier, B., Teiller, A.: Multistage knapsack. In: Proceedings of the MFCS, pp. 22:1–22:14 (2019)
9. Bienstock, D., Goemans, M.X., Simchi-Levi, D., Williamson, D.P.: A note on the prize collecting traveling salesman problem. Math. Prog. **59**, 413–420 (1993)
10. Buchbinder, N., Chen, S., Naor, J.: Competitive analysis via regularization. In: Proceedings of the SODA, pp. 436–444 (2014)
11. Buchbinder, N., Chen, S., Naor, J., Shamir, O.: Unified algorithms for online learning and competitive analysis. Math. Oper. Res. **41**(2), 612–625 (2016)
12. Christofides, N.: Worst-case analysis of a new heuristics for the traveling salesman problem. In: Report 388, Graduate School of Industrial Administration, Carnegie-Mellon University (1976)
13. Eisenstat, D., Mathieu, C., Schabanel, N.: Facility location in evolving metrics. In: Esparza, J., Fraigniaud, P., Husfeldt, T., Koutsoupias, E. (eds.) ICALP 2014, Part II. LNCS, vol. 8573, pp. 459–470. Springer, Heidelberg (2014). https://doi.org/10.1007/978-3-662-43951-7_39
14. Fluschnik, T., Niedermeier, R., Rohm, V., Zschoche, P.: Multistage vertex cover. In: Proceedings of the IPEC, pp. 14:1–14:14 (2019)
15. Fotakis, D., Kavouras, L., Kostopanagiotis, P., Lazos, P., Skoulakis, S., Zarifis, N.: Reallocating multiple facilities on the line. In: Proceedings of the IJCAI, pp. 273–279 (2019)
16. Goemans, M.X., Williamson, D.P.: A general approximation technique for constrained forest problems. SIAM J. Comput. **24**(2), 296–317 (1995)
17. Gu, A., Gupta, A., Kumar, A.: The power of deferral: maintaining a constant-competitive Steiner tree online. SIAM J. Comput. **45**(1), 1–28 (2016)
18. Gupta, A., Talwar, K., Wieder, U.: Changing bases: multistage optimization for matroids and matchings. In: Esparza, J., Fraigniaud, P., Husfeldt, T., Koutsoupias, E. (eds.) ICALP 2014, Part I. LNCS, vol. 8572, pp. 563–575. Springer, Heidelberg (2014). https://doi.org/10.1007/978-3-662-43948-7_47
19. Held, M., Karp, R.M.: The travelling salesman problem and minimum spanning trees. Oper. Res. **18**, 1138–1162 (1970)
20. Hochbaum, D.S.: Solving integer programs over monotone inequalities in three variables: a framework for half integrality and good approximations. Eur. J. Oper. Res. **140**(2), 291–321 (2002)
21. Joseph, V., de Veciana, G.: Jointly optimizing multi-user rate adaptation for video transport over wireless systems: mean-fairness-variability tradeoffs. In: Proceedings of the IEEE INFOCOM, pp. 567–575 (2012)
22. Megow, N., Skutella, M., Verschae, J., Wiese, A.: The power of recourse for online MST and TSP. SIAM J. Comput. **45**(3), 859–880 (2016)
23. Nagarajan, C., Williamson, D.P.: Offline and online facility leasing. Discret. Optim. **10**(4), 361–370 (2013)

24. Olver, N., Pruhs, K., Schewior, K., Sitters, R., Stougie, L.: The itinerant list update problem. In: Epstein, L., Erlebach, T. (eds.) WAOA 2018. LNCS, vol. 11312, pp. 310–326. Springer, Cham (2018). https://doi.org/10.1007/978-3-030-04693-4_19
25. Rottner, C.: Combinatorial Aspects of the Unit Commitment Problem. Ph.D. thesis, Sorbonne Université (2018)
26. Shmoys, D., Williamson, D.: Analyzing the held-Karp TSP bound: a monotonicity property with applications. Inf. Process. Lett. **35**(6), 281–285 (1988)
27. Sleator, D.D., Tarjan, R.E.: Amortized efficiency of list update and paging rules. Commun. ACM **28**(2), 202–208 (1985)
28. Williamson, D.P., Shmoys, D.B.: The Design of Approximation Algorithms, 1st edn. Cambridge University Press, New York (2011)
29. Wolsey, L.A.: Heuristic analysis, linear programming and branch and bound. Math. Program. Study **13**, 121–134 (1980)

A Faster FPTAS for Knapsack Problem with Cardinality Constraint

Wenxin Li[1(✉)], Joohyun Lee[2], and Ness Shroff[3]

[1] Department of ECE, The Ohio State University, Columbus, USA
li.7328@osu.edu

[2] Division of Electrical Engineering, Hanyang University, Seoul, South Korea
joohyunlee@hanyang.ac.kr

[3] Department of ECE and CSE, The Ohio State University, Columbus, USA
shroff.11@osu.edu

Abstract. We study the K-item knapsack problem (*i.e.*, 1.5-dimensional knapsack problem), a generalization of the famous 0–1 knapsack problem (*i.e.*, 1-dimensional knapsack problem) in which an upper bound K is imposed on the number of items selected. This problem is of fundamental importance and is known to have a broad range of applications in various fields. It is well known that, there is no *fully polynomial time approximation scheme* (FPTAS) for the d-dimensional knapsack problem when $d \geq 2$, unless P = NP. While the K-item knapsack problem is known to admit an FPTAS, the complexity of all existing FPTASs have a high dependency on the cardinality bound K and approximation error ε, which could result in inefficiencies especially when K and ε^{-1} increase. The current best results are due to [Mastrolilli and Hutter, 2006], in which two schemes are presented exhibiting a space-time tradeoff–one scheme with time complexity $O(n + Kz^2/\varepsilon^2)$ and space complexity $O(n + z^3/\varepsilon)$, and another scheme that requires a run-time of $O(n + (Kz^2 + z^4)/\varepsilon^2)$ but only needs $O(n + z^2/\varepsilon)$ space, where $z = \min\{K, 1/\varepsilon\}$.

In this paper we close the space-time tradeoff exhibited in [Mastrolilli and Hutter, 2006] by designing a new FPTAS with a running time of $\widetilde{O}(n + z^2/\varepsilon^2)$, while simultaneously reaching a space complexity ($\widetilde{O}$ notation hides terms poly-logarithmic in n and $1/\varepsilon$) of $O(n + z^2/\varepsilon)$. Our scheme provides $\widetilde{O}(K)$ and $O(z)$ improvements on the state-of-the-art algorithms in time and space complexity respectively, and is the *first* scheme that achieves a running time that is *independent* of the cardinality bound K (up to logarithmic factors) under fixed ε. Another salient feature of our algorithm is that it is the *first* FPTAS that achieves better time and space complexity bounds than the very first standard FPTAS *over all parameter regimes*.

1 Introduction

The famous *0–1 knapsack problem* (0–1 KP), also known as the *binary knapsack problem* (BKP), is a classical combinatorial optimization problem which often

C. Kaklamanis and A. Levin (Eds.): WAOA 2020, LNCS 12806, pp. 16–29, 2021.
https://doi.org/10.1007/978-3-030-80879-2_2

arises when there are resources to be allocated within a budget. In addition, the 0–1 knapsack problem can be also viewed as the most fundamental non-trivial *integer linear programming* (ILP) problem, and can be formally formulated as follows:

$$\max \sum_{i \in E} p_i x_i, \tag{1}$$

$$s.t. \sum_{i \in E} w_i x_i \leq W \text{ and } x_i \in \{0, 1\}. \tag{2}$$

The value and size of each item i is called *profit* (p_i) and *weight* (w_i) respectively. For any positive integer m, let $[m] = \{1, 2, \dots, m\}$, we use set $E = [n]$ to denote the ground set, which includes all possible items. Our goal is to make a binary choice for each item i to maximize the overall profit subject to a budget constraint W. Beyond this basic model, there are several interesting practical extensions and variations of 0–1 KP, readers are referred to [9] for details.

In this paper, we study the *K-item knapsack problem* (KKP), a well known generalization of the famous 0–1 KP that can be formulated as (1)–(2) with the additional constraint $\sum_{i \in E} x_i \leq K$, which means that the number of items in any feasible solutions is upper bounded by K. The KKP can be cast as a special case of the *two-dimensional knapsack problem*, which is a knapsack problem with two different packing constraints. Hence KKP problem can also be interpreted as 1.5-*dimensional knapsack problem* (1.5-KP) [9, p. 269]. Another closely related problem is the *exact K-item knapsack problem* (E-KKP), for which the results in this paper still hold and discussions are included in [18].

The KKP (and E-KKP) represents many practical applications in various fields ranging from *assortment planning* [5] to *multiprocessor task scheduling* [3], and *crowdsourcing* [19]. For example, the worker selection problem in crowdsourcing systems [7], *i.e.*, maximizing opinion diversity in constructing a wise crowd, can be reduced to E-KKP. On the other hand, KKP also appears as a key subproblem in the solutions of several more complicated problems [1,2,6,8,12]. For example, in the *bin packing* problem [6], to apply the ellipsoid algorithm to approximately solve the linear program, the (approximation) algorithm to the KKP is utilized to construct a polynomial time (approximate) separation oracle. In many such practical and theoretical applications, e.g., the *single-sink capacitated K-facility location problem* [1] and the *resource constrained scheduling problem* [8], the subroutine utilized to solve KKP frequently appears to be one of the main complexity bottleneck. These observations and facts motivate our study of designing a faster algorithm for KKP.

Complexity of Knapsack Problems. An FPTAS is highly desirable for NP-hard problems. Unfortunately, it has been shown that there exists no FPTAS for d-dimensional knapsack problem for $d \geq 2$, unless P =NP [11].

1.1 Theoretical Motivations and Contributions

Known Results of KKP. In this paper we focus on FPTAS for KKP (and E-KKP). The first FPTAS for KKP was proposed in [3], by utilizing standard dynamic programming and profit scaling techniques, which runs in $O(nK^2/\varepsilon)$ time and requires $O(n+K^3/\varepsilon)$ space. This algorithm was later improved by [13]. Based on the *hybrid rounding* technique, two alternative FPTASs (denoted by Scheme A and Scheme B) were presented, which significantly accelerate the dynamic programming procedure while exhibiting a space-time tradeoff. More specifically, Scheme A achieves a time complexity of $O(n + Kz^2/\varepsilon^2)$ and space complexity of $O(n + z^3/\varepsilon)$, Scheme B needs $O(n + z^2/\varepsilon)$ space but requires a run-time of $O(n + Kz^2 + z^4/\varepsilon^2)$. We remark that [10] also investigated this problem, under an additional assumption that item profits follow an underlying distribution. This assumption enables the design of a fast algorithm via rounding the item profits adaptively according to the profit distribution.

The current fastest FPTAS (Scheme A) sacrifices its space complexity in order to improve run-time performance. This may not be desirable as the space requirement is often a more serious bottleneck for practical applications than running time [9, p. 168]. Despite the recent widespread applications of the KKP problem [2,5,6,16,17,19], the state-of-the-art complexity results established in [13] have not been improved since then. This lack of progress brings us to our first question: *Is it possible to design a more efficient FPTAS with lower time and/or space complexity to enhance practicality?*

Moreover, while the two schemes in [13] achieve substantial improvements compared with [3], it is worth noting that there exists a hard parameter regime $\mathcal{H} = \{(n, K, \varepsilon)|K = \Theta(n), \varepsilon^{-1} = \Omega(n)\}$, in which existing FPTASs in the literature fail to surpass both the time and space complexity barriers guaranteed by the standard scheme in [3]. For example, the run-time of Scheme B is higher than that of [3]. Hence from a theoretical point of view, it is natural to ask: *Can we design a new FPTAS that has lower time complexity or space complexity than the standard FPTAS [3] over all parameter regimes?*

Table 1. Comparisons between different FPTASs

Reference	Time complexity	Space complexity
[3]	$O(\frac{nK^2}{\varepsilon})$	$O(n+\frac{K^3}{\varepsilon})$
[13] (Scheme A)	$O(n+\frac{Kz^2}{\varepsilon^2})$	$O(n+\frac{z^3}{\varepsilon})$
[13] (Scheme B)	$O(n+\frac{Kz^2+z^4}{\varepsilon^2})$	$O(n+\frac{z^2}{\varepsilon})$
This paper [2]	$\widetilde{O}(n+\frac{z^2}{\varepsilon^2})$	$O(n+\frac{z^2}{\varepsilon})$

[1] $z = \min\{K, \varepsilon^{-1}\}$;
[2] [2] As shown in Theorem 2, our time complexity can be refined to $\widetilde{O}(n + z^4 + \frac{z^2}{\varepsilon} \cdot \min\{n, \varepsilon^{-1}\})$.

Our Contributions. As summarized in Table 1, we break the longstanding barrier and answer the aforementioned questions in the affirmative. In particular, we present a new FPTAS with $\widetilde{O}(n + z^2/\varepsilon^2)$ running time and $O(n + z^2/\varepsilon)$ space requirement, which offers $\widetilde{O}(K)$ and $O(z)$ improvements in time and space complexity respectively. Our FPTAS is the **first** to achieve time complexity that is independent of K (up to logarithmic factors, for a given ε). According to Theorem 2, the time complexity of our algorithm can be indeed refined to $\widetilde{O}(n + z^4 + \frac{z^2}{\varepsilon} \cdot \min\{n, \varepsilon^{-1}\})$. From this refined bound, it can be seen that even in the hard regime $\mathcal{H}$, our algorithm has the same time complexity (up to log factors) as the standard FPTAS [3], while improving its space complexity by a factor of n. This implies that our algorithm is also the **first** FPTAS that outperforms the standard FPTAS [3] over all parameter regimes, thus answering the second question in the affirmative.

Our new scheme also helps to improve the state-of-the-art complexity results of several problems in other fields, owing to the widespread applications of KKP. In [18], we take the resource constrained scheduling problem [8] as an illustrative example.

1.2 Technique Overview

Different from the *hybrid rounding* technique proposed in [13], which simplifies the structure of the input instance and approximately guarantees the objective value, we show that it is possible to achieve a better complexity result solely via the *geometric rounding* in the preprocessing phase. We divide items into two classes according to their profits and present distinct methods for each class of items. To solve the subproblem for items with low profit, we present a continuous relaxation function, using the natural linear programming relaxation and other alternatives based on structured weights and scaled budget constraint. The carefully designed relaxation function well approximates the optimal objective value of the subproblem and allows us to exploit the redundancy among various input. For every new input parameters, the relaxation can be computed in $O(z/\varepsilon)$ time on average. As for items with large profit, our treatment mainly follows from the novel "functional" approximation approach and point of view, which was recently proposed in [4]. As a straightforward generalization of the 0–1 KP, a two dimensional convolution operator is defined. We perform the convolution procedure in parallel planes to reduce the running time. The fact that there are at most z elements with large profits helps us to bound the discretization precision via parameter z, instead of the number of profit functions. Here we adopt a slightly different but rather (unnecessary) sophisticated and tedious presentation via the lens of numerical discretization. We hope that this presentation helps to make the approach more clear (in the context of KKP). Finally, an approximate solution is obtained by appropriately putting these two modules together.

2 Item Preprocessing

Definition 1 (Item Partition). *Let $\mathcal{L}$ and $\mathcal{S}$ denote the set of large and small items, respectively. Item $e \in E$ is called a* small *item if its profit is no more than*

εOPT, *otherwise it is called a* large *item*[1], i.e., $\mathcal{S} = \{e \in E | \varepsilon\text{OPT}/K \leq p_e \leq \varepsilon\text{OPT}\}$ *and* $\mathcal{L} = \{e \in E | p_e \in \Xi\}$, *where* $\Xi = [\varepsilon\text{OPT}, \text{OPT}]$. *We further divide* $\mathcal{L}$ *and* $\mathcal{S}$ *into different classes,* $\{\mathcal{L}_i^\dagger\}_{i \in [r_\mathcal{L}]}$ *and* $\{\mathcal{S}_i^\dagger\}_{i \in [r_\mathcal{S}]}$, *where* $\mathcal{L}_i^\dagger = \{e \in \mathcal{L} | p_e \in (\varepsilon(1+\varepsilon)^{i-1}\text{OPT},\ \varepsilon(1+\varepsilon)^i\text{OPT}]\}$ $(i \in [r_\mathcal{L}])$ *and* $\mathcal{S}_i^\dagger = \{e \in \mathcal{S} | p_e \in (\varepsilon(1+\varepsilon)^{-i}\text{OPT},\ \varepsilon(1+\varepsilon)^{-i+1}\text{OPT}]\}$ $(i \in [r_\mathcal{S}])$. *Let* r *denote the number of non-empty classes in* E, *as shown in [18], we have*

$$r = O(\min\{r_\mathcal{L} + r_\mathcal{S}, n\}) = O(\min\{\log(K/\varepsilon)/\varepsilon, n\}) = \widetilde{O}(\min\{1/\varepsilon, n\}). \quad (3)$$

Definition 2 (Geometric Rounding). *Without loss of generality, we can assume that elements in the same class have the same profit value. More specifically, we let* $p_e = p_i^\dagger = \varepsilon(1+\varepsilon)^i\text{OPT}$ $(\forall e \in \mathcal{L}_i)$ *and* $p_e = p_i^\ddagger = \varepsilon(1+\varepsilon)^{-i}\text{OPT}$ $(\forall e \in \mathcal{S}_i)$.

The simplification in Definition 2 does not hurt the solution since it will incur a loss of $O(\varepsilon\text{OPT})$ in the objective value. Let O^* denote the optimal solution, exploiting the simple structure of item profits after item partition and profit rounding, we are able to derive the following more fine-grained bound on the size of $|O^* \cap \mathcal{L}|$ and $\mathcal{S}$. Its proof is deferred to [18].

Proposition 1. *There are no more than* $|O^* \cap \mathcal{L}| \leq z$ *large items in the optimal solution set* O^*. *Without loss of generality, we can assume that the number of small items* $|\mathcal{S}| = O(\min\{K \cdot \log(K/\varepsilon)/\varepsilon, n\}) = \widetilde{O}(\min\{K/\varepsilon, n\})$.

3 Algorithm for Large Items

To approximately solve the K-item knapsack problem on ground set E, the first step of our approach is to divide this problem into two smaller KKP problems, which are defined on the large item set $\mathcal{L}$ and small item set $\mathcal{S}$ respectively. In this section we study the subproblem on $\mathcal{L}$, which is the same as the original problem, except that the ground set is substituted by $\mathcal{L}$ and the cardinality upper bound k must be no less than z.

3.1 An Abstract Algorithm Based on Convolution

We first define the profit function $\varphi_{(\cdot)}(\cdot,\cdot) : 2^\mathcal{L} \times \mathbb{R}^+ \times [z] \to \mathbb{R}^+$. From the definition we can see that $\varphi_\mathcal{L}(\omega, k)$ is equal to the optimal objective value of the subproblem considered in this section.

Definition 3 (Profit function [4]**).** *For any given set* $T \subseteq E$, *real number* ω, *and integer* k, $\varphi_T(p, k)$ *is given by* $\varphi_T(\omega, k) = \max\{\sum_{e \in T'} p_e | \sum_{e \in T'} w_e \leq \omega, |T'| \leq k, T' \subseteq T \subseteq E\}$, *which denotes the optimal objective value of the* K*-item knapsack problem that is defined on set* T, *while the budget and cardinality are* ω, k *respectively.*

[1] We discuss the method of obtaining OPT in [18].

Our objective is to approximately compute matrix $\mathbf{A}_{\mathcal{L}} = \{\varphi_{\mathcal{L}}(\omega, k)\}_{\omega \in X, k \in [z]}$, in which the value of X will be specified in Sect. 3.3. This matrix plays an important role in our final item combination procedure, as we will show later in Sect. 5. To compute the profit function efficiently, we introduce the following *inverse weight function* $\phi_{(\cdot)}(\cdot, \cdot) : 2^{\mathcal{L}} \times \Xi \times [z] \to \mathbb{R}^+$, which is one of the key ingredients in computing the profit function.

Definition 4 (Inverse weight function). *For any given set $T \subseteq E$, real number p and integer k, $\phi_T(p, k)$ is given by $\phi_T(p, k) = \min\{\sum_{e \in T'} w_e | \sum_{e \in T'} p_e \geq p, |T'| \leq k, T' \subseteq T\}$, which characterizes the minimum possible total weights under which there exists a subset of T with total profit being no less than p and cardinality no more than k.*

An immediate consequence of Definitions 3 and 4 is that we can easily obtain the value of $\varphi_{\mathcal{L}}(\omega, k)$ based on ϕ, *i.e.*, via equation $\varphi_{\mathcal{L}}(\omega, k) = \sup\{p \in \mathbb{R}^+ | \phi_{\mathcal{L}}(p, k) \leq \omega\}$. Therefore it suffices to derive the inverse weight function $\phi_{\mathcal{L}}(\cdot, \cdot)$ to compute $\mathbf{A}_{\mathcal{L}}$.

Algorithm for Computing $\phi_{\mathcal{L}}$. If we partition the large item set $\mathcal{L}$ into ℓ disjoint subsets as $\mathcal{L} = \cup_{i=1}^{\ell} \mathcal{L}^{(i)}$, then $\phi_{\mathcal{L}}$ can be computed by performing convolution operations sequentially. We specify the details of the algorithm in [18]. The *convolution operation* $\otimes$ is defined as follows.

Definition 5 (Two Dimensional Convolution Operator $\otimes$). *For any two disjoint sets $S_1, S_2 \subseteq E$, we use $(\phi_{S_1} \otimes \phi_{S_2})(\cdot, \cdot)$ to denote the convolution of functions $\phi_{S_1}(\cdot, \cdot)$ and $\phi_{S_2}(\cdot, \cdot)$, then it can be represented as,*

$$(\phi_{S_1} \otimes \phi_{S_2})(p, k) = \min\left\{\phi_{S_1}(p_1, k_1) + \phi_{S_2}(p_2, k_2) \middle| k_1 + k_2 \leq k, p_1 + p_2 \geq p\right\}$$
$$\equiv \phi_{S_1 \cup S_2}(p, k).$$

Under this notation, function $\phi_{\mathcal{L}}(\cdot, \cdot)$ defined on $\mathcal{L}$ can be represented as $\phi_{\mathcal{L}}(p, k) = (\otimes_{i=1}^{\ell} \phi_{\mathcal{L}^{(i)}})(p, k)$. It is important to remark that the algorithm is a rather general description of the convolution procedure, and the partition scheme should be further specified. Generally speaking, different partition schemes will induce different complexity results. For example, if we partition $\mathcal{L}$ into singletons, *i.e.*, $\mathcal{L}^{(i)} = \{e_i\}$ and $\ell = |\mathcal{L}|$, then $\phi_{\mathcal{L}}(p, K) = (\otimes_{i=1}^{|\mathcal{L}|} \phi_{\{e_i\}})(p, K)$. In this case, the algorithm is equivalent to the standard dynamic programming paradigm. In each stage we are in charge of making the decision of whether to include item e_i or not. Here we divide $\mathcal{L}$ in the same way as Definition 1, *i.e.*, $\mathcal{L}^{(i)} = \mathcal{L}_i^{\dagger}, \forall i \in [r_{\mathcal{L}}]$.

3.2 Discretizing the Function Domain

At the current stage, it is worth pointing out that in the convolution operation between inverse weight functions, the profit variable p appears as a decision variable that varies continuously in Ξ. In addition, we are not able to obtain the closed form solution of the convolution operation analytically. The solution is to transform the problem into a computationally tractable one via discretization, then compute an (approximate) solution utilizing the computable version.

Discretizing the Profit Space. To implement the convolution in polynomial time, we discretize the interval Ξ with the points $\{x_i\}_{i\in[m]}$ as $X = \{x_i : \varepsilon\text{OPT} = x_1 < x_2 < \ldots < x_{m-1} < x_m = \text{OPT}\} \subseteq \Xi$. We denote the *discretization parameter* of X by *discretization parameter* $\delta_X = \max_{1\leq i\leq m-1}\{x_{i+1} - x_i\}$. To tackle the computational challenge induced by the continuity of profit p, we execute the convolution operation over the discrete functions that are defined on $X \times [z]$, *i.e.*,

$$(\phi_{S_1} \otimes \phi_{S_2})^X(p,k) = \min_{p_1,p_2\in X}\left\{\phi^X_{S_1}(p_1,k_1) + \phi^X_{S_2}(p_2,k_2)\Big| k_1 + k_2 \leq k, p_1 + p_2 \geq p\right\}.$$

More specifically, we start with functions $\phi^X_{\mathcal{L}^{(i)}}$, and compute $\phi^X_{\cup_{j=1}^{i}\mathcal{L}^{(j)}}$ iteratively until $\phi^X_{\mathcal{L}}$ is obtained. In general, function $\phi^X_{\cup_{i\in I}\mathcal{L}^{(i)}}(\cdot,\cdot) \equiv (\otimes_{i\in I}\phi^X_{\mathcal{L}^{(i)}})(\cdot,\cdot)$ for any $I \subseteq [\ell]$. The discrete profit function $\varphi^X_S(\cdot,\cdot)$ can also be recovered by its relation with the inverse weight function, *i.e.*, $\varphi^X_S(\omega,k) = \max\{p \in X : \phi^X_S(p,k) \leq \omega\}, \forall S \subseteq E$.

Convergence Behaviour of $\varphi^X(\cdot,\cdot)$. We first show point-wise convergence of $\{\varphi^X_{(\cdot)}(\cdot,\cdot)\}_X$ towards $\varphi_{(\cdot)}(\cdot,\cdot)$ when δ_X goes to zero. The proof of Lemma 1 is deferred to [18]. It is worth pointing out that the straightforward intuition that convergence occurs if discretization is small, may not always hold. Indeed we can verify that the weight function ϕ^X may not converge to ϕ through the example in [18].

Lemma 1. *For any finite index set* I *and* ω, k, *we have* $\lim_{\delta_X\to 0}\varphi^X_{\cup_{i\in I}\mathcal{L}^{(i)}}(\omega,k) = \varphi_{\cup_{i\in I}\mathcal{L}^{(i)}}(\omega,k)$ *for fixed* ω, k.

The theoretical convergence of $\varphi^X(\cdot,\cdot)$ ensures the near-optimality of the solution obtained by discretization, as long as X is dense enough in Ξ. However, what matters greatly is the *order of the accuracy*, which refers to how rapidly the error decreases in the limit as the discretization parameter tends to zero. The formal definition of the convergence speed of discretization methods is given in [18]. This speed is directly related to the complexity of our algorithm. From the following lemma, we can conclude that the method of discretizing X by a uniform grid set converges with order 1, as $\delta_X = O(1/|X|)$ for uniform grid set X. The proof of Lemma 2 is deferred to [18].

Lemma 2. *Let* $\phi^X_{\mathcal{L}}$ *be the weight function, then for any given budget* $\omega \leq W$, *cardinality upper bound* $k \leq z$, *and discretization set* X, *we have* $|\varphi^X_{\mathcal{L}}(\omega,k) - \varphi_{\mathcal{L}}(\omega,k)| \leq C\delta_X$, *where the coefficient* $C = z+1$. *As a consequence,* $|X|$ *must be of order* $\Omega(z/\varepsilon)$ *to ensure an error of order* $O(\varepsilon\text{OPT})$.

3.3 Fast Convolution Algorithm

In this subsection we settle the problem of designing a fast convolution algorithm, which is the last remaining issue that has a critical impact on the efficiency of the algorithm for large items. To this end, we show an inherent connection between convolution results under different inputs p and k, which is formally described in Lemma 3. Owing to this observation, we are able to remove a large amount of redundant calculations when facing new input parameters. To start with, we first sort items in each $\mathcal{L}_i^\dagger$ in non-increasing order of weights, which takes $O(z \log z)$ time. We define the optimum index function as follows.

Definition 6 (Optimum index function). $\psi : X \times [K] \to [K]$ *is defined as,*

$$\psi(p,k) = \operatorname{argmin}\Big\{\theta \in [k] \Big| \phi^X_{\mathcal{L}_a^\dagger}(\max\{x \in X : x \le \theta \cdot p_a^\dagger\}, \theta) + \phi^X_S(\max\{x \in X : x \le p - \theta \cdot p_a^\dagger\}, k - \theta)\Big\} \ (p \in X). \quad (4)$$

Here (4) benefits from the partition in which all items in the same set $\mathcal{L}_i^\dagger$ have equal profit value. Specifically, when we derive the result of $(\phi^X_{\mathcal{L}_a^\dagger} \otimes \phi^X_S)(p,k)$, there is indeed only one decision variable θ, *i.e.*, the number of elements selected from $\mathcal{L}_i^\dagger$, that should be figured out. Hence, we denote the optimal value of θ by the index function ψ. Our primary objective is then reduced to figure out all the indices $\{\psi(p,k)\}_{p \in X, k \in [z]}$, for which we give a graphic illustration in Fig. 1(a). It can be regarded as finding *column minimums* in the cube, here column minimum refers to the optimal indices defined in Definition 6.

Consider the Problem in Parallel Slices. We divide the cube into parallel slices. Consider slice

$$H = \Big\{(p,k) \Big| p = p_0 + \zeta\lambda_a, k = k_0 + \zeta\Big\} \bigcap \Big(\Xi \times [0, z]\Big), \quad (5)$$

as shown in Fig. 1(b), where (p_0, k_0) denotes the boundary point of slice H, hence $p_0 k_0 = 0$, and ζ represents the drift of point (p,k) from boundary. It can be seen that the angle between slice H and the frontal plane is equal to $\arctan \lambda_a^{-1}$, and there are $O(|X|) = O(z/\varepsilon)$ such parallel slices in the cube. On the other hand, plugging (5) into (4), the index function can be simplified to

$$\chi_H(\zeta) = \operatorname{argmin}\Big\{\theta \in [z] \Big| \phi^X_{\mathcal{L}_a^\dagger}(\lambda_a\theta, \theta) + \phi^X_S(p_0 + \lambda_a[\zeta - \theta], k_0 + [\zeta - \theta])\Big\}.$$

Bounded "Gradient" of χ_H. Without loss of generality we could assume that there exists an integer $\tau_a \in \mathbb{Z}^+$ such that $p_a^\dagger = \tau_a \cdot \frac{\varepsilon \mathrm{OPT}}{z}$, otherwise we can always modify $p_a^\dagger$ by an $O(\frac{\varepsilon \mathrm{OPT}}{z})$ additive factor to meet this criteria while inducing a $O(\varepsilon \mathrm{OPT})$ loss in the objective function. Consequently we have $\lambda_a = \tau_a \varepsilon \mathrm{OPT}$. We consider the case when Ξ is discretized by the uniform grid set $X = \{i \cdot \frac{\varepsilon \mathrm{OPT}}{z} | i \in [z/\varepsilon]\}$. Then the following key observation about the distribution of column minima in slice H holds. The proof of this lemma is deferred to [18].

Lemma 3. *Consider two columns in H that are indexed by ζ_1 and ζ_2. We have* $\frac{\chi_H(\zeta_2) - \chi_H(\zeta_1)}{\zeta_2 - \zeta_1} \le 1$.

Divide-and-Conquer on Slice H. In Lemma 3, we establish an upper bound on the growth rate of the index function. Taking advantage of this lemma, we are able to reduce the size of the searching space in one column, given that we have figured out the optimum indices at some other columns in the slice H. More specifically, consider columns indexed by $\zeta_1 \leq \zeta_2 \leq \zeta_3$, the information of $\chi_H(\zeta_1)$ and $\chi_H(\zeta_3)$ indeed provide two cutting planes to help us locate $\chi_H(\zeta_2)$ in a smaller interval $[\chi_H(\zeta_3) + \zeta_2 - \zeta_3, \chi_H(\zeta_1) + \zeta_2 - \zeta_1]$.

Inspired by this observation, for any slice in the form of (5), we design a *divide-and-conquer* procedure to compute the optimum indices efficiently. We start with a recursive call to determine the optimum indices of all the even-indexed columns. Here a column is called even (odd) column if and only if its corresponding ζ value in (5) is even (odd). Then for each odd column $\chi_H(2i)$, it can be computed by enumerating the interval $[\chi_H(2i+1) - 1, \chi_H(2i-1) + 1]$. The details are specified in [18].

The time complexity of computing the index function for a single slice is summarized in the following proposition, whose proof is presented in [18].

Proposition 2. *It takes $O(z \log z) = \widetilde{O}(z)$ time to compute $\chi_H(\cdot)$.*

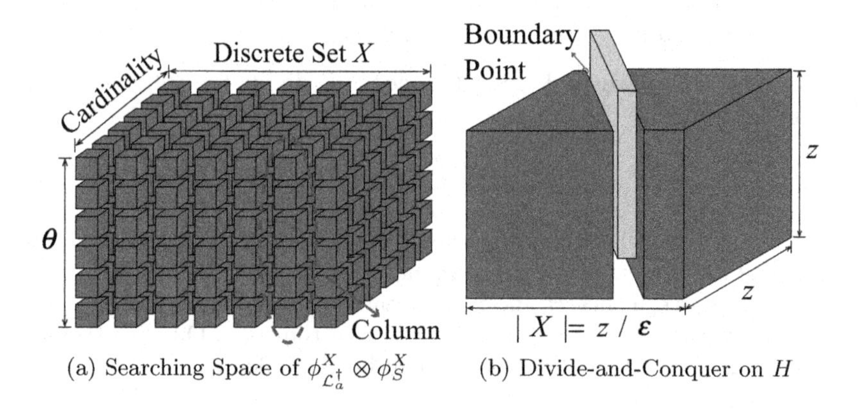

(a) Searching Space of $\phi^X_{\mathcal{C}^\dagger_a} \otimes \phi^X_S$ (b) Divide-and-Conquer on H

Fig. 1. Graphic illustrations of the convolution operation

Fast Convolution Operation. The details of the convolution operation are specified in [18]. The following lemma summarizes the complexity of our algorithm, the proof is presented in [18].

Lemma 4. *It takes $O(n)$ space and*

$$O(n + (z^2 \log z/\varepsilon) \cdot \min\{\log(1/\varepsilon)/\varepsilon, n\}) = O(n) + \widetilde{O}(\min\{z^2/\varepsilon^2, nz^2/\varepsilon\})$$

time to complete the convolution operation.

Generally speaking, for given $p \in X$ and $k \in \mathbb{Z}^+$, it requires $O(z \cdot |X|)$ arithmetic operations to compute $(\phi_{S_1} \otimes \phi_{S_2})(p,k)$ if we enumerate all possible pairs of (p_1, k_1), which further results in a total complexity of $O(z^2|X|^2)$ for operator $\otimes$. Compared with our Algorithm, this is unnecessarily inefficient, since it restarts all the arithmetic operations when the input parameters varies.

4 Continuous Relaxation for Small Items

In Sect. 3 we have shown how to approximately select the most profitable large items under any given budget and cardinality constraints. One important task left is to solve the subproblem with only small items involved. In this section we show how to approximately solve this subproblem efficiently. Similar to Definition 4, the profit function of small items, $\varphi_{\mathcal{S}}(\cdot,\cdot) : \mathbb{R}^+ \times [K] \to \mathbb{R}^+$, is given by $\varphi_{\mathcal{S}}(\omega,k) = \max\{\sum_{e\in\mathcal{S}} p_e x_e | \sum_{e\in\mathcal{S}} x_e \le k, \sum_{e\in\mathcal{S}} w_e x_e \le \omega, x_e \in \{0,1\}\}$. The main spirit of our approach for small items is similar to that of Sect. 3, *i.e.*, find a new function $\widetilde{\varphi}_{\mathcal{S}}$, which is a good approximation of $\varphi_{\mathcal{S}}$ and is economical in computations. To this end, our main result in this subsection is formally stated in the following lemma. We leave the proof of this lemma in [18], which relies on our analysis in the following two subsections.

Lemma 5. *There exists a relaxation* $\widetilde{\varphi}_{\mathcal{S}}(\cdot,\cdot) : \mathbb{R}^+ \times [K] \to \mathbb{R}^+$ *that satisfies* $|\widetilde{\varphi}_{\mathcal{S}} - \varphi_{\mathcal{S}}| = O(\varepsilon\mathrm{OPT})$, *and the corresponding matrix* $\mathbf{A}_{\mathcal{S}} = \{\widetilde{\varphi}_{\mathcal{S}}(\omega,k) | \omega \in \mathcal{W}, k \in \mathcal{K}\}$ *can be computed within* $\widetilde{O}(n + z^4 + \min\{\frac{z^2}{\varepsilon^2}, \frac{nz}{\varepsilon}\})$ *time when* $|\mathcal{W}| = O(\varepsilon^{-1})$ *and* $|\mathcal{K}| = O(z)$, *while requiring* $O(z/\varepsilon)$ *space.*

One question that may arise is the following: can the methods in Sect. 3 still work for the small item set $\mathcal{S}$, *i.e.*, can we apply the algorithm for large items over $\mathcal{S}$ and use the output discrete function as an approximation of $\varphi_{\mathcal{S}}$? Unfortunately it can be verified that $O(n) + \widetilde{O}(K^2/\varepsilon^2)$ time is required, which is significantly high especially when K is large, and fails to provide the desired complexity result. This is because there could be many more small items than large items, which will result in a larger searching space.

To construct the new function $\widetilde{\varphi}_{\mathcal{S}}$, we turn to the continuous relaxation of the subproblem, as the continuous problem is much easier to deal with. More importantly, the boundness of small item profits will ensure that the gap between the optimal values of the two problems is small.

4.1 Continuous Relaxation Design and Error Analysis

Designing $\widetilde{\varphi}_{\mathcal{S}}$. In our algorithm, we let

$$\widetilde{\varphi}_{\mathcal{S}}(\omega,k) = \widetilde{\varphi}_{\mathcal{S}}^{(1)}(\omega,k) \cdot \mathbb{1}_{\{K \le \varepsilon^{-1}\}} + \widetilde{\varphi}_{\mathcal{S}}^{(2)}(\omega,k) \cdot \mathbb{1}_{\{K > \varepsilon^{-1}\}},$$

in which the two building block functions $\widetilde{\varphi}_{\mathcal{S}}^{(i)}$ $(i = 1,2)$ are specified in the following definition. The first function $\widetilde{\varphi}_{\mathcal{S}}^{(1)}$ is the most natural linear programming relaxation of $\widetilde{\varphi}_{\mathcal{S}}$, in which all the integer variables are relaxed to real numbers in $[0,1]$. In the second function $\widetilde{\varphi}_{\mathcal{S}}^{(2)}$, we only relax variables corresponding to elements in $\mathcal{S}_\omega$, while the element weights are rounded to an integer power of $(1+\varepsilon)$, and the budget is given by $(1-\varepsilon)\omega$ instead of ω.

Definition 7 (Definition of $\widetilde{\varphi}_{\mathcal{S}}^{(1)}, \widetilde{\varphi}_{\mathcal{S}}^{(2)}$). *Functions $\widetilde{\varphi}_{\mathcal{S}}^{(i)}(\cdot,\cdot) : \mathbb{R}^+ \times [K] \rightarrow \mathbb{R}^+$ $(i=1,2)$ are constructed as*

$$\widetilde{\varphi}_{\mathcal{S}}^{(1)}(\omega,k) = \max\Big\{\sum_{e\in\mathcal{S}} p_e x_e \Big| \sum_{e\in\mathcal{S}} x_e \le k, \sum_{e\in\mathcal{S}} w_e x_e \le \omega, x_e \in [0,1]\Big\},$$

and

$$\widetilde{\varphi}_{\mathcal{S}}^{(2)}(\omega,k) = \max_{0\le t\le k}\Big\{\widetilde{\varphi}_{\mathcal{S}_\omega}^{(2)}(\omega,k-t) + \widetilde{\varphi}_{\bar{\mathcal{S}}_\omega}^{(2)}(t)\Big\}.$$

Here set $\bar{\mathcal{S}}_\omega = \{e \in \mathcal{S} | w_e \le \varepsilon\omega/K\}$ represents the set of elements in $\mathcal{S}$ with weight less than a threshold $\varepsilon\omega/K$, and $\mathcal{S}_\omega = (\mathcal{S}\backslash\bar{\mathcal{S}}_\omega)\backslash\{e\in\mathcal{S}|w_e > \omega\}$. Function $\widetilde{\varphi}_{\bar{\mathcal{S}}_\omega}^{(2)}(t) = \max\{\sum_{e\in T} p_e | T \subseteq \bar{\mathcal{S}}_\omega, |T| \le t\}$ denotes the total profits of the top t elements in $\bar{\mathcal{S}}_\omega$. In addition, $\widetilde{\varphi}_{\mathcal{S}_\omega}^{(2)}(\omega,t)$ is given by

$$\widetilde{\varphi}_{\mathcal{S}_\omega}^{(2)}(\omega,t) = \max_{x_e\in[0,1]}\Big\{\sum_{e\in\mathcal{S}_\omega} p_e x_e \Big| \sum_{e\in\mathcal{S}_\omega} x_e \le t, \sum_{e\in\mathcal{S}_\omega} w'_e x_e \le (1-\varepsilon)\omega\Big\} \quad (6)$$

where $w'_e = (\omega(1+\varepsilon)^{\lceil \log_{(1+\varepsilon)}(\frac{\varepsilon K w_e}{\omega})\rceil})/(K\varepsilon)$ and $\lceil\cdot\rceil$ refers to the ceiling function.

Error of Approximation. We show that $\widetilde{\varphi}_{\mathcal{S}}$ provides a good approximation of $\varphi_{\mathcal{S}}$ in the following lemma. The proof of Lemma 6 is presented in [18].

Lemma 6. *The differences between functions $\widetilde{\varphi}_{\mathcal{S}}$ and $\varphi_{\mathcal{S}}$ is bounded as $|\widetilde{\varphi}_{\mathcal{S}}(\omega,k) - \varphi_{\mathcal{S}}(\omega,k)| \le 4\varepsilon\mathrm{OPT}$.*

Obtain the Final Solution Set. Recall that our ultimate objective is to retrieve an solution set that has near optimal objective function value. To this end, for the subproblem of small items, we can solve the continuous problem and return the corresponding integer components $S = \{e | x_e^* = 1\}$ as an approximate solution, where $\mathbf{x}^*$ denotes the optimal fractional solution.

4.2 Computing $\widetilde{\varphi}_{\mathcal{S}}$ Efficiently

In this subsection, we consider how to efficiently compute the function $\widetilde{\varphi}_{\mathcal{S}}(\cdot,\cdot)$. More specifically, our objective is to compute set $\{\widetilde{\varphi}_{\mathcal{S}}(\omega,k) | \omega \in \mathcal{W}, k \in \mathcal{K}\}$, for given $\mathcal{K} \in \mathbb{Z}^{|\mathcal{K}|}$ and $\mathcal{W} \in \mathbb{R}^{|\mathcal{W}|}$.

Computing Relaxation $\widetilde{\varphi}_{\mathcal{S}}^{(1)}(\cdot,\cdot)$. One straightforward approach is to utilize the linear time algorithm [3,14,15] to solve $\widetilde{\varphi}_{\mathcal{S}}^{(1)}$ under distinct parameters in $\mathcal{W}, \mathcal{K}$ separately, which will result in a total complexity of $O(|\mathcal{S}| \cdot |\mathcal{K}| \cdot |\mathcal{W}|) =$

$O(\frac{z}{\varepsilon} \cdot \min\{K/\varepsilon, n\})$. Note that under this approach, the complexity has a high dependence on the parameter K.

Computing Relaxation $\widetilde{\varphi}_{\mathcal{S}}^{(2)}(\cdot, \cdot)$. Let $f_t(\omega, k) = \widetilde{\varphi}_{\mathcal{S}_\omega}^{(2)}(\omega, k-t) + \widetilde{\varphi}_{\bar{\mathcal{S}}_\omega}^{(2)}(t)$, then $\widetilde{\varphi}_{\mathcal{S}}^{(2)}(\omega, k) = \max_{0 \le t \le k} f_t(\omega, k)$ according to Definition 7. We first claim the following key observation with regard to $\{f_t(\omega, k)\}_{0 \le t \le k}$. Basically this concavity property enables us to compute $\widetilde{\varphi}_{\mathcal{S}}^{(2)}$ using $O(\log k)$ calls to the subroutine of computing $f_t(\omega, k)$. The proof is presented in [18].

Theorem 1 (Concavity of f_t). *The sequence $\{f_t(\omega, k)\}_{t \in [k]}$ is a concave sequence with respect to t. As a result, $\widetilde{\varphi}_{\mathcal{S}}^{(2)}(\omega, k)$ can be computed in $O(\mathcal{T}_f \log k) = \widetilde{O}(\mathcal{T}_f)$ time, where $\mathcal{T}_f$ represents the worst case time complexity for computing $f_t(\omega, k)$ under fixed values of t, ω, k.*

At the current stage, the problem of computing $\widetilde{\varphi}_{\mathcal{S}}^{(2)}(\omega, k)$ has been shown to have the same time complexity (up to a factor of $O(\log k)$) as computing $f_t(\omega, k)$, which is further determined by the following two subroutines–calculating $\widetilde{\varphi}_{\mathcal{S}_\omega}^{(2)}(\omega, t)$ and $\widetilde{\varphi}_{\bar{\mathcal{S}}_\omega}^{(2)}(t)$. We dualize the budget constraint as

$$L(\mu, \omega, t) = \max_{x_e \in [0,1]} \Big\{ \sum_{e \in \mathcal{S}_\omega} (p_e - \mu w_e) x_e + \mu\omega \Big| \sum_{e \in \mathcal{S}_\omega} x_e \le t \Big\},$$

which helps to figure out the first function under multiple input parameters. Indeed we can always apply binary search on set

$$\mathcal{B}' = \Big\{ (1+\varepsilon)^b \cdot \frac{(1+\varepsilon)^c - 1}{(1+\varepsilon)^d - 1} \Big| |b|, |c|, |d| \le \log(K/\varepsilon)/\varepsilon, \text{ and } b, c, d \in \mathbb{Z} \Big\}$$

to figure out the optimal multiplier μ^*, owing to the convexity of the Lagrange function. As for function $\widetilde{\varphi}_{\bar{\mathcal{S}}_\omega}^{(2)}(t)$, we take advantage of the fact that $\{\widetilde{\varphi}_{\bar{\mathcal{S}}_\omega}^{(2)}(t)\}_{t \in [|\mathcal{S}|]}$ can be computed together to reduce running time, under the same budget ω. We present the details and complexity analysis in [18].

5 Putting the Pieces Together–Combining Small and Large Items

In our main algorithm, we utilize our two algorithms established in Sect. 3 and 4 as two basic building blocks, to approximately enumerate all the possible profit allocations among $\mathcal{L}$ and $\mathcal{S}$. The details are specified in [18] and performance guarantee is given by Theorem 2. We remark that set X' in the algorithm is not equal to X but a subset of X, and is given by $X' = \{i\varepsilon\mathrm{OPT} | i \in [1/\varepsilon]\}$. The proof of theorem 2 is deferred to [18].

Theorem 2. *The total profits of items in set S_o given in the main algorithm is no less than $p(S_o) \ge (1 - O(\varepsilon))\mathrm{OPT}$, while requires $\widetilde{O}(n + z^4 + \frac{z^2}{\varepsilon} \cdot \min\{n, \varepsilon^{-1}\})$ time, which is within the order of $\widetilde{O}(n + \frac{z^2}{\varepsilon^2})$.*

6 Conclusion

In this paper we proposed a new FPTAS for the *K-item knapsack problem* (and *Exactly K-item knapsack problem*) that exhibits $\widetilde{O}(K)$ and $O(z)$ improvements in time and space complexity respectively, compared with the state-of-the-art [13]. More importantly, our result suggests that for a fixed value of ε, an $(1-\varepsilon)$-approximation solution of KKP can be computed in time asymptotically independent of cardinality bound K. Our scheme is also the first FPTAS that achieves better time and space complexity (up to logarithmic factors) than the standard dynamic programming scheme in [3] over all parameter regimes.

References

1. Aardal, K., van den Berg, P.L., Gijswijt, D., Li, S.: Approximation algorithms for hard capacitated k-facility location problems. Eur. J. Oper. Res. **242**(2), 358–368 (2015)
2. Ahuja, R.K., Orlin, J.B., Pallottino, S., Scaparra, M.P., Scutellà, M.G.: A multi-exchange heuristic for the single-source capacitated facility location problem. Manage. Sci. **50**(6), 749–760 (2004)
3. Caprara, A., Kellerer, H., Pferschy, U., Pisinger, D.: Approximation algorithms for knapsack problems with cardinality constraints. Eur. J. Oper. Res. **123**(2), 333–345 (2000)
4. Chan, T.M.: Approximation schemes for 0–1 knapsack. In: SOSA (2018)
5. Désir, A., Goyal, V., Segev, D.: Assortment optimization under a random swap based distribution over permutations model. In: EC, pp. 341–342 (2016)
6. Epstein, L., Levin, A.: Bin packing with general cost structures. Math. Program. **132**(1–2), 355–391 (2012)
7. Gong, X., Shroff, N.: Incentivizing truthful data quality for quality-aware mobile data crowdsourcing. In: MobiHoc (2018)
8. Jansen, K., Porkolab, L.: On preemptiveresource constrained scheduling: polynomial-time approximation schemes. SIAM J. Discrete Math. **20**(3), 545–563 (2006)
9. Kellerer, H., Pferschy, U., Pisinger, D.: Knapsack Problems. Springer, Berlin (2004). http://dx.doi.org/10.1007/978-3-540-24777-7
10. Krishnan, B.K.: A multiscale approximation algorithm for the cardinality constrained knapsack problem. Ph.D. thesis, Massachusetts Institute of Technology (2006)
11. Magazine, M.J., Chern, M.S.: A note on approximation schemes for multidimensional knapsack problems. Math. Oper. Res. **9**(2), 244–247 (1984)
12. Martello, S., Pisinger, D., Toth, P.: Dynamic programming and strong bounds for the 0–1 knapsack problem. Manage. Sci. **45**(3), 414–424 (1999)
13. Mastrolilli, M., Hutter, M.: Hybrid rounding techniques for knapsack problems. Discrete Appl. Math. **154**(4), 640–649 (2006)
14. Megiddo, N.: Linear programming in linear time when the dimension is fixed. J. ACM **31**(1), 114–127 (1984)
15. Megiddo, N., Tamir, A.: Linear time algorithms for some separable quadratic programming problems. Oper. Res. Lett. **13**(4), 203–211 (1993)

16. Nobibon, F.T., Leus, R., Spieksma, F.C.: Optimization models for targeted offers in direct marketing: exact and heuristic algorithms. Eur. J. Oper. Res. **210**(3), 670–683 (2011)
17. Soldo, F., Argyraki, K., Markopoulou, A.: Optimal source-based filtering of malicious traffic. IEEE/ACM Trans. Networking **20**(2), 381–395 (2012)
18. Wenxin, L., Joohyun Lee, N.S.: A faster fptas for knapsack problem with cardinality constraint. arXiv:1902.00919
19. Wu, T., Chen, L., Hui, P., Zhang, C.J., Li, W.: Hear the whole story: towards the diversity of opinion in crowdsourcing markets. Proc. VLDB Endowment **8**(5), 485–496 (2015)

Distributed Algorithms for Matching in Hypergraphs

Oussama Hanguir(✉) and Clifford Stein

Columbia University, New York, NY 10027, USA
{oh2204,cs2035}@columbia.edu

Abstract. We study the d-Uniform Hypergraph Matching (d-UHM) problem: given an n-vertex hypergraph G where every hyperedge is of size d, find a maximum cardinality set of disjoint hyperedges. For $d \geq 3$, the problem of finding the maximum matching is NP-complete, and was one of Karp's 21 NP-complete problems. In this paper we are interested in the problem of finding matchings in hypergraphs in the massively parallel computation (MPC) model that is a common abstraction of MapReduce-style computation. In this model, we present the first three parallel algorithms for d-Uniform Hypergraph Matching, and we analyse them in terms of resources such as memory usage, rounds of communication needed, and approximation ratio. The highlights include:

- A $O(\log n)$-round d-approximation algorithm that uses $O(nd)$ space per machine.
- A 3-round, $O(d^2)$-approximation algorithm that uses $\tilde{O}(\sqrt{nm})$ space per machine.
- A 3-round algorithm that computes a subgraph containing a $(d-1+\frac{1}{d})^2$-approximation, using $\tilde{O}(\sqrt{nm})$ space per machine for linear hypergraphs, and $\tilde{O}(n\sqrt{nm})$ in general.

1 Introduction

As massive graphs become more ubiquitous, the need for scalable parallel and distributed algorithms that solve graph problems grows as well. In recent years, we have seen progress in many graph problems (e.g. spanning trees, connectivity, shortest paths [4,5]) and, most relevant to this work, matchings [21,25]. A natural generalization of matchings in graphs is to matchings in *hypergraphs*. Hypergraph Matching is an important problem with many applications such as capital budgeting, crew scheduling, facility location, scheduling airline flights [49], forming a coalition structure in multi-agent systems [46] and determining the winners in combinatorial auctions [45] (see [51] for a partial survey). Although matching problems in graphs are one of the most well-studied problems in algorithms and optimization, the NP-hard problem of finding a maximum matching in a hypergraph is not as well understood.

C. Stein—Research partially supported by NSF grants CCF-1714818 and CCF-1822809. A full version of the paper is available at http://arxiv.org/abs/2009.09605.

C. Kaklamanis and A. Levin (Eds.): WAOA 2020, LNCS 12806, pp. 30–46, 2021.
https://doi.org/10.1007/978-3-030-80879-2_3

In this work, we are interested in the problem of finding matchings in very large hypergraphs, large enough that we cannot solve the problem on one computer. We develop, analyze and experimentally evaluate three parallel algorithms for hypergraph matchings in the MPC model. Two of the algorithms are generalizations of parallel algorithms for matchings in graphs. The third algorithm develops new machinery which we call a *hyper-edge degree constrained subgraph* (HEDCS), generalizing the notion of an edge-degree constrained subgraph (EDCS). The EDCS has been recently used in parallel and dynamic algorithms for graph matching problems [6,15,16]. We show a range of algorithm trade-offs between approximation ratio, rounds, memory and computation, evaluated both as worst case bounds, and via computational experiments.

More formally, a *hypergraph* G is a pair $G = (V, E)$ where V is the set of vertices and E is the set of hyperedges. A *hyperedge* $e \in E$ is a nonempty subset of the vertices. The cardinality of a hyperedge is the number of vertices it contains. When every hyperedge has the same cardinality d, the hypergraph is said to be *d-uniform*. A hypergraph is *linear* if the intersection of any two hyperedges has at most one vertex. A *hypergraph matching* is a subset of the hyperedges $M \subseteq E$ such that every vertex is covered at most once, i.e. the hyperedges are mutually disjoint. This notion generalizes matchings in graphs. The cardinality of a matching is the number of hyperedges it contains. A matching is called maximum if it has the largest cardinality of all possible matchings, and maximal if it is not contained in any other matching. In the d-Uniform Hypergraph Matching Problem (also referred to as Set Packing or d-Set Packing), a d-uniform hypergraph is given and one needs to find the maximum cardinality matching.

We adopt the most restrictive MapReduce-like model of modern parallel computation among [4,11,27,36], the Massively Parallel Computation (MPC) model of [11]. This model is widely used to solve different graph problems such as matching, vertex cover [2,6,7,25,40], independent set [25,30], as well as many other algorithmic problems. In this model, we have k machines (processors) each with space s. N is the size of the input and our algorithms will satisfy $k \cdot s = \tilde{O}(N)$, which means that the total space in the system is only a polylogarithmic factor more than the input size. The computation proceeds in rounds. At the beginning of each round, the data (e.g. vertices and edges) is distributed across the machines. In each round, a machine performs local computation on its data (of size s), and then sends messages to other machines for the next round. Crucially, the total amount of communication sent or received by a machine is bounded by s, its space. Each machine treats the received messages as the input for the next round. Our model limits the number of machines and the memory per machine to be substantially sublinear in the size of the input. On the other hand, no restrictions are placed on the computational power of any individual machine. The main complexity measure is therefore the memory per machine and the number of rounds R required to solve a problem. For the rest of the paper, $G(V, E)$ is a d-uniform hypergraph with n vertices and m hyperedges, and when the context is clear, we will simply refer to G as a graph and to its hyperedges as edges. $MM(G)$ denotes the maximum matching in G, and $\mu(G) = |MM(G)|$

is the size of that matching. We define $d_G(v)$ to be the degree of a vertex v (the number of hyperedges that v belongs to) in G. When the context is clear, we omit indexing by G and simply denote it $d(v)$.

2 Our Contribution and Results

We design and implement algorithms for the d-UHM in the MPC model. We give three different algorithms, demonstrating different trade-offs between the model's parameters. Due to space constraints, the proofs are are deferred to the full version. Our algorithms are inspired by methods to find maximum matchings in graphs, but require developing significant new tools to address hypergraphs. We are not aware of previous algorithms for hypergraph matching in the MPC model. First we generalize the randomized coreset algorithm of [7] which finds an 3-rounds $O(1)$-approximation for matching in graphs. Our algorithm partitions the graph into random pieces across the machines, and then simply picks a maximum matching of each machine's subgraph. We show that this natural approach results in a $O(d^2)$-approximation.

Theorem 1 (Restated). *There exists an MPC algorithm that with high probability computes a $(3d(d-1)+3+\epsilon)$-approximation for the d-UHM problem in 3 MPC rounds on machines of memory $s = \tilde{O}(\sqrt{nm})$.*

Our second result concerns the MPC model with per-machine memory $O(d{\cdot}n)$. We adapt the sampling technique and post-processing strategy of [40] to construct maximal matchings in hypergraphs, and are able to show that in d-uniform hypergraphs, this technique yields a maximal matching, and thus a d-approximation to the d-UHM problem in $O(\log n)$ rounds.

Theorem 2 (Restated). *There exists an MPC algorithm that given a d-uniform hypergraph $G(V,E)$ with high probability computes a maximal matching in G in $O(\log n)$ MPC rounds on machines of memory $s = \Theta(d \cdot n)$.*

Our third result generalizes the edge degree constrained subgraphs (EDCS), originally introduced by Bernstein and Stein [15] for maintaining large matchings in dynamic graphs, and later used for several other problems including matching and vertex cover in the streaming and MPC model [6,16]. We call these generalized subgraphs hyper-edge degree constrained subgraphs (HEDCS). We show that they exist for specific parameters, and that they contain a good approximation for the d-UHM problem. We prove that an HEDCS of a hypergraph G, with well chosen parameters, contain a fractional matching with a value at least $\frac{d}{d^2-d+1}\mu(G)$, and that the underlying fractional matching is special in the sense that for each hyperedge, it either assigns a value of 1 or a value less than some chosen ϵ. We call such a fractional matching an ϵ-restricted fractional matching. For Theorem 4, we compute an HEDCS of the hypergraph in a distributed fashion. This procedure relies on the robustness properties that we prove for the HEDCSs under sampling. Table 1 summarizes our results.

Theorem 3 (Restated Informally). *Let G be a d-uniform hypergraph and $0 < \epsilon < 1$. There exists an HEDCS that contains an ϵ-restricted fractional matching M_f^H with total value at least $\mu(G)\left(\frac{d}{d^2-d+1} - \epsilon\right)$.*

Theorem 4. *There exists an MPC algorithm that given a d-uniform hypergraph $G(V, E)$, where $|V| = n$ and $|E| = m$, can construct an HEDCS of G in 2 MPC rounds on machines of memory $s = \tilde{O}(n\sqrt{nm})$ in general and $s = \tilde{O}(\sqrt{nm})$ for linear hypergraphs.*

Corollary 4. *There exists an MPC algorithm that with high probability achieves a $d(d-1+1/d)^2$-approximation to the d-Uniform Hypergraph Matching in 3 rounds.*

Table 1. Our parallel algorithms for the d-uniform hypergraph matching problem.

Approximation ratio	Rounds	Memory per machine	Computation per round
$3d(d-1)+3$	3	$\tilde{O}(\sqrt{nm})$	Exponential
d	$O(\log n)$	$O(dn)$	Polynomial
$d(d-1+1/d)^2$	3	$\tilde{O}(n\sqrt{nm})$ in general $\tilde{O}(\sqrt{nm})$ for linear hypergraphs	Polynomial

Experimental Results. We implement our algorithms in a simulated MPC model environment and test them on random and real-world instances. Our experimental results are consistent with the theoretical bounds on most instances, and show a trade-off between the extent to which the algorithms use the power of parallelism and the quality of the approximations. This trade-off is illustrated by comparing the number of rounds and the performance of the algorithms on machines with the same memory size. See Sect. 7 for more details.

Our Techniques. For Theorem 1 and Theorem 4, we use the concept of composable coresets, which has been employed in several distributed optimization models such as the streaming and MapReduce models [1,8–10,33,43]. Informally, the main idea behind this technique is as follows: first partition the data into smaller parts. Then compute a representative solution, referred to as a coreset, from each part. Finally, obtain a solution by solving the optimization problem over the union of coresets. We use a randomized variant of composable coresets, first introduced in [42], where the above idea is applied on a random clustering of the data. This randomized variant has been used after for Graph Matching and Vertex Cover [6,7], as well as Column Subset Selection [17]. Our algorithm for Theorem 1 is similar to previous works, but the analysis requires new techniques for handling hypergraphs. Theorem 2 is a relatively straightforward generalization of the corresponding result for matching in graphs.

The majority of our technical innovation is contained in Theorem 3 and 4. Our general approach is to construct, in parallel, an HEDCS that will contain a good approximation of the maximum matching in the original graph and that will fit on one machine. Then we can run an approximation algorithm on the resultant HEDCS to come up with a good approximation to the maximum matching in this HEDCS and hence in the original graph. In order to make this approach work, we need to generalize much of the known EDCS machinery [15,16] to hypergraphs. This endeavor is quite involved, as almost all the proofs do not generalize easily and, as the results show, the resulting bounds are weaker than those for graphs. We first show that HEDCSs exist, and that they contain large fractional matchings. We then show that they are useful as a coreset, which amounts to showing that even though there can be many different HEDCSs of some fixed hypergraph $G(V, E)$, the degree distributions of every HEDCS (for the same parameters) are almost identical. We show also that HEDCS are robust under edge sampling, in the sense that edge sampling from a HEDCS yields another HEDCS. These properties allow to use HEDCS in our coresets and parallel algorithm in the rest of the paper.

3 Related Work

Hypergraph Matching. The problem of finding a maximum matching in d-uniform hypergraphs is NP-hard for any $d \geq 3$ [37,44], and APX-hard for $d = 3$ [35]. The most natural approach for an approximation algorithm for the hypergraph matching problem is the greedy algorithm: repeatedly add an edge that doesn't intersect any edges already in the matching. This solution is clearly within a factor of d from optimal: from the edges removed in each iteration, the optimal solution can contain at most d edges (at most one for each element of the chosen edge). It is also easy to construct examples showing that this analysis is tight for the greedy approach. All the best known approximation algorithms for the Hypergraph Matching Problem in d-uniform hypergraphs are based on local search methods [13,14,19,28,32]. The first such result by Hurkens and Schrijver [32] gave a $(\frac{d}{2}+\epsilon)$-approximation algorithm. Halldorsson [28] presented a quasi-polynomial $(\frac{d+2}{3})$-approximation algorithm for the unweighted d-UHM. Sviridenko and Ward [50] established a polynomial time $(\frac{d+2}{3})$-approximation algorithm that is the first polynomial time improvement over the $(\frac{d}{2}+\epsilon)$ result from [32]. Cygan [20] and Furer and Yu [24] both provide a $(\frac{d+1}{3}+\epsilon)$ polynomial time approximation algorithm, which is the best approximation guarantee known so far. On the other hand, Hazan, Safra and Schwartz [31] proved that it is hard to approximate d-UHM problem within a factor of $\Omega(d/\log d)$.

Matching in Parallel Computation Models. The study of the graph maximum matching problem in parallel computation models can be traced back to PRAM algorithms of 1980s [3,34,41]. Since then it has been studied in the LOCAL and MPC models, and $(1+\epsilon)$- approximation can be achieved in $O(\log\log n)$ rounds using a space of $O(n)$ [6,21,25]. The question of finding a maximal matching in a small number of rounds has also been considered in [26,40]. Recently, Behnezhad *et al.* [12] presented a $O(\log\log \Delta)$ round algorithm for maximal matching with $O(n)$ memory per machine. While we are not aware of previous work on hypergraph matching in the MPC model, finding a maximal hypergraph matching has been considered in the LOCAL model. Both Fischer *et al.* [22] and Harris [29] provide a deterministic distributed algorithm that computes $O(d)$-approximation to the d-UHM.

4 A 3-Round $O(d^2)$-Approximation

In this section, we generalize the randomized composable coreset algorithm of Assadi and Khanna [7]. They used a maximum matching as a coreset and obtained a $O(1)$-approximation. We use a hypergraph maximum matching and we obtain a $O(d^2)$-approximation. We first define a k-partitioning and then present our greedy approach.

Definition 1 (Random k-partitioning). *Let E be an edge-set of a hypergraph $G(V, E)$. We say that a collection of edges $E^{(1)}, \ldots, E^{(k)}$ is a random k-partition of E if the sets are constructed by assigning each edge $e \in E$ to some $E^{(i)}$ chosen uniformly at random. A random k-partition of E results in partitioning the graph G into k subgraphs $G^{(1)}, \ldots, G^{(k)}$ where $G^{(i)} := G(V, E^{(i)})$ for all $i \in [k]$.*

Let $G(V, E)$ be any d-uniform hypergraph and $G^{(1)}, \ldots, G^{(k)}$ be a random k-partitioning of G. We describe a simple greedy process for combining the maximum matchings of $G^{(i)}$, and prove that this process results in a $O(d^2)$-approximation of the maximum matching of G.

Algorithm 1: Greedy

1 Construct a random k-partitioning of G across the k machines. Let $M^{(0)} := \emptyset$.
2 **for** $i = 1$ *to* k **do**
3 | Set $MM(G^{(i)})$ to be an arbitrary hypergraph maximum matching of $G^{(i)}$.
4 | Let $M^{(i)}$ be a maximal matching obtained by adding to $M^{(i-1)}$ the edges of $MM(G^{(i)})$ that do not violate the matching property.
5 **end**
6 **return** $M := M^{(k)}$.

Theorem 1. *Greedy computes, with high probability, a $(3d(d-1) + 3 + \epsilon)$-approximation for the d-UHM problem in 3 MPC rounds on machines of memory $s = \tilde{O}(\sqrt{nm})$.*

In the rest of the section, we present a proof sketch for Theorem 1. Let $c = \frac{1}{3d(d-1)+3}$, we show that $M^{(k)} \geq c \cdot \mu(G)$ w.h.p, where $M^{(k)}$ is the output of *Greedy*. The randomness stems from the fact that the matchings $M^{(i)}$ (for $i \in \{1, \ldots, k\}$) constructed by *Greedy* are random variables depending on the random k-partitioning. We adapt the general approach in [7] for d-uniform hypergraphs, with $d \geq 3$. Suppose at the beginning of the i-th step of *Greedy*, the matching $M^{(i-1)}$ is of size $o(\mu(G)/d^2)$. One can see that in this case, there is a matching of size $\Omega(\mu(G))$ in G that is entirely incident on vertices of G that are not matched by $M^{(i-1)}$. We can show that in fact $\Omega(\mu(G)/(d^2k))$ edges of this matching are appearing in $G^{(i)}$, even when we condition on the assignment of the edges in the first $(i-1)$ graphs. Next we argue that the existence of these edges forces any maximum matching of $G^{(i)}$ to match $\Omega(\mu(G)/(d^2k))$ edges in $G^{(i)}$ between the vertices that are not matched by $M^{(i-1)}$. These edges can always be added to the matching $M^{(i-1)}$ to form $M^{(i)}$. Therefore, while the maximal matching in *Greedy* is of size $o(\mu(G))$, we can increase its size by $\Omega(\mu(G)/(d^2k))$ edges in each of the first $k/3$ steps, hence obtaining a matching of size $\Omega(\mu(G)/d^2)$ at the end. The following Lemma 1 formalizes this argument.

Lemma 1. *For any $i \in [k/3]$, if $M^{(i-1)} \leq c \cdot \mu(G)$, then, w.p. $1 - O(1/n)$,*

$$M^{(i)} \geq M^{(i-1)} + \frac{1 - 3d(d-1)c - o(1)}{k} \cdot \mu(G) \ .$$

We conclude this section by stating that computing a maximum matching on every machine is only required for the analysis, i.e., to show that there exists a large matching in the union of coresets. In practice, we can use an approximation algorithm to obtain a large matching from the coresets. In our experiments, we use a maximal matching instead of maximum.

5 A $O(\log n)$-Rounds d-Approximation Algorithm

In this section, we show how to compute a maximal matching in $O(\log n)$ MPC rounds, by generalizing the algorithm in [40] to d-uniform hypergraphs. The algorithm provided by Lattanzi *el al.* [40] computes a maximal matching in graphs in $O(\log n)$ if $s = \Theta(n)$, and in at most $\lfloor c/\epsilon \rfloor$ iterations when $s = \Theta(n^{1+\epsilon})$, where $0 < \epsilon < c$ is a fixed constant. Harvey *et al.* [30] show a similar result.

The algorithm first samples $O(s)$ edges and finds a maximal matching M_1 on the resulting subgraph (we will specify a bound on the memory s later). Given this matching, we can safely remove edges that are in conflict (i.e. those incident on nodes in M_1) from the original graph G. If the resulting filtered graph H is small enough to fit onto a single machine, the algorithm augments M_1 with a matching found on H. Otherwise, we augment M_1 with the matching found by recursing on H. Since the size of the graph reduces from round to round, the effective sampling probability increases, resulting in a larger sample of the remaining graph.

Theorem 2. *Given a d-uniform hypergraph $G(V, E)$, Iterated-Sampling computes a maximal matching in G with high probability in $O(\log n)$ MPC rounds on machines of memory $s = \Theta(d \cdot n)$.*

We show that after sampling edges, the size of the graph $G[I]$ induced by unmatched vertices decreases exponentially with high probability, hence the algorithm terminates in $O(\log n)$ rounds.

Algorithm 2: Iterated-Sampling

1 Set $M := \emptyset$ and $\mathcal{S} = E$.
2 Sample every edge $e \in \mathcal{S}$ uniformly with probability $p = \frac{s}{5|\mathcal{S}|d}$ to form E'.
3 If $|E'| > s$ the algorithm fails. Otherwise give the graph $G(V, E')$ as input to a single machine and compute a maximal matching M' on it. Set $M = M \cup M'$.
4 Let I be the unmatched vertices in G and $G[I]$ the induced subgraph with edges $E[I]$. If $E[I] > s$, set $\mathcal{S} := E[I]$ and return to step 2.
5 Compute a maximal matching M'' on $G[I]$ and output $M = M \cup M''$.

6 A 3-Round $O(d^3)$-Approximation Using HEDCSs

In graphs, edge degree constrained subgraphs (EDCS) [15,16] have been used as a local condition for identifying large matchings, and leading to good dynamic and parallel algorithms [6,15,16]. These papers showed that an EDCS exists, that it contains a large matching, that it is robust to sampling and composition, and that it can be used as a coreset. In this section, we present a generalization of EDCS for hypergraphs. We prove that an HEDCS exists, that it contains a large matching, that it is robust to sampling and composition, and that it can be used as a coreset. The proofs and algorithms, however, require significant developments beyond the graph case. We first present definitions of a fractional matching and HEDCS.

Definition 2. *[Fractional matching]*
In a hypergraph $G(V, E)$, a fractional matching is a mapping from $y : E \mapsto [0, 1]$ such that for every edge e we have $0 \leq y_e \leq 1$ and for every vertex v : $\sum_{e \in v} y_e \leq 1$. The value of such fractional matching is equal to $\sum_{e \in E} y_e$. For $\epsilon > 0$, a fractional matching is an ϵ-restricted fractional matching if for every edge e: $y_e = 1$ or $y_e \in [0, \epsilon]$.

Definition 3. *For any hypergraph $G(V, E)$ and integers $\beta \geq \beta^- \geq 0$ a hyperedge degree constraint subgraph (HEDCS)(H, β, β^-) is a subgraph $H := (V, E_H)$ of G satisfying:*

- *(P1): For any hyperedge $e \in E_H$: $\sum_{v \in e} d_H(v) \leq \beta$.*
- *(P2): For any hyperedge $e \notin E_H$: $\sum_{v \in e} d_H(v) \geq \beta^-$.*

We show via a constructive proof that a hypergraph contains an HEDCS when the parameters of this HEDCS satisfy the inequality $\beta - \beta^- \geq d - 1$.

Lemma 2. *Any hypergraph G contains an HEDCS(G, β, β^-) for any parameters $\beta - \beta^- \geq d - 1$.*

We prove that an HEDCS of a graph contains a large ϵ-restricted fractional matching that approximates the maximum matching by a factor less than d.

Theorem 3. *Let G be a d-uniform hypergraph and $0 \leq \epsilon < 1$. Let H be a hyperedge degree constrained subgraph$(G, \beta, \beta(1 - \lambda))$ with $\lambda = \frac{\epsilon}{6}$ and $\beta \geq \frac{8d^2}{d-1}$.*

λ^{-3}. *Then* H *contains an* ϵ*-restricted fractional matching* M_f^H *with total value at least* $\mu(G)(\frac{d}{d^2-d+1} - \frac{\epsilon}{d-1})$.

6.1 Sampling and Constructing an HEDCS in the MPC Model

Our results in this section are general and applicable to every computation model. We prove structural properties about the HEDCS that will help us construct HEDCS in the MPC model. We show that the degree distributions of every HEDCS (for the same parameters β and λ) are almost identical. In other words, the degree of any vertex v is almost the same in every HEDCS of the same hypergraph G. We show also that HEDCS are robust under edge sampling, i.e. that edge sampling from and HEDCS yields another HEDCS, and that the degree distributions of any two HEDCS for two different edge sampled subgraphs of G is almost the same no matter how the two HEDCS are selected. In the following lemma, we argue that any two HEDCS of a graph G (for the same parameters β, β^-) are "somehow identical" in that their degree distributions are close to each other. In the rest of this section, we fix the parameters β, β^- and two subgraphs A and B that are both $HEDCS(G, \beta, \beta^-)$.

Lemma 3. *(Degree Distribution Lemma). Fix a d-uniform hypergraph* $G(V, E)$ *and parameters* β, $\beta^- = (1-\lambda)\cdot\beta$ *(for* λ *small enough). For any two subgraphs A and B that are* $HEDCS(G, \beta, \beta^-)$*, and any vertex* $v \in V$*, then* $|d_A(v) - d_B(v)| = O(\sqrt{n})\lambda^{1/2}\beta$.

Corollary 1. *For d-uniform linear hypergraphs, the degree distribution is tighter, and* $|d_A(v) - d_B(v)| = O(\log n)\lambda\beta$.

Next we prove two lemmas regarding the structure of different HEDCS across sampled subgraphs. The first lemma shows that edge sampling an HEDCS results in another HEDCS for the sampled subgraph. The second lemma shows that the degree distributions of any two HEDCS for two different edge sampled subgraphs of G is almost the same.

Lemma 4. *Let* H *be a HEDCS*(G, β_H, β_H^-) *for parameters* $\beta_H := (1 - \frac{\lambda}{\alpha}) \cdot \frac{\beta}{p}$ *and* $\beta_H^- := \beta_H - (d-1)$*, and* $\beta \geq 15d(\alpha d)^2 \cdot \lambda^{-2} \cdot \log n$ *such that* $p < 1 - \frac{2}{\alpha}$*. Suppose* $G_p := G_p^E(V, E_p)$ *is an edge sampled subgraph of* G *and* $H_p := H \cap G_p$*; then, with high probability:*

1. *For any vertex* $v \in V$: $|d_{H_p}(v) - p \cdot d_H(v)| \leq \frac{\lambda}{\alpha d}\beta$
2. H_p *is a HEDCS of* G_p *with parameters* $(\beta, (1-\lambda)\cdot\beta)$.

Lemma 5. *(HEDCS in Edge Sampled Subgraph). Fix any hypergraph* $G(V, E)$ *and* $p \in (0, 1)$*. Let* G_1 *and* G_2 *be two edge sampled subgraphs of* G *with probability* p *(chosen not necessarily independently). Let* H_1 *and* H_2 *be arbitrary HEDCSs of* G_1 *and* G_2 *with parameters* $(\beta, (1-\lambda)\cdot\beta)$*. Suppose* $\beta \geq 15d(\alpha d)^2 \cdot \lambda^{-2} \cdot \log n$*, then, with probability* $1 - \frac{4}{n^{5d-1}}$*, simultaneously for all* $v \in V$: $|d_{H_1}(v) - d_{H_2}(v)| = O(n^{1/2})\lambda^{1/2}\beta$.

Corollary 2. *If G is linear, then* $|d_{H_1}(v) - d_{H_2}(v)| = O(\log n)\lambda\beta$.

We are now ready to present a parallel algorithm that will use the HEDCS subgraph. We first compute an HEDCS in parallel via edge sampling. Let $G^{(1)}, \ldots, G^{(k)}$ be a random k-partition of a graph G. We show that if we compute an arbitrary HEDCS of each graph $G^{(i)}$ and combine them together, we obtain a HEDCS for the original graph G. We then store this HEDCS in one machine and compute a maximal matching on it . We present our algorithm for all range of memory $s = n^{\Omega(1)}$. Lemma 6 and Corollary 3 serve as a proof to Theorem 4.

Algorithm 3: HEDCS-Matching

---

1 Define $k := \frac{m}{s \log n}$, $\lambda := \frac{1}{2n \log n}$ and $\beta := 500 \cdot d^3 \cdot n^2 \cdot \log^3 n$.
2 $G^{(1)}, \ldots, G^{(k)}$:= random k-partition of G.
3 **for** $i = 1$ *to* k, *in parallel* **do**
4 | Compute $C^{(i)} = HEDCS(G^{(i)}, \beta, (1 - \lambda) \cdot \beta)$ on machine i.
5 **end**
6 Define the multi-graph $C(V, E_C)$ with $E_C := \cup_{i=1}^{k} C^{(i)}$.
7 Compute and output a maximal matching on C.

Lemma 6. *Suppose $k \leq \sqrt{m}$. Then with high probability*

1. *The subgraph C is a $HEDCS(G, \beta_C, \beta_C^-)$ for parameters: $\lambda_C = O(n^{1/2})\lambda^{1/2}$, $\beta_C = (1 + d \cdot \lambda_C) \cdot k \cdot \beta$ and $\beta_C^- = (1 - \lambda - d \cdot \lambda_C) \cdot k \cdot \beta$.*
2. *The total number of edges in each subgraph $G^{(i)}$ of G is $\tilde{O}(s)$.*
3. *If $s = \tilde{O}(n\sqrt{nm})$, then the graph C can fit in the memory of one machine.*

Corollary 3. *If G is linear, then by choosing $\lambda := \frac{1}{2 \log^2 n}$ and $\beta := 500 \cdot d^3 \cdot \log^4 n$ in Algorithm 3 we have:*

1. *With high probability, the subgraph C is a $HEDCS(G, \beta_C, \beta_C^-)$ for parameters: $\lambda_C = O(\log n)\lambda$, $\beta_C = (1 + d \cdot \lambda_C) \cdot k \cdot \beta$ and $\beta_C^- = (1 - \lambda - d \cdot \lambda_C) \cdot k \cdot \beta$.*
2. *If $s = \tilde{O}(\sqrt{nm})$ then C can fit on the memory of one machine.*

Theorem 4. *HEDCS-Matching constructs a HEDCS of G in 2 MPC rounds on machines of memory $s = \tilde{O}(n\sqrt{nm})$ in general and $s = \tilde{O}(\sqrt{nm})$ for linear hypergraphs.*

Corollary 4. *HEDCS-Matching achieves a $d(d - 1 + 1/d)^2$-approximation to the d-Uniform Hypergraph Matching in 3 rounds with high probability.*

Proof of Corollary 4. We show that with high probability, C verifies the assumptions of Theorem 3. From Lemma 6, we get that with high probability, the subgraph C is a $HEDCS(G, \beta_C, \beta_C^-)$ for parameters: $\lambda_C = O(n^{1/2})\lambda^{1/2}$, $\beta_C = (1 + d \cdot \lambda_C) \cdot k \cdot \beta$ and $\beta_C^- = (1 - \lambda - d \cdot \lambda_C) \cdot k \cdot \beta$. We can see that $\beta_c \geq \frac{8d^2}{d-1} \cdot \lambda_c^{-3}$.

Therefore by Theorem 3, C contains a $(d-1+\frac{1}{d})$-approximate ϵ-restricted matching. Since the integrality gap of the d-UHM is at most $d-1+\frac{1}{d}$ (see [18] for details), then C contains a $(d-1+\frac{1}{d})^2$-approximate matching. Taking a maximal matching in C multiplies the approximation factor by at most d. Therefore, any maximal matching in C is a $d(d-1+\frac{1}{d})^2$-approximation. □

7 Computational Experiments

We conduct a wide variety of experiments on both random and real-life data. We implement the three algorithms Greedy, Iterated-Sampling and HEDCS-Matching using Python, and more specifically relying on the class *pygraph.hypergraph* to construct and perform operations on hypergraphs. We simulate the MPC model by computing a k-partitioning and splitting it into k different inputs. Parallel computations on different parts of the k-partitioning are handled through the use of the *multiprocessing* library in Python. We compute the optimal matching through an Integer Program for small instances of random hypergraphs ($d \leq 10$) as well as geometric hypergraphs. The datasets differ in their number of vertices, hyperedges, vertex degree and hyperedge cardinality. In the following tables, n and m denote the number of vertices and number of hyperedges respectively, and d is the size of hyperedges. k is the number of machines used to distribute the hypergraph initially. We limit the number of edges that a machine can store to $\frac{2m}{k}$. The columns Gr, IS and HEDCS contain the average ratio between the size of the matching computed by the algorithms Greedy, Iterated-Sampling and HEDCS-Matching respectively, and the size of a benchmark. This ratio is computed by the percentage $\frac{ALG}{BENCHMARK}$, where ALG is the output of the algorithm, and $BENCHMARK$ denotes the size of the benchmark. The benchmarks include the size of optimal solution when it is possible to compute, or the size of a maximal matching computed sequentially. $\#I$ denotes the number of generated instances for fixed n, m and d. β and β^- are the parameters used to construct the HEDCS subgraphs in HEDCS-Matching. These subgraphs are constructed using the procedure in the proof of Lemma 2.

7.1 Experiments with Random Uniform Hypergraphs

We present the results experiments on random uniform hypergraphs. The full version contains results on random geometric hypergraphs as well. For a fixed n, m and d, each potential hyperedge is sampled independently and uniformly at random from the set of vertices. In Table 2, we use the size of a perfect matching $\frac{n}{d}$ as a benchmark, because a perfect matching in random graphs exists with probability $1-o(1)$ under some conditions on m, n and d. If $d(n,m) = \frac{m \cdot d}{n}$ is the expected degree of a random uniform hypergraph, Frieze and Janson [23] showed that $\frac{d(n,m)}{n^{1/3}} \to \infty$ is a sufficient condition for the hypergraph to have a perfect matching with high probability. Kim [38] further weakened this condition to $\frac{d(n,m)}{n^{1/(5+2/(d-1))}} \to \infty$. We empirically verify, by solving the IP, that for $d = 3, 5$ and for small instances of $d = 10$, our random graphs contain a perfect matching. In terms of solution quality HEDCS-Matching performs consistently better than Greedy, and Iterated-Sampling performs significantly better than the other two. The performance of the algorithms decreases as d grows, which is theoretically expected since their approximations ratio are both proportional to d. The number of rounds for Iterated-Sampling grows slowly with n, which is consistent with $O(\log n)$ bound. The number of rounds of Iterated-Sampling still grows with n, confirming the theoretical bound.

Table 2. Comparison on random instances with perfect matching benchmark.

n	m	d	k	#I	Gr	IS	HEDCS	β	β^-	Rounds IS
50	800	5	6	500	66.0%	76.2%	67.0%	16	11	4.89
100	2,800		10		68.0%	79.6%	69.8%	16	11	4.74
300	4,000		10		62.2%	75.1%	65.5%	10	5	6.62
500	8,000		16		63.3%	76.4%	65.6%	10	5	7.62
500	15,000	10	16	500	44.9%	58.3%	53.9%	20	10	6.69
1,000	50,000		20		47.3%	61.3%	50.5%	20	10	8.25
2,500	100,000		20		45.6%	59.9%	48.2%	20	10	8.11
5,000	200,000		20		45.0%	59.7%	47.8%	20	10	7.89
1,000	50,000	25	25	100	27.5%	34.9%	30.8%	75	50	8.10
2,500	100,000		25		26.9%	34.0%	27.0%	75	50	8.26
5,000	250,000		30		26.7%	33.8%	28.8%	75	50	8.23
10,000	500,000		30		26.6%	34.1%	28.2%	75	50	8.46

7.2 Experiments with Real Data

PubMed and Cora Datasets. We employ two citation network datasets, the Cora and Pubmed datasets [47,52]. These datasets are represented with a graph, with vertices being publications, and edges being citation links from one article to another. We construct the hypergraphs in two steps 1) each article is a vertex; 2) each article is taken as a centroid and forms a hyperedge to connect those articles which have citation links to it (either citing it or being cited). The Cora hypergraph has an average edge size of 3.0 ± 1.1, while the average in the Pubmed hypergraph is 4.3 ± 5.7. The number of edges in both is significantly smaller than the number of vertices, therefore we allow each machine to store only $\frac{m}{k} + \frac{1}{4}\frac{m}{k}$ edges. We randomly split the edges on each machine, and compute the optimal matchings on each machine, as well as on the whole hypergraphs. We perform ten runs of each algorithm with different random k-partitioning and take the maximum cardinality obtained. Table 3 shows that none of the algorithms is able to retrieve the optimal matching. This behaviour can be explained by the loss of information that using parallel machines implies. We see that Iterated-Sampling, like in previous experiments, outperforms the other algorithms due to highly sequential design. HEDCS-Matching particularly performs worse than the other algorithms, mainly because it fails to construct sufficiently large HEDCSs.

Social Network Communities. We include two larger real-world datasets, orkut-groups and LiveJournal, from the Koblenz Network Collection [39]. We use two hypergraphs that were constructed from these datasets by Shun [48]. Vertices represent individual users, and hyperedges represent communities. We use the size of a maximal matching as a benchmark. Table 3 shows that Iterated-Sampling still provides the best approximation. HEDCS-Sampling performs worse than Greedy on Livejournal, mainly because the ratio $\frac{m}{n}$ is not big enough to construct an HEDCS with a large matching (Table 3).

Table 3. Comparison on co-citation and social network hypergraphs.

Name	n	m	k	Gr	IS	HEDCS	Rounds IS
Cora	2,708	1,579	2	75.0%	83.2%	63.9%	6
PubMed	19,717	7,963	3	72.0%	86.6%	62.4%	9
Orkut	2,32 $\times 10^6$	1,53 $\times 10^7$	10	55.6%	66.1%	58.1%	11
Livejournal	3,20 $\times 10^6$	7,49 $\times 10^6$	3	44.0%	55.2%	43.3%	10

7.3 Experimental Conclusions

In the majority of our experiments, Iterated-Sampling provides the best approximation to the d-UHM problem, which is consistent with its theoretical superiority. On random graphs, HEDCS-Matching performs consistently better than Greedy. We suspect it is because the $O(d^3)$-approximation bound on HEDCS-Matching is loose. We conjecture that rounding an $\epsilon-$restricted matching can be done efficiently, which would improve the approximation ratio. The performance of the three algorithms decreases as d grows. The number of rounds of Iterated-Sampling is also consistent with the theoretical bound. However, due to its sequential design, and by centralizing the computation on one single machine while using the other machines simply to coordinate, Iterated-Sampling not only takes more rounds than the other two algorithms, but is also slower when we account for the absolute runtime as well as the runtime per round. In the full version, we compare the runtimes of our three algorithms on a set of random uniform hypergraphs. We show that the absolute and per round run-times of Iterated-Sampling grow considerably faster with the size of the hypergraphs. We also see that HEDCS-Sampling is slower than Greedy, since the former performs heavier computation on each machine.

8 Conclusion

We have presented the first algorithms for the d-UHM problem in the MPC model. Our theoretical and experimental results highlight the trade-off between the approximation ratio, the necessary memory per machine and the number of rounds it takes to run the algorithms. We have also introduced the notion of

HEDCS subgraphs, and have shown that an HEDCS contains a good approximation for the maximum matching and that they can be constructed in few rounds in the MPC model. We believe better approximation algorithms should be possible, especially if we can give better rounding algorithms for a ϵ-restricted fractional hypergraph matching. For future work, it would be interesting to explore whether we can achieve better-than-d approximation in the MPC model in a polylogarithmic number of rounds.

References

1. Abbar, S., Amer-Yahia, S., Indyk, P., Mahabadi, S., Varadarajan, K.R.: Diverse near neighbor problem. In: Proceedings of the Twenty-Ninth Annual Symposium on Computational Geometry, pp. 207–214. ACM (2013)
2. Ahn, K.J., Guha, S.: Access to data and number of iterations: dual primal algorithms for maximum matching under resource constraints. ACM Trans. Parallel Comput. (TOPC) **4**(4), 17 (2018)
3. Alon, N., Babai, L., Itai, A.: A fast and simple randomized parallel algorithm for the maximal independent set problem. J. Algor. **7**(4), 567–583 (1986)
4. Andoni, A., Nikolov, A., Onak, K., Yaroslavtsev, G.: Parallel algorithms for geometric graph problems. In: Proceedings of the Forty-Sixth Annual ACM Symposium on Theory of Computing, pp. 574–583. ACM (2014)
5. Andoni, A., Song, Z., Stein, C., Wang, Z., Zhong, P.: Parallel graph connectivity in log diameter rounds. In: 2018 IEEE 59th Annual Symposium on Foundations of Computer Science (FOCS), pp. 674–685. IEEE (2018)
6. Assadi, S., Bateni, M., Bernstein, A., Mirrokni, V., Stein, C.: Coresets meet edcs: algorithms for matching and vertex cover on massive graphs. In: Proceedings of the Thirtieth Annual ACM-SIAM Symposium on Discrete Algorithms, pp. 1616–1635. SIAM (2019)
7. Assadi, S., Khanna, S.: Randomized composable coresets for matching and vertex cover. In: Proceedings of the 29th ACM Symposium on Parallelism in Algorithms and Architectures, pp. 3–12 (2017)
8. Badanidiyuru, A., Mirzasoleiman, B., Karbasi, A., Krause, A.: Streaming submodular maximization: Massive data summarization on the fly. In: Proceedings of the 20th ACM SIGKDD International Conference on Knowledge Discovery and Data Mining, pp. 671–680. ACM (2014)
9. Balcan, M.F.F., Ehrlich, S., Liang, Y.: Distributed k-means and k-median clustering on general topologies. In: Advances in Neural Information Processing Systems, pp. 1995–2003 (2013)
10. Bateni, M., Bhashkara, A., Lattanzi, S., Mirrokni, V.: Mapping core-sets for balanced clustering. Adv. Neural Inf. Process. Syst. **26**, 5–8 (2014)
11. Beame, P., Koutris, P., Suciu, D.: Communication steps for parallel query processing. In: Proceedings of the 32nd ACM SIGMOD-SIGACT-SIGAI Symposium on Principles of Database Systems, pp. 273–284. ACM (2013)
12. Behnezhad, S., Hajiaghayi, M., Harris, D.G.: Exponentially faster massively parallel maximal matching. In: 2019 IEEE 60th Annual Symposium on Foundations of Computer Science (FOCS), pp. 1637–1649. IEEE (2019)
13. Berman, P.: A d/2 approximation for maximum weight independent set in d-claw free graphs. In: SWAT 2000. LNCS, vol. 1851, pp. 214–219. Springer, Heidelberg (2000). https://doi.org/10.1007/3-540-44985-X_19

14. Berman, P., Krysta, P.: Optimizing misdirection. In: Proceedings of the Fourteenth Annual ACM-SIAM Symposium on Discrete Algorithms, pp. 192–201. Society for Industrial and Applied Mathematics (2003)
15. Bernstein, A., Stein, C.: Fully dynamic matching in bipartite graphs. In: Halldórsson, M.M., Iwama, K., Kobayashi, N., Speckmann, B. (eds.) ICALP 2015. LNCS, vol. 9134, pp. 167–179. Springer, Heidelberg (2015). https://doi.org/10.1007/978-3-662-47672-7_14
16. Bernstein, A., Stein, C.: Faster fully dynamic matchings with small approximation ratios. In: Proceedings of the Twenty-Seventh Annual ACM-SIAM Symposium on Discrete Algorithms, pp. 692–711. Society for Industrial and Applied Mathematics (2016)
17. Bhaskara, A., Rostamizadeh, A., Altschuler, J., Zadimoghaddam, M., Fu, T., Mirrokni, V.: Greedy column subset selection: new bounds and distributed algorithms (2016)
18. Chan, Y.H., Lau, L.C.: On linear and semidefinite programming relaxations for hypergraph matching. Math. Program. **135**(1–2), 123–148 (2012)
19. Chandra, B., Halldórsson, M.M.: Greedy local improvement and weighted set packing approximation. J. Algor. **39**(2), 223–240 (2001)
20. Cygan, M.: Improved approximation for 3-dimensional matching via bounded pathwidth local search. In: 2013 IEEE 54th Annual Symposium on Foundations of Computer Science (FOCS), pp. 509–518. IEEE (2013)
21. Czumaj, A., Łacki, J., Madry, A., Mitrovic, S., Onak, K., Sankowski, P.: Round compression for parallel matching algorithms. SIAM J. Comput. (0), STOC18-1 (2019)
22. Fischer, M., Ghaffari, M., Kuhn, F.: Deterministic distributed edge-coloring via hypergraph maximal matching. In: 2017 IEEE 58th Annual Symposium on Foundations of Computer Science (FOCS), pp. 180–191. IEEE (2017)
23. Frieze, A., Janson, S.: Perfect matchings in random s-uniform hypergraphs. Random Struct. Algor. **7**(1), 41–57 (1995)
24. Furer, M., Yu, H.: Approximate the k-set packing problem by local improvements (2013). arXiv preprint arXiv:1307.2262
25. Ghaffari, M., Gouleakis, T., Konrad, C., Mitrović, S., Rubinfeld, R.: Improved massively parallel computation algorithms for mis, matching, and vertex cover. In: Proceedings of the 2018 ACM Symposium on Principles of Distributed Computing, pp. 129–138 (2018)
26. Ghaffari, M., Uitto, J.: Sparsifying distributed algorithms with ramifications in massively parallel computation and centralized local computation. In: Proceedings of the Thirtieth Annual ACM-SIAM Symposium on Discrete Algorithms, pp. 1636–1653. SIAM (2019)
27. Goodrich, M.T., Sitchinava, N., Zhang, Q.: Sorting, searching, and simulation in the mapreduce framework. In: Asano, T., Nakano, S., Okamoto, Y., Watanabe, O. (eds.) ISAAC 2011. LNCS, vol. 7074, pp. 374–383. Springer, Heidelberg (2011). https://doi.org/10.1007/978-3-642-25591-5_39
28. Halldórsson, M.M.: Approximating discrete collections via local improvements. SODA **95**, 160–169 (1995)
29. Harris, D.G.: Distributed approximation algorithms for maximum matching in graphs and hypergraphs (2018). arXiv preprint arXiv:1807.07645
30. Harvey, N.J., Liaw, C., Liu, P.: Greedy and local ratio algorithms in the mapreduce model. In: Proceedings of the 30th on Symposium on Parallelism in Algorithms and Architectures, pp. 43–52 (2018)

31. Hazan, E., Safra, S., Schwartz, O.: On the complexity of approximating k-set packing. Comput. Compl. **15**(1), 20–39 (2006)
32. Hurkens, C.A.J., Schrijver, A.: On the size of systems of sets every t of which have an sdr, with an application to the worst-case ratio of heuristics for packing problems. SIAM J. Disc. Math. **2**(1), 68–72 (1989)
33. Indyk, P., Mahabadi, S., Mahdian, M., Mirrokni, V.S.: Composable core-sets for diversity and coverage maximization. In: Proceedings of the 33rd ACM SIGMOD-SIGACT-SIGART Symposium on Principles of Database Systems, pp. 100–108. ACM (2014)
34. Israeli, A., Itai, A.: A fast and simple randomized parallel algorithm for maximal matching. Inf. Process. Lett. **22**(2), 77–80 (1986)
35. Kann, V.: Maximum bounded 3-dimensional matching in max snp-complete. Inf. Process. Lett. **37**(1), 27–35 (1991)
36. Karloff, H., Suri, S., Vassilvitskii, S.: A model of computation for mapreduce. In: Proceedings of the Twenty-First Annual ACM-SIAM Symposium on Discrete Algorithms, pp. 938–948. SIAM (2010)
37. Karp, R.M.: Reducibility among combinatorial problems. In: Complexity of Computer Computations, pp. 85–103. Springer, Heidelberg (1972). https://doi.org/10.1007/978-1-4684-2001-2_9
38. Kim, J.H.: Perfect matchings in random uniform hypergraphs. Random Struct. Algor. **23**(2), 111–132 (2003)
39. Kunegis, J.: Konect: the koblenz network collection. In: Proceedings of the 22nd International Conference on World Wide Web, pp. 1343–1350 (2013)
40. Lattanzi, S., Moseley, B., Suri, S., Vassilvitskii, S.: Filtering: a method for solving graph problems in mapreduce. In: Proceedings of the Twenty-Third Annual ACM Symposium on Parallelism in Algorithms and Architectures, pp. 85–94. ACM (2011)
41. Luby, M.: A simple parallel algorithm for the maximal independent set problem. SIAM J. Comput. **15**(4), 1036–1053 (1986)
42. Mirrokni, V., Zadimoghaddam, M.: Randomized composable core-sets for distributed submodular maximization. In: Proceedings of the Forty-Seventh Annual ACM Symposium on Theory of Computing, pp. 153–162. ACM (2015)
43. Mirzasoleiman, B., Karbasi, A., Sarkar, R., Krause, A.: Distributed submodular maximization: Identifying representative elements in massive data. In: Advances in Neural Information Processing Systems, pp. 2049–2057 (2013)
44. Papadimitriou, C.H.: Computational Complexity. John Wiley and Sons Ltd., Hoboken (2003)
45. Sandholm, T.: Algorithm for optimal winner determination in combinatorial auctions. Artif Intelligence **135**(1–2), 1–54 (2002)
46. Sandholm, T., Larson, K., Andersson, M., Shehory, O., Tohmé, F.: Coalition structure generation with worst case guarantees. Artif. Intell. **111**(1–2), 209–238 (1999)
47. Sen, P., Namata, G., Bilgic, M., Getoor, L., Galligher, B., Eliassi-Rad, T.: Collective classification in network data. AI Mag. **29**(3), 93–93 (2008)
48. Shun, J.: Practical parallel hypergraph algorithms. In: Proceedings of the 25th ACM SIGPLAN Symposium on Principles and Practice of Parallel Programming, pp. 232–249 (2020)
49. Skiena, S.S.: The Algorithm Design Manual, vol. 1. Springer, Heidelberg (1998). https://doi.org/10.1007/978-3-030-54256-6

50. Sviridenko, M., Ward, J.: Large neighborhood local search for the maximum set packing problem. In: Fomin, F.V., Freivalds, R., Kwiatkowska, M., Peleg, D. (eds.) ICALP 2013. LNCS, vol. 7965, pp. 792–803. Springer, Heidelberg (2013). https://doi.org/10.1007/978-3-642-39206-1_67
51. Vemuganti, R.: Applications of set covering, set packing and set partitioning models: A survey. In: Handbook of Combinatorial Optimization, pp. 573–746. Springer, Heidelberg(1998). https://doi.org/10.1007/978-1-4613-0303-9_9
52. Yang, J., Leskovec, J.: Defining and evaluating network communities based on ground-truth. Knowl. Inf. Syst. **42**(1), 181–213 (2015)

Online Coloring and a New Type of Adversary for Online Graph Problems

Yaqiao Li[1], Vishnu V. Narayan[2(✉)], and Denis Pankratov[3]

[1] Université de Montréal, Montreal, Canada
yaqiao.li@umontreal.ca
[2] McGill University, Montreal, Canada
vishnu.narayan@mail.mcgill.ca
[3] Concordia University, Montreal, Canada
denis.pankratov@concordia.ca

Abstract. We introduce a new type of adversary for online graph problems. The new adversary is parameterized by a single integer κ, which upper bounds the number of connected components that the adversary can use at any time during the presentation of the online graph G. We call this adversary "κ components bounded", or κ-CB for short. On one hand, this adversary is restricted compared to the classical adversary because of the κ-CB constraint. On the other hand, we seek competitive ratios parameterized *only* by κ with no dependence on the input length n, thereby giving the new adversary power to use arbitrarily large inputs.

We study online coloring under the κ-CB adversary. We obtain finer analysis of the existing algorithms *FirstFit* and *CBIP* by computing their competitive ratios on trees and bipartite graphs under the new adversary. Surprisingly, *FirstFit* outperforms *CBIP* on trees. When it comes to bipartite graphs *FirstFit* is no longer competitive under the new adversary, while *CBIP* uses at most 2κ colors. We also study several well known classes of graphs, such as 3-colorable, C_k-free, d-inductive, planar, and bounded-treewidth, with respect to online coloring under the κ-CB adversary. We demonstrate that the extra adversarial power of unbounded input length outweighs the restriction on the number of connected components leading to non-existence of competitive algorithms for these classes.

Keywords: Online algorithms · Online coloring · Parameterized analysis · Connected components · First fit · CBIP

1 Introduction

In online graph problems the input graph is not known in advance, but is rather revealed one item at a time. In this paper we are concerned with the so-called

Y. Li—Acknowledges postdoctoral support via NSERC discovery grant RGPIN-04500.
D. Pankratov—Research is supported by Natural Sciences and Engineering Research Council of Canada (NSERC).

C. Kaklamanis and A. Levin (Eds.): WAOA 2020, LNCS 12806, pp. 47–62, 2021.
https://doi.org/10.1007/978-3-030-80879-2_4

vertex arrival model, where the graph is revealed one vertex at a time. When a new vertex is revealed, an online algorithm learns the identity of the vertex as well as its neighborhood restricted to the already revealed vertices. Note that the algorithm gets no information about future vertices. Many graph problems do not admit any non-trivial online algorithms in the adversarial vertex-arrival model. Be that as it may, online graph problems often arise in real life applications in computer networks, public transit networks, electrical grids, and so on. Recently, interest in online and "online-like"[1] graph models and algorithms has increased, sparked by the proliferation of online social networks. Thus, it is necessary to introduce various restrictions of the basic adversarial model that allow nontrivial algorithms while capturing interesting real-life scenarios.

One obtains a plethora of restricted adversaries simply by insisting that the adversary generates a graph belonging to a particular family of graphs, such as χ-colorable, planar, d-inductive, etc. Another way to relax the classical adversarial model is to consider distributions on graphs and perform average-case analysis. One of the most studied distributions is, of course, the Erdös-Rényi random graph. While it is mathematically appealing, real life graphs rarely follow this distribution. For example, one of the early empirical observations was that distributions on degrees of nodes in real social networks are most accurately modeled by power-law distributions [12], whereas the Erdös-Rényi model induces a binomial distribution on degrees of nodes. Thus, new models of random graphs have been introduced in an attempt to approximate power-law distributions on degrees. Many of these new generative models are inherently offline. A notable exception is the preferential attachment model [2], which perfectly fits within the vertex arrival model. The formal definition is technical, but at a high level this model works as follows. When a new vertex v arrives its neighborhood is generated by connecting v to an already existing vertex u with probability proportional to the current degree of u. This model has a natural motivation: consider a person signing up for some social network, where people can "follow" each other. This person is signing up not because they want to be left alone (i.e. form a new connected component), but because they already have a list of people in mind who they will follow (i.e., join existing connected component(s), potentially merging some components together). It is more likely that the person is going to follow well known people, e.g. celebrities, who in turn have some of the highest numbers of followers in the network. This is akin to a new node in vertex arrival model likely being connected to existing nodes of high degree.

The starting point of our work is the observation that when a social network graph is generated via the preferential attachment process, there are very few connected components in the online graph at any point in time. Formalizing this observation in the adversarial setting, we investigate a new type of adversary that is restricted to using at most κ connected components at any point in

[1] Some examples of "online-like" models of computation are dynamic graph algorithms, temporal graph algorithms, streaming graph algorithms, and priority graph algorithms.

time during the generation of the online input graph. We call such adversary κ components bounded, or κ-CB for short.

In this paper we focus on the online coloring problem under the κ-CB adversary. Indeed, another motivation for considering the κ-CB adversary is to extend our understanding of lower bound techniques for online coloring. Most of the past research uses the following methodology: the adversary creates a collection of disjoint components with some properties, then the adversary merges some of these components by creating a vertex appropriately connected to the components. The aim of this technique is to allow the adversary to observe the coloring of each component chosen by the algorithm, and then choose a "correct" coloring of the components that differs from the one chosen by the algorithm. The adversary then connects the components together, forcing the algorithm to use extra colors (since the algorithm's coloring is incorrect inside at least one component). By iterating this process, the adversary tries to force the online algorithm to perform badly. Some variant of this technique has been used, for example, in [1,4,6,7,14]. A notable exception is [8], where this create-and-merge components technique is not directly involved. Usually, this type of construction involves a large number of disjoint components, typically logarithmic in the number of vertices – see, for example, [6,7]. Our goal is to formally analyze the power of this technique and the extent of dependence of existing lower bounds on this technique. Specifically, we ask, what happens if the adversary in the online coloring problem is κ-CB? In this work we investigate this question, while allowing the adversary to use an unlimited number of vertices to compensate for a limited number of components.

Our first set of results gives a finer understanding of the $FirstFit$ and $CBIP$ algorithms, which are well known in the graph coloring community. We show that, perhaps surprisingly, $FirstFit$ outperforms $CBIP$ on trees with respect to the κ-CB adversary. For general bipartite graphs, we show that $CBIP$ uses at most 2κ colors against the κ-CB adversary. This result is particularly interesting in the context of existing lower bounds on the performance of $CBIP$ on bipartite graphs. In a series of works [4,6,7] it is shown that any online algorithm must use at least roughly $2\log n$ colors where n is the number of vertices. The construction for this lower bound uses $\log n$ disjoint components. Our result shows that this is necessary. One often measures the performance of an online algorithm by its competitive ratio – the worst-case ratio between the objective value achieved by an algorithm and the offline optimum. In the case of nontrivial bipartite graphs the offline optimum is 2, thus the difference between the absolute number of colors used by $CBIP$ and its competitive ratio is just a factor of 2. But this difference has a philosophical significance: our result shows that the competitive ratio of $CBIP$ on bipartite graphs is simply κ: the number of components that the adversary is allowed to use. Our second set of results shows that for several classes of graphs, including χ-colorable graphs, the κ-CB adversary equipped with an unlimited number of vertices is powerful enough to rule out competitive algorithms even when $\kappa = 1$. These two sets of results provide another contrast between bipartite graphs and other classes of graphs.

The rest of the paper is organized as follows. In Sect. 2 we go over some preliminaries. The new adversarial model is introduced in Sect. 3. The $FirstFit$ algorithm is analyzed in Sect. 4, while $CBIP$ is analyzed in Sect. 5. The analysis of various classes of graphs can be found in Sect. 6. We finish with some discussion and open problems in Sect. 7. Due to length restrictions, some proofs are omitted. These are present in the full version of the paper[2].

2 Preliminaries

In online coloring, an adversary creates a simple undirected graph[3] $G = (V, E)$ and a presentation order of vertices[4] $\sigma : [n] \to V$. The graph is then presented online in the *vertex arrival model*: at time i vertex $v = \sigma(i)$ arrives, and we learn all its neighbors among already appearing vertices. An online algorithm must declare how to color the new vertex $c(v)$ prior to the arrival of the next vertex $\sigma(i+1)$. A priori, an online algorithm does not know V or even n. Alternatively, we can view the online input as a sequence of induced subgraphs:

$$G \cap \sigma([1]), G \cap \sigma([2]), \ldots, G \cap \sigma([n]).$$

We call the set of neighbors of v that an online algorithm learns about at the time of arrival of v as the *pre-neighborhood* of v, denoted by $N^-(v)$.

A natural greedy algorithm is called $FirstFit$ (see, for example, [7,9,13]): when a vertex v arrives, $FirstFit$ colors it with the first color that does not appear in the pre-neighborhood of v.

Another famous algorithm due to [11] for online coloring of *bipartite graphs* is called $CBIP$: when a vertex $v = \sigma(i)$ arrives, $CBIP$ computes an entire connected component CC to which v belongs in the partial graph known so far. Since we assume that the input graph is bipartite, the connected component CC can be partitioned into two sets of vertices A and B such that all edges go between A and B only. Suppose that $v \in A$, then v is colored with the first color that is not present among vertices in B.

In the adversarial arguments presented in this work we often need to control the chromatic number of constructed instances. These instances can get quite complicated and computing their chromatic number exactly might be rather difficult. The following technique is widely used in the online coloring community, see e.g., [8]. The adversary is not only going to construct an online instance, but it will also maintain a valid coloring of that instance. Thus, when specifying the adversary we need to define not only how the next input item is generated, but also how it is colored. The key idea is that since the adversary knows and controls how future input items will be generated, it can anticipate its own moves and create a much better coloring than what an online algorithm can achieve without this knowledge. Unless explicitly stated otherwise, by an "algorithm" we always mean a *deterministic algorithm.*

[2] The full paper is at https://arxiv.org/abs/2005.10852.

[3] We will always use n to refer to $|V|$ and m to refer to $|E|$.

[4] Notation $[n]$ stands for $\{1, 2, \ldots, n\}$; notation $[k, n]$ stands for $\{k, k+1, \ldots, n\}$.

2.1 Bins Vs. Colors

We adopt the terminology introduced in [8]: when there is a possibility of ambiguity we say that an online algorithm colors with *bins* and the adversary with *colors* in order to distinguish the two. Let v be a vertex. We use the notation $b(v)$ to denote the bin that is assigned to v by an online algorithm, and $c(v)$ to denote the color that is assigned to v by the adversary. The functions b and c naturally extend to be defined over sets of vertices. Let A be a set of vertices. Define $b(A) = \{b(v) : v \in A\}$ and $c(A) = \{c(v) : v \in A\}$. Sometimes, we say that bin b contains v to mean that $b(v) = b$.

2.2 Saturated Bins

We define a notion that is inspired by several previous works [6,8,14].

Definition 1. *Suppose that the adversary is constructing a χ-colorable graph. A bin b is said to be* p-saturated *if there are p vertices $v_1, \ldots, v_p$ such that*

$$b(v_1) = \cdots = b(v_p) = b; \quad |\{c(v_1), c(v_2), \ldots, c(v_p)\}| = p. \tag{1}$$

A bin b is said to be perfectly p-saturated *if bin b is p-saturated and it contains exactly p vertices. When a bin b is χ*-saturated, *we simply say bin b is* saturated.

The following simple fact demonstrates why this notion might be interesting.

Fact 1. *If t bins are all saturated, then every color class contains t vertices in distinct bins. By connecting a new vertex to these t vertices, the algorithm is forced to use a new bin.*

The notion of saturated bins is already implicit in some previous works. For example, the so-called *two-sided colors*[5] in [6] are saturated bins when $\chi = 2$. The construction in [6] forces many two-sided colors, i.e., many saturated bins. The proof of a lower bound in [8], which we mentioned earlier as an example that does *not* use the common create-and-merge components strategy directly, could be summarized as follows. The adversary has a strategy to force any algorithm to use perfectly p-saturated bins for $1 \leq p \leq \chi/2$. In [14], the lower bound construction does not necessarily force p-saturated bins, but seeks to create a situation where a weaker form of the Fact 1 is bound to appear.

In Sect. 4, we show that κ-CB adversary can successively force saturated bins on $FirstFit$ for χ-colorable graphs. This leads to the algorithm being noncompetitive. The construction of forcing saturated bins on $FirstFit$ is generalized in Sect. 6 to work for all algorithms.

[5] What is called "colors" in [6], as in many other works, is what we call bins.

3 A New Type of Adversary

Let $cc(G)$ denote the number of connected components of graph G.

Definition 2. *An adversary is said to be* κ components bounded, *or* κ*-CB, if the input graph* G *and the presentation order* σ *satisfy*

$$\forall\, i \in [n] \;\; cc(G \cap \sigma([i])) \leq \kappa. \tag{2}$$

Let $\mathcal{A}$ *denote a deterministic online coloring algorithm. Define* $\beta(\mathcal{A}, \kappa, G)$ *to be the maximal number of bins* $\mathcal{A}$ *has to use when a* κ*-CB adversary constructs the graph* G*. Let* $\mathcal{G}$ *denote a class of graphs. Define*

$$\beta(\mathcal{A}, \kappa, \mathcal{G}) = \sup_{G \in \mathcal{G}} \beta(\mathcal{A}, \kappa, G), \tag{3}$$

and

$$\beta(\kappa, \mathcal{G}) = \inf_{\mathcal{A}} \beta(\mathcal{A}, \kappa, \mathcal{G}). \tag{4}$$

Let $\chi \in \mathbb{N}$. By identifying χ with the class of graphs that are χ-colorable, the notation $\beta(\mathcal{A}, \kappa, \chi)$ and $\beta(\kappa, \chi)$ are defined via (3) and (4), respectively. Let TREE denote the class of graphs that are trees.

Different from traditional online coloring models, a feature of the κ-CB adversary model is that the number of vertices of a graph is *not necessarily* a parameter in the model. In this work, the κ-CB adversary is allowed to construct graphs with arbitrarily many vertices. We will be interested in understanding what the power and limitations are for an adversary who can use unlimited number of vertices but is κ-CB.

Lemma 1. *Let* Ω *denote the set of all possible graphs and let* $\mathcal{A}$ *be an arbitrary algorithm. Then,*

$$\beta(\mathcal{A}, 1, \Omega) = \infty.$$

Proof. For every $n \in \mathbb{N}$, let K_n denote the complete graph on n vertices. Obviously, an adversary can present K_n in any presentation order while maintaining a single connected component. Thus, $\beta(\mathcal{A}, 1, K_n) \geq n$. As n is arbitrary, the result follows. □

Hence, the κ-CB adversary model becomes interesting when we consider special classes of graphs. For example, $\beta(FirstFit, \kappa, 2)$ denotes the maximal number of bins that the κ-CB adversary can force $FirstFit$ to use by constructing a bipartite graph.

4 *FirstFit* on χ-colorable Graphs, Triangle-Free Graphs, and Trees

In this section, we characterize the performance of $FirstFit$ on χ-colorable graphs, triangle-free graphs, and trees for κ-CB adversaries. We begin with the following theorem, which completely determines $\beta(FirstFit, \kappa, \chi)$ for all $\kappa, \chi \in \mathbb{N}$.

Theorem 2. *Let* $\kappa \in \mathbb{N}, \chi \in \mathbb{N}$.

(1) $\beta(FirstFit, \kappa, 1) = 1$ *for every* $\kappa \geq 1$;
(2) $\beta(FirstFit, 1, 2) = 2$;
(3) $\beta(FirstFit, 2, 2) = \infty$
(4) $\beta(FirstFit, \kappa, \chi) = \infty$ *for every* $\kappa \geq 1$ *and* $\chi \geq 3$.

The construction in part (4) of Theorem 2 results in a triangle-free graph. Let TRIANGLE-FREE denote the class of triangle-free graphs. Thus, we immediately obtain the following.

Corollary 1. $\beta(FirstFit, 1, \texttt{TRIANGLE-FREE}) = \infty$.

Part (4) of Theorem 2 and Corollary 1 will be generalized in Sect. 6.

We conclude this section by giving a complete analysis of $FirstFit$ on trees with respect to κ-CB adversary.

Theorem 3. $\beta(FirstFit, \kappa, \texttt{TREE}) = \kappa + 1$.

Proof. The lower bound $\beta(FirstFit, \kappa, \texttt{TREE}) \geq \kappa + 1$ is witnessed by the so-called forest construction due to Bean [3] (also independently discovered in [7]). We claim that a κ-CB adversary can construct a forest consisting of κ trees $T_1, T_2, \ldots, T_\kappa$ with the property that for each i the $FirstFit$ algorithm uses color i on some vertex v_i belonging to the tree T_i. We first prove this claim and later see how it implies the lower bound.

The construction is recursive and so we prove the above statement by induction on κ. Base case is trivial: when $\kappa = 1$ the adversary can give a single isolated vertex. Assume that the statement is true for κ and we wish to establish it for $\kappa + 1$. The adversary begins by invoking induction and creating $T_1, T_2, \ldots, T_\kappa$ such that for $i \in [\kappa]$ there is $v_i \in T_i$ such that $b(v_i) = i$ is assigned by $FirstFit$. Then, the adversary creates a new vertex u connected to all the v_i. This process, merges all existing trees into a single tree, which we call $T'_{\kappa+1}$. Moreover, this forces $FirstFit$ to assign $b(u) = \kappa + 1$. This tree is set aside, and to satisfy the claim for $\kappa + 1$, the adversary invokes the induction again to create another set of trees $T'_1, \ldots, T'_\kappa$ with $v'_i \in T'_i$ such that $b(v'_i) = i$. Note that creating T'_i requires at most κ components, so the adversary is $(\kappa + 1)$-CB (remember that we have an additional component $T'_{\kappa+1}$ set aside during the second invocation of induction). Moreover, note that $b(u) = \kappa + 1$, so the trees $T'_1, \ldots, T'_{\kappa+1}$ satisfy the claim.

This claim implies the lower bound since the κ-CB adversary can present $T_1, \ldots, T_\kappa$ with $b(v_i) = i$ for some $v_i \in T_i$. In the last step, the adversary presents u connected to each v_i. This process does not increase the number of components and forces $b(u) = \kappa + 1$. An example of this construction is shown in Fig. 1.

Next, we show the upper bound $\beta(FirstFit, \kappa, \texttt{TREE}) \leq \kappa + 1$. Suppose that $FirstFit$ uses $m+1$ bins for some m. Let v be the first vertex which is placed into bin $m+1$ by $FirstFit$. By the definition of $FirstFit$, there are m vertices in the pre-neighborhood $N^-(v)$ that have been previously assigned to bins $1, 2, \ldots, m$.

Since the adversary is constructing a tree, there can be no cycle. Hence, these m vertices must be in distinct components, i.e., there are at least m distinct components. □

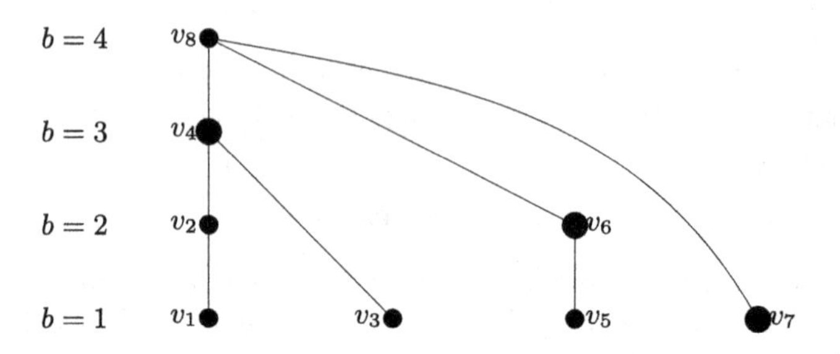

Fig. 1. An example of the forest construction used in the proof of Theorem 3. The adversary presents the vertices in order: $v_1, v_2, v_3, \ldots, v_8$. The *FirstFit* uses 4 bins while the adversary uses 3 connected components during this construction.

5 *CBIP* on Bipartite Graphs

This section contains the most technical result of this paper. We establish the tight bound of 2κ on the number of bins used by $CBIP$ with respect to a κ-CB adversary on bipartite graphs and trees. This provides a finer understanding of the performance of $CBIP$ and is particularly interesting in light of previous lower bounds. Gutowski et al. [6] proved that any online algorithm has to use at least $2\log n - 10$ bins for coloring bipartite graphs with n vertices, which matches the upper bound on $CBIP$ from [11] up to the additive constant -10. The construction in [6] applied to $CBIP$ (or even $FirstFit$) uses $\log n$ disjoint connected components to force $2\log n$ bins. In particular, the main result of this section, which we state next, demonstrates that this is a necessary feature of their construction.

Theorem 4. $\beta(CBIP, \kappa, \texttt{TREE}) = \beta(CBIP, \kappa, 2) = 2\kappa.$

Proof. Since $\beta(CBIP, \kappa, 2) \geq \beta(CBIP, \kappa, \texttt{TREE})$, the lower bound follows from Lemma 2 and the upper bound follows from Lemma 3. □

Observe that Theorem 4 implies that, in the class of bipartite graphs, worst case input already appears in TREE for $CBIP$.

We begin by stating the lower bound used in the above theorem. The main idea behind the proof is the adversarial construction of a sequence of trees T_i under the κ-CB constraint, such that $CBIP$ uses 2κ bins for the final tree $T_{2\kappa}$.

Lemma 2. $\beta(CBIP, \kappa, \texttt{TREE}) \geq 2\kappa.$ □

We finish this section with a matching upper bound for the class of bipartite graphs.

Lemma 3. $\beta(CBIP, \kappa, 2) \leq 2\kappa$.

Proof. Consider a certain point in execution of $CBIP$ on the input graph. As usual, let $b(v)$ denote the bin to which v is assigned by $CBIP$. We say that a connected component CC is of Type $1[\ell]$ if CC can be partitioned in two blocks A and B such that

- all edges go between A and B
- $b(A) = [\ell - 2]$
- $b(B) = [\ell - 1]$

Similarly, we say that a connected component CC is of Type $2[\ell]$ if CC can be partitioned in two blocks A and B such that

- all edges go between A and B
- $b(A) = [\ell - 2] \cup \{\ell\}$
- $b(B) = [\ell - 1]$

Figure 2 shows an example construction with two connected components of Type 2[4] and Type 1[5].

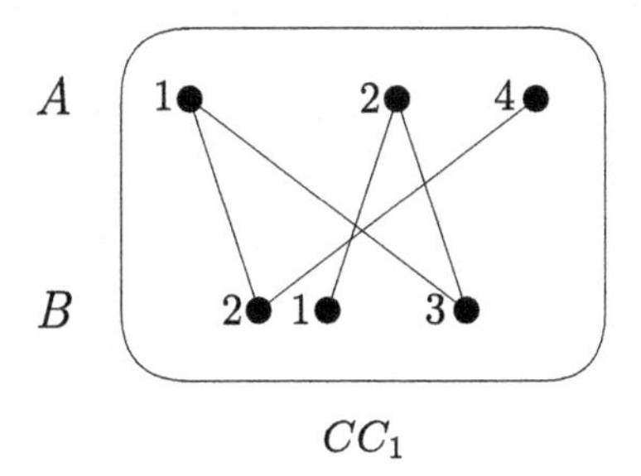

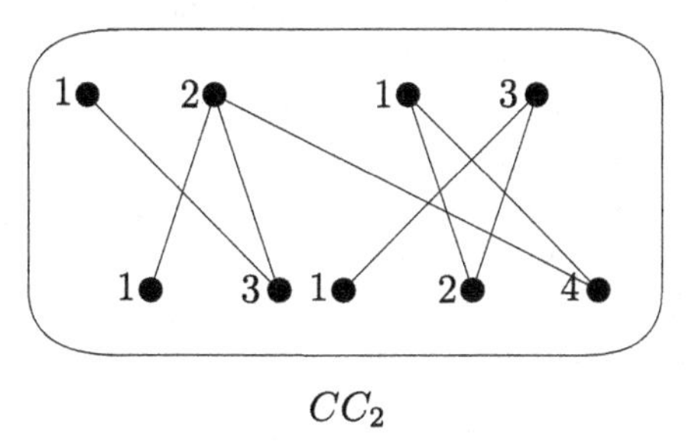

Fig. 2. A snapshot of an execution of $CBIP$ on an input instance, with vertices presented in left-to-right order. At this step in the presentation, the graph contains two connected components CC_1 and CC_2 of Type 2[4] and Type 1[5] respectively. Labels indicate bins used by $CBIP$.

The high level idea is that the κ-CB adversary can only force components that are either of Type $1[\ell]$ or of Type $2[\ell]$ for some $\ell \leq 2\kappa$. Before we prove it formally, we observe that when a new vertex v arrives, it can either (1) be an isolated vertex (taken to be of Type 1[2]), (2) be added to an existing component, or (3) be used to merge two or more existing components. Formally, we say that components $CC_1, CC_2, \ldots, CC_j$ get *merged* at time t if vertex $v = \sigma(t)$ satisfies $N^-(v) \cap CC_i \neq \emptyset$ for $i \in [j]$, and CC_i were distinct connected components at time $t - 1$.

We record what happens to types of components after each of the above operations (1), (2) and (3). During operation (1), a new vertex of Type $1[\ell]$ for $\ell = 2$ is added. Clearly, $\ell \leq 2\kappa$ for any $\kappa \geq 1$. Next, we consider operation (2), i.e., when a new vertex v gets added to a component CC. We assume that the two blocks of vertices of CC are A and B and that they satisfy the conditions of Type 1 or 2. The resulting component is called CC'. The changes to types after vertex v is presented are recorded in Table 1.

Table 1. Type changes for case (1), i.e., when v is added to an existing component.

v is connected to	Type of CC	Type of CC'
A	$1[\ell]$	$1[\ell]$
A	$2[\ell]$	$2[\ell]$
B	$1[\ell]$	$2[\ell]$
B	$2[\ell]$	$2[\ell]$

Finally, we consider what happens when a vertex v is used to merge two or more components. We distinguish four types of components, the numbers of which are denoted by k_1, k_2, k_3, and k_4, respectively:

1. CC_i^A of Type $1[\ell_i^A]$ for $i \in [k_1]$. Vertex v has a neighbor on the A-side of such components.
2. CC_i^B of Type $1[\ell_i^B]$ for $i \in [k_2]$. Vertex v has a neighbor on the B-side of such components.
3. CC'^A_i of Type $2[\ell'^A_i]$ for $i \in [k_3]$. Vertex v has a neighbor on the A-side of such components.
4. CC'^B_i of Type $2[\ell'^B_i]$ for $i \in [k_4]$. Vertex v has a neighbor on the B-side of such components.

Let $m = \max\{\ell_{i_1}^A, \ell_{i_2}^B, \ell'^A_{i_3}, \ell'^B_{i_4} : i_1 \in [k_1], i_2 \in [k_2], i_3 \in [k_3], i_4 \in [k_4]\}$. We call m the *type parameter* of the partially constructed input graph.

We say that a block A or B of a particular component being merged is on the opposite side of v if v has a neighbor among the vertices of the block. Otherwise, we say that the block is on the same side as v. For example, block A of CC_i^A component is on the opposite side of v, whereas block B of the same component is on the same side as v. Let S_{-v} denote the set of bins already used for the vertices of blocks on the opposite side of v, and let S_v denote the set of bins already used for the vertices of blocks on the same side as v. By the definitions of Type 1 and 2 components as well as m, it is easy to see that each of S_v, S_{-v} can be only one of the following four options: $[m-2], [m-1], [m-2] \cup \{m\}, [m]$. This reduces the problem of computing the type of the merged component to analyzing 16 cases. For example, if $S_{-v} = [m-2]$ and $S_v = [m-2]$ then vertex v will be assigned bin $m-1$ and the merged component will be of Type $1[m]$ since it will have one side with bins $[m-2]$ and the opposite side with bins $[m-1]$.

Table 2. Type changes for case (2), i.e., when v is used to merge some existing components.

S_{-v}	S_v	Possible?	Bin of v	Type of CC'
$[m-2]$	$[m-2]$	yes	$m-1$	$1[m]$
$[m-2]$	$[m-1]$	yes	$m-1$	$1[m]$
$[m-2]$	$[m-2] \cup \{m\}$	no	NA	NA
$[m-2]$	$[m]$	no	NA	NA
$[m-1]$	$[m-2]$	yes	m	$2[m]$
$[m-1]$	$[m-1]$	yes	m	$1[m+1]$
$[m-1]$	$[m-2] \cup \{m\}$	yes	m	$2[m]$
$[m-1]$	$[m]$	yes	m	$1[m+1]$
$[m-2] \cup \{m\}$	$[m-2]$	no	NA	NA
$[m-2] \cup \{m\}$	$[m-1]$	yes	$m-1$	$2[m]$
$[m-2] \cup \{m\}$	$[m-2] \cup \{m\}$	no	NA	NA
$[m-2] \cup \{m\}$	$[m]$	no	NA	NA
$[m]$	$[m-2]$	no	NA	NA
$[m]$	$[m-1]$	yes	$m+1$	$2[m+1]$
$[m]$	$[m-2] \cup \{m\}$	no	NA	NA
$[m]$	$[m]$	yes	$m+1$	$1[m+2]$

We denote the merged component by CC' and Table 2 summarizes all of the 16 cases.

The reason that certain combinations in Table 2 are impossible is that if one side is colored with bins $[m]$ then the opposite side must use bin $m-1$ (because of how $CBIP$ works).

Observe that from Table 2, the type parameter of CC' can either stay the same, increase by additive 1, or increase by additive 2. Furthermore, it can be directly verified from the table that an increase is possible only if there are at least two components having type parameters not less than $m-1$. We refer to this property as the *continuity of the type parameter.*

Assume that the input graph G and the presentation order σ satisfy the κ-CB condition, then the above observations imply the following statements:

(i) $\forall\ i$ we have $G \cap \sigma([i])$ consists of Type $1/2[\ell]$ components for $\ell \leq 2\kappa$;
(ii) $\forall\ i$ there can be at most one component that is of one of the following four types: Type $1[2\kappa-1]$, Type $1[2\kappa]$, Type $2[2\kappa-1]$, Type $2[2\kappa]$.

Note that (i) immediately implies the statement of this lemma.

These statements can be proved by induction on κ. The base case $\kappa = 1$ is easy to verify. Indeed, if $i = 1$, i.e., there is just a single vertex, then it is of Type $1[2]$. Consider $i \geq 2$. Observe that for $\kappa = 1$ the algorithms $CBIP$ and $FirstFit$ have identical behavior. Therefore, (2) in Theorem 2 shows (i) is true, and the single component is of Type $2[2]$. Hence, (ii) is true.

We proceed to the induction step. Assume (i) and (ii) are true for κ, we consider the case $\kappa+1$. First, we show (ii). Let CC_1 be the first component that is of one of the following types: Type $1[2\kappa+1]$, Type $1[2\kappa+2]$, Type $2[2\kappa+1]$, Type $2[2\kappa+2]$. Since the adversary is $(\kappa+1)$-CB, the existence of CC_1 implies that the adversary becomes κ-CB when creating any new component that is disjoint from CC_1. By the induction assumption, the adversary can only create components that are of Type $1/2[\ell]$ for $\ell \leq 2\kappa$. This proves (ii). Next we show (i). By the *continuity of the type parameter*, in order for the type parameter to go beyond $2\kappa+2$ there need to be at least two components both having type parameters not less than $2\kappa+1$. By (ii), this is impossible, so (i) is true. □

6 Lower Bounds for Several Graph Classes

In this section we establish non-existence of competitive algorithms against κ-CB adversaries for various classes of graphs. We begin by establishing a strong non-competitiveness result for χ-colorable graphs for $\chi \geq 3$. This generalizes part (4) of Theorem 2 to arbitrary algorithms.

Theorem 5. *$\beta(1,\chi)=\infty$ for every $\chi \geq 3$.*

Proof. Since $\beta(1,\chi)$ is non-decreasing in χ, it suffices to prove that $\beta(1,3)=\infty$. Fix an arbitrary coloring algorithm $\mathcal{A}$. We show that for every $t \in \mathbb{N}$, a 1-CB adversary can construct a 3-colorable graph G so that $\mathcal{A}$ uses at least t different bins to color vertices in G. It may be helpful to consult Fig. 3 while reading this proof.

The construction of G proceeds in layers, which we denote by $L_1, L_2, \ldots, L_{t-1}$. Vertices (and their pre-neighborhoods) in L_1 are presented first, followed by L_2, and so on. The construction stops as soon as $\mathcal{A}$ uses t distinct bins, which may happen before L_{t-1} and will be guaranteed to happen in L_{t-1}.

Each layer consists of "sufficiently many" vertices, meaning that there should be enough vertices in lower layers to guarantee that the construction of higher layers goes through. Initially, we don't quantify "sufficiently many," although we shall give some estimates on sizes of layers at the end of this proof.

Layer L_1 is simply a path P of sufficiently large length ℓ_1. The adversary presents the vertices in P in the order in which they appear on the path. There are two possibilities: (i) $\mathcal{A}$ already uses at least t bins to color P; (ii) $\mathcal{A}$ uses fewer than t bins to color P. In case (i) the construction is over and the adversary has achieved its goal.

Next, we handle case (ii). Observe that $\mathcal{A}$ has to use at least two different bins to color P correctly. Consider two bins b_1 and b_2 with the most number of vertices assigned to them by $\mathcal{A}$. Let the sets of nodes assigned to those bins be B_1 and B_2, respectively, with $|B_1| \geq |B_2|$. The definition of case (ii) implies that $|B_1| \geq \ell_1/t$. Since P is a path, no bin can contain more than $\ell_1/2+1$ nodes. Thus, the number of nodes not in B_1 is at least $\ell_1/2-1$. Since they are partitioned among at most $t-1$ bins, including B_2, and B_2 is most populous then $|B_2| \geq (\ell_1/2 - 1)/(t-1) = (\ell_1 - 2)/(2t-2)$. Next, we select subsets $B_1' \subseteq B_1$ and $B_2' \subseteq B_2$ so that all the nodes in $B_1' \cup B_2'$ are non-adjacent in P and $|B_1'| = |B_2'| = \ell_1/(10t)$. This can be done as follows: alternatively pick a node from B_1 or B_2 to include in B_1' or B_2', respectively, and remove its neighbors from B_2 or B_1, respectively. Each pair of such steps includes one vertex into B_1' and one vertex into B_2' removing at most 3 vertices from each B_1 and B_2 from future considerations. Thus, this can go on for at least $|B_2|/3 \geq \ell_1/(10t)$ rounds. In conclusion, we end up with sets of nodes B_1' and B_2' such that

- all nodes in B_i' are placed in bin b_i by $\mathcal{A}$, where $i \in \{1,2\}$;
- $|B_1'| = |B_2'| = \ell_1/(10t)$.

In particular, the second item implies that $|B_1'|$ and $|B_2'|$ can be assumed to be sufficiently large.

Construction of each following layer L_i for $i \geq 2$ either terminates early because $\mathcal{A}$ used at least t different bins or forces $\mathcal{A}$ to assign sufficiently many vertices to bin b_{i+1}. We shall denote the set of such vertices[6] B_{i+1}' for layer L_i. Assuming that the construction hasn't terminated in layer L_{i-1}, the next layer L_i is constructed by the adversary by repeating the following steps sufficiently many times:

(1) the adversary chooses vertices $u_j \in B_j'$ for all $j \leq i$ arbitrarily;
(2) the adversary presents a new vertex v with pre-neighborhood $\{u_1, \ldots, u_i\}$;
(3) the adversary updates $B_j' \leftarrow B_j' \setminus \{u_j\}$ for all $j \leq i$.

Due to step (3) we say that v *consumes* nodes u_j from B_j' for $j \in [i]$. Observe that step (2) guarantees that $\mathcal{A}$ has to assign v to a bin other than $b_1, \ldots, b_i$. Just as for layer L_1, if $\mathcal{A}$ uses t different bins in this layer then we are done. Otherwise, let b_{i+1} be the bin that has the most number of vertices assigned to it in layer L_i. If the adversary presents ℓ_i vertices in layer L_i then the number of vertices assigned to b_{i+1} is at least ℓ_i/t. We let B_{i+1}' be an arbitrary subset of such vertices of size exactly ℓ_i/t.

This construction continues until layer L_{t-1} where the adversary can present a single node according to the above scheme forcing $\mathcal{A}$ to assign it to a new bin b_t. Overall, $\mathcal{A}$ then uses t different bins, namely, $b_1, \ldots, b_t$.

To guarantee that step (1) in the above construction always works, we need to make sure that all sets B_j' are sufficiently large for this construction to reach layer L_{t-1}. This is possible provided that for $i \geq 2$ we have $|B_{i+1}'| \geq \sum_{j=i+1}^{t-1} \ell_j$ since each node in a layer above i consumes one node from B_i' (step (3) of the above construction). We also need a similar condition for layer 1, namely, that $|B_2'| = |B_1'| \geq \sum_{j=2}^{t-1} \ell_j$. Thus, we end up with the following system of inequalities:

- $\ell_{t-1} = 1$;
- $\ell_i/t \geq \sum_{j=i+1}^{t-1} \ell_j$ for $i \in \{2, 3, \ldots, t-2\}$;

[6] Note that the index of B_{i+1}' is off by one with respect to the index of layer L_i with which it is associated. This happens for $i \geq 2$ since layer L_1 has two sets B_1' and B_2' associated with it.

– $\ell_1/(10t) \geq \sum_{j=2}^{t-1} \ell_j$.

It is straightforward to check that $\ell_{t-1} = 1$, $\ell_i = t(t+1)^{t-i-2}$ for $i \in [2, t-2]$ and $\ell_1 = 10t(t+1)^{t-3}$ is a valid solution to the above system. Thus, a feasible construction can be carried out by the adversary. The total number of nodes in this construction is at most $20t(t+1)^{t-3}$.

Observe that the construction clearly satisfies the 1-CB constraint, since layer L_1 is presented as a single connected component and every vertex in a higher layer is adjacent to a vertex in layer L_1.

We also note that the construction creates an *almost-forest*. More specifically, call the vertex v in step (2) of the above construction the *parent* of the corresponding u_j for $j \in [i]$ chosen in step (1). Observe that due to step (3), each vertex has at most one parent. Therefore, the only thing preventing this construction from being a forest is layer L_1, which can be thought of as a path going through all the leaves of the forest. This implies that the constructed graph is 3-colorable. Consider the subgraph obtained by removing all edges in layer L_1 along with all vertices that do not have parents. Since it is a forest, it is 2-colorable. Moreover, any valid 2-coloring of this subgraph is also a partial valid 2-coloring of the entire graph since we chose B'_1 and B'_2 to be non-adjacent. We can then extend this partial coloring to a complete 3-coloring of the entire graph by using a greedy strategy. Note that uncolored vertices in L_1 have degree at most 2, so a greedy coloring would use at most 3 colors. □

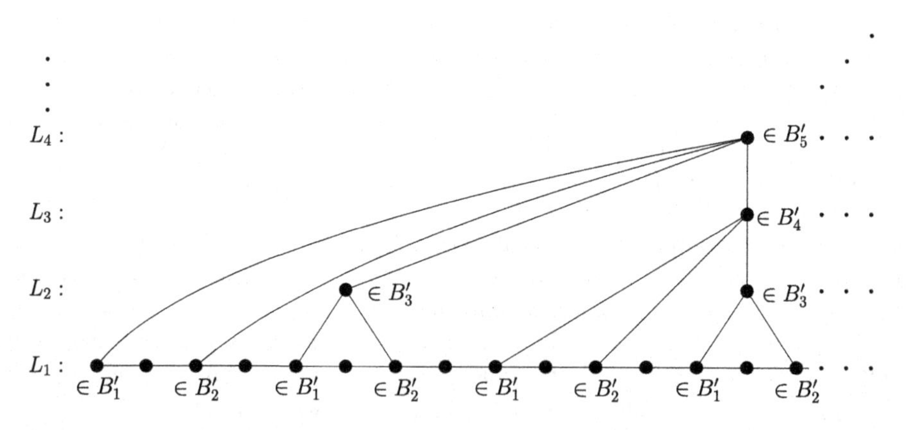

Fig. 3. Example of the construction used in Theorem 5. This is a hypothetical example for some $\mathcal{A}$ that assigns bins to vertices according to the figure.

The above construction can be modified so that either $\mathcal{A}$ uses t bins or the adversary can successively force saturated bins. For example, the adversary can extend the level L_1 and repeat the construction on the extended part. The adversary can do this sufficiently many times and recolor each copy so that saturated bins are forced.

The rough estimates on sufficient lengths of layers presented in the above proof immediately lead to the following quantitative version of the result.

Corollary 2. *The 1-CB adversary can construct a 3-colorable graph on n vertices so that any online coloring algorithm uses at least $\Omega(\log n / \log \log n)$ bins.*

Next, we note that the construction from Theorem 5 is quite robust. It can be modified in various ways to obtain similar non-competitiveness results for other classes of graphs. We first define the relevant classes.

C_k-FREE: the class of graphs that do not contain cycle of length k as a (not necessarily induced) subgraph.
d-INDUCTIVE: the class of d-inductive graphs, i.e., those graphs whose vertices can be numbered so that each vertex has at most d adjacent vertices among higher numbered vertices.
PLANAR: the class of planar graphs.
TREEWIDTH-k: the class of graphs of treewidth at most k.

We are now ready to state the following corollary of the construction from Theorem 5. Parts 1 and 2 are straightforward, while 3 follows from a modification of the above construction. Part 4 follows from the modified construction being 2-outerplanar, and a result of Bodlaender [5].

Corollary 3. *1.* $\beta(1, C_k\text{-FREE}) = \infty$ *for every* $k \geq 3$.
2. $\beta(1, d\text{-INDUCTIVE}) = \infty$ *for every* $d \geq 2$.
3. $\beta(1, \text{PLANAR}) = \infty$.
4. $\beta(1, \text{TREEWIDTH-}k) = \infty$ *for every* $k \geq 5$.

7 Conclusion and Open Problems

We have introduced a new type of adversary for online graph problems and studied online coloring with respect to this adversary. This led to an improved understanding of the properties of the two widely studied online coloring algorithms $FirstFit$ and $CBIP$. Furthermore, when the adversary is κ-CB for $\kappa = O(1)$, Theorems 4 and 5 show a sharp contrast between bipartite graphs, for which $CBIP$ uses only $O(1)$ bins, and 3-colorable graphs for which any algorithm has to use infinitely many bins. While our work suggests many directions for future research, we find the following questions particularly intriguing:

1. What is $\beta(1, \text{TREEWIDTH-}k)$ for $k \in \{2, 3, 4\}$?
2. We allow the adversary to use an unlimited number of vertices. A natural extension of our work is to study the dependency on n while the adversary is κ-CB. Corollary 2 is a step in that direction. Can the lower bound in Corollary 2 be improved to $\Omega(\log^2 n)$, matching the best known lower bound in [14] for 3-colorable graphs when the adversary is unconstrained? On the other hand, we discusssed in the introduction that in practice it seems the case that κ is often $O(1)$, hence it would be very interesting to see whether it is possible to improve the existing upper bounds (such as those in [14] and [10]) under the $O(1)$-CB adversary assumption.

3. For a graph G and presentation order σ define $\kappa(G, \sigma) = \max_i cc(G \cap \sigma([i]))$. What is the behaviour of $\kappa(G, \sigma)$ in real-world instances? As we mention in the introduction, it is expected that $\kappa(G, \sigma)$ is "small" for social networks. How "small" is it actually? What are typical values of $\kappa(G, \sigma)$? For a class of real-world instances for a particular application (such as transportation networks, social networks, or electrical networks), do $\kappa(G, \sigma)$ values follow some well-defined distribution?
4. One can study the power and limitations of the κ-CB adversary in other related models, e.g., online algorithms with advice, streaming algorithms, temporal or dynamic graphs algorithms. Interactions between various features of those models and the κ-CB constraint might lead to new algorithms or finer understanding of existing algorithms.
5. Finally, it would be interesting to study other online graph problems besides coloring under the κ-CB adversary.

References

1. Albers, S., Schraink, S.: Tight bounds for online coloring of basic graph classes. In: 25th Annual European Symposium on Algorithms (ESA 2017). Schloss Dagstuhl-Leibniz-Zentrum fuer Informatik (2017)
2. Barabási, A.-L., Albert, R.: Emergence of scaling in random networks. Science **286**(5439), 509–512 (1999)
3. Bean, D.R.: Effective coloration. J. Symbolic Logic **41**(2), 469–480 (1976)
4. Bianchi, M.P., Böckenhauer, H.-J., Hromkovič, J., Keller, L.: Online coloring of bipartite graphs with and without advice. Algorithmica **70**(1), 92–111 (2013). https://doi.org/10.1007/s00453-013-9819-7
5. Bodlaender, H.L.: A partial k-arboretum of graphs with bounded treewidth. Theor. Comput. Sci. **209**(1), 1–45 (1998)
6. Gutowski, G., Kozik, J., Micek, P., Zhu, X.: Lower bounds for on-line graph colorings. In: Ahn, H.-K., Shin, C.-S. (eds.) ISAAC 2014. LNCS, vol. 8889, pp. 507–515. Springer, Cham (2014). https://doi.org/10.1007/978-3-319-13075-0_40
7. Gyárfás, A., Lehel, J.: On-line and first fit colorings of graphs. J. Graph Theor. **12**(2), 217–227 (1988)
8. Halldórsson, M.M., Szegedy, M.: Lower bounds for on-line graph coloring. In: Proceedings of the third annual ACM-SIAM symposium on Discrete algorithms, pp. 211–216. Society for Industrial and Applied Mathematics (1992)
9. Kierstead, H.A.: Coloring graphs on-line. In: Fiat, A., Woeginger, G.J. (eds.) Online Algorithms. LNCS, vol. 1442, pp. 281–305. Springer, Heidelberg (1998). https://doi.org/10.1007/BFb0029574
10. Kierstead, H.A.: On-line coloring k-colorable graphs. Israel J. Math. **105**(1), 93–104 (1998) https://doi.org/10.1007/BF02780324
11. Lovász, L., Saks, M., Trotter, W.T.: An on-line graph coloring algorithm with sublinear performance ratio. Discrete Math. **75**(1–3), 319–325 (1989)
12. Newman, M.: Networks: An Introduction. Oxford University Press Inc, USA (2010)
13. Steffen, B.C.: Advice complexity of online graph problems. PhD thesis, ETH Zurich (2014)
14. Vishwanathan, S.: Randomized online graph coloring. In: Proceedings 1990 31st Annual Symposium on Foundations of Computer Science, pp. 464–469. IEEE (1990)

Maximum Coverage with Cluster Constraints: An LP-Based Approximation Technique

Guido Schäfer[1,2] and Bernard G. Zweers[1(✉)]

[1] Centrum Wiskunde and Informatica (CWI), Amsterdam, The Netherlands
{g.schaefer,b.g.zweers}@cwi.nl
[2] Vrije Universiteit, Amsterdam, The Netherlands

Abstract. Packing problems constitute an important class of optimization problems. However, despite the large number of variants that have been studied in the literature, most packing problems encompass a single tier of capacity restrictions only. For example, in the *Multiple Knapsack Problem*, we want to assign a selection of items to multiple knapsacks such that their capacities are not exceeded. But what if these knapsacks are partitioned into *clusters*, each imposing an additional (aggregated) capacity restriction on the knapsacks contained in that cluster?

In this paper, we study the *Maximum Coverage Problem with Cluster Constraints (MCPC)*, which generalizes the *Maximum Coverage Problem with Knapsack Constraints (MCPK)* by incorporating such cluster constraints. Our main contribution is a general LP-based technique to derive approximation algorithms for such cluster capacitated problems. Our technique basically allows us to reduce the cluster capacitated problem to the respective original packing problem. By using an LP-based approximation algorithm for the original problem, we can then obtain an effective rounding scheme for the problem, which only loses a small fraction in the approximation guarantee.

We apply our technique to derive approximation algorithms for MCPC. To this aim, we develop an LP-based $\frac{1}{2}(1-\frac{1}{e})$-approximation algorithm for MCPK by adapting the *pipage rounding technique*. Combined with our reduction technique, we obtain a $\frac{1}{3}(1-\frac{1}{e})$-approximation algorithm for MCPC. We also derive improved results for a special case of MCPC, the *Multiple Knapsack Problem with Cluster Constraints (MKPC)*. Based on a simple greedy algorithm, our approach yields a $\frac{1}{3}$-approximation algorithm. By combining our technique with a more sophisticated iterative rounding approach, we obtain a $\frac{1}{2}$-approximation algorithm for certain special cases of MKPC.

Keywords: Budgeted maximum coverage problem · Multiple Knapsack Problem · Pipage rounding · Iterative rounding

C. Kaklamanis and A. Levin (Eds.): WAOA 2020, LNCS 12806, pp. 63–80, 2021.
https://doi.org/10.1007/978-3-030-80879-2_5

1 Introduction

Many optimization problems encountered in real-life can be modeled as a *packing problem*, where a set of diverse objects (or items, elements) need to be packed into a limited number of containers (or knapsacks, bins) such that certain feasibility constraints are satisfied. Given their high practical relevance, there is a large variety of packing problems that have been studied in the literature. For example, the monograph by Kellerer et al. [13] discusses more than 20 variants of the classical *Knapsack Problem (KP)* alone.

However, despite the large number of variants that exist, most packing problems encompass a *single* tier of capacity restrictions only. For the sake of concreteness, consider the *Multiple Knapsack Problem (MKP)*: In this problem, the goal is to find a most profitable selection of items that can be assigned to the (multiple) knapsacks without violating their capacities, i.e., there is a single capacity constraint per knapsack that needs to be satisfied. But what about the setting where these knapsacks are partitioned into *clusters*, each imposing an additional (aggregated) capacity restriction on all knapsacks contained in it?

We believe that the study of such problems is well motivated and important, both because of their practical relevance and theoretical appeal. For example, a potential application of the *Multiple Knapsack Problem with Cluster Constraints (MKPC)* (as outlined above) is when items have to be packed into containers, obeying some weight capacities, and these boxes have to be loaded onto some weight-capacitated vehicles (e.g., ships). Another example is a situation in which customers have to be assigned to facilities, which have a certain capacity, but these facilities are also served by larger warehouses that again have a restricted capacity.

Maximum Coverage with Cluster Constraints. In this paper, we propose and initiate the study of such extensions for packing problems. More specifically, we consider the following packing problem, which we term the *Maximum Coverage Problem with Cluster Constraints (MCPC)* (see Sect. 2 for formal definition): Basically, in this problem we are given a collection of subsets of items, where each subset is associated with some cost and each item has a profit, and a set of knapsacks with individual capacities. In addition, the knapsacks are partitioned into clusters which impose additional capacity restrictions on the total cost of the subsets assigned to the knapsacks in each cluster. The goal is to determine a feasible assignment of a selection of the subsets to the knapsacks such that both the knapsack and the cluster capacities are not exceeded, and the total profit of all items covered by the selected subsets is maximized.

As we detail in the related work section below, MCPC is related to several other packing problems but has not been studied in the literature before (to the best of our knowledge). It generalizes several fundamental packing problems such as the *Maximum Coverage Problem with Knapsack Constraints (MCPK)* (see, e.g., [1,14]), which in turn is a generalization of the Multiple Knapsack Problem (MKP). Another important special case of MCPC is what we term the *Multiple Knapsack Problem with Cluster Constraints (MKPC)* (as outlined above). Also this problem has not been addressed in the literature before and will be considered in this paper.

All problems considered in this paper are (strongly) NP-hard (as they contain MKP as a special case), which rules out the existence of a polynomial-time algorithm to compute an optimal solution to the problem (unless NP = P). We therefore resort to the development of approximation algorithms for these problems. Recall that an *α-approximation algorithm* computes a feasible solution recovering at least an α-fraction ($0 \leq \alpha \leq 1$) of the optimal solution in polynomial time. Note that the *Maximum Coverage Problem* is a special case of MCPC and this problem is known to be $(1 - \frac{1}{e})$-inapproximable (see [10]). As a consequence, we cannot expect to achieve an approximation factor better than this for MCPC.

We remark that while our focus here is mostly on the extensions of MCPK and MKP, the idea of imposing a second tier of capacity restrictions is generic and can be applied to other problems as well (not necessarily packing problems only).[1]

Our Contributions: We present a general technique to derive approximation algorithms for packing problems with two tiers of capacity restrictions. Basically, our idea is to extend the natural integer linear programming (ILP) formulation of the original packing problem (i.e., without cluster capacities) by incorporating the respective cluster capacity constraints. But, crucially, these new constraints are set up in such a way that an optimal solution to the LP relaxation of this formulation defines some *reduced capacities* for each knapsack individually. This enables us to reduce the cluster capacitated problem to the respective original packing problem with knapsack constraints only. We then use an LP-based approximation algorithm for the original problem to round the optimal LP-solution. This rounding requires some care because of the reduced capacities (see below for details).

Here we apply our technique to derive approximation algorithms for MCPC and MKPC. As mentioned, to this aim we need LP-based approximation algorithms for the respective problems without cluster capacities, namely MCPK and MKP, respectively. While for the latter a simple greedy algorithm gives a $\frac{1}{2}$-approximation, we need more sophisticated techniques to derive an LP-based approximation algorithm for MCPK. In particular, we adapt the *pipage rounding technique* [1] to obtain an LP-based $\frac{1}{2}(1 - \frac{1}{e})$-approximation for this problem (Sect. 3), which might be of independent interest; this is also one of our main technical contributions in this paper. Based on these algorithms, we then obtain a $\frac{1}{3}(1 - \frac{1}{e})$-approximation algorithm for MCPC (Sect. 4) and a $\frac{1}{3}$-approximation algorithm for MKPC (Sect. 5). Finally, we show that by combining our technique with a more sophisticated iterative rounding approach, we can obtain an

[1] In fact, some preliminary results (omitted from this paper) show that our technique can also be applied to derive an approximation algorithm for the capacitated facility location problem with cluster constraints.

improved $\frac{1}{2}$-approximation algorithm for a special case of the MKPC (Sect. 5); this part is another main technical contribution of the paper. Our iterative rounding approach applies whenever the clusters satisfy a certain *isolation property* (which is guaranteed, for example, if the clusters can be disentangled in a natural way).

Related Literature: The coverage objective that we consider in this paper is a special case of a submodular function and there is a vast literature concerning the problem of maximizing a (monotone) submodular function. Nemhauser and Wolsey [17] were the first to study this problem under a cardinality constraint.

The authors propose a natural greedy algorithm to solve this problem and showed that it achieves a $(1 - \frac{1}{e})$-approximation guarantee. Later, Feige [10] shows that the factor of $(1 - \frac{1}{e})$ is best possible for this problem (unless P = NP).

Khuller et al. [14] use a modification of the greedy algorithm of [17] in combination with partial enumeration to obtain a $(1 - \frac{1}{e})$-approximation algorithm for the Maximum Coverage Problem with a budget constraint. Ageev and Sviridenko [1] introduce the technique of *pipage rounding* and also derive a $(1 - \frac{1}{e})$-approximation algorithm for this problem. Sviridenko [21] observed that the algorithm of [14] can be applied to submodular functions and achieves a $(1 - \frac{1}{e})$-approximation ratio. Ene and Nguyen [7] build upon the approach of Badanidiyuru and Vondrák [2] to derive a $(1 - \frac{1}{e} - \varepsilon)$-approximation algorithm for maximizing a submodular function given a knapsack constraint. For non-monotone submodular functions, a $(1 - \frac{1}{e} - \varepsilon)$-approximation algorithm is given by Kulik et al. [15].

Further, the problem of maximizing a submodular function subject to multiple knapsack constraints has also been studied (see [4,15,16]): Kulik et al. [15] obtain a $(1 - \frac{1}{e} - \varepsilon)$-approximation algorithm (for any $\varepsilon > 0$) for the Maximum Coverage Problem with a d-dimensional knapsack constraint. The technique also extends to non-monotone submodular maximization with a d-dimensional knapsack constraint. Lee et al. [16] give a $(\frac{1}{5} - \varepsilon)$-approximation algorithm for non-monotone submodular maximization with a d-dimensional knapsack. The randomized rounding technique that is used in [16] is similar to that of [15], but the algorithm to solve the fractional relaxation is different. However, in all three works [4,15,16], the interpretation of multiple knapsack constraints is different from the one we use here. In particular, in their setting the costs of the sets and the capacity of a single knapsack are d-dimensional and a feasible assignment needs to satisfy the capacity in every dimension. On the other hand, in our definition there are multiple knapsacks, but their capacity is only one-dimensional. Hence, the techniques in [4,15,16] do not apply to our problem.

Simultaneously to our work, two papers have appeared that also study MCPK (see [8,20]). Fairstein et al. [8] present a randomized $(1 - \frac{1}{e} - \varepsilon)$-approximation algorithm for MCPK in which a greedy approach for submodular maximization is combined with a partitioning of knapsacks in groups of approximately the same size. In Sun et al. [20] a deterministic greedy $(1 - \frac{1}{e} - \varepsilon)$-approximation algorithm is given for the case in which all knapsacks have the same size. For the general case, their deterministic algorithm has an approximation ratio of $\frac{1}{2} - \varepsilon$, and they present a randomized $\left(1 - \frac{1}{e} - \varepsilon\right)$-approximation algorithm.

Another problem that is related to our MCPC is the Cost-Restricted Maximum Coverage Problem with Group Budget Constraints (CMCG) introduced by Chekuri and Kumar [5]. In this problem, there are predefined groups that are a union of sets. Each group has a budget that must not be exceeded by its assigned sets, and there is a single knapsack constraint for all selected sets. The authors give a greedy algorithm that achieves an approximation ratio of $\frac{1}{12}$. Besides this cost-restricted version, Chekuri and Kumar [5] also study the cardinality-restricted version for the number of sets assigned to a group. For this problem, they obtain a $\frac{1}{\alpha+1}$-approximation algorithm that uses an oracle that, given a current solution, returns a set which contribution to the current solution is within a factor α of the optimal contribution of a single set. For the cost-restricted version, Farbstein and Levin [9] obtain a $\frac{1}{5}$-approximation. Guo et al. [12] present a pseudo-polynomial algorithm for CMCG whose approximation ratio is $(1 - \frac{1}{e})$. The MCPC with a single cluster is a special case CMCG, in which each group correspond to a knapsack and each set has a copy for each feasible knapsack. If the solution for a single cluster is seen as a new set, then the MCPC with multiple clusters can be seen as the cardinality restricted version in which each cluster correspond to a group to which at most one set can be assigned. Combining the approaches of Farbstein and Levin [9] and Chekuri and Kumar [5] for the cost-restricted and cardinalty-restricted version, respectively, we obtain a $\frac{1}{6}$-approximation algorithm. In this paper, we improve this ratio to $\frac{1}{3}\left(1 - \frac{1}{e}\right)$.

Finally, we elaborate on the relationship between MKPC and closely related problems. We focus on three knapsack problems that look similar to MKPC but are still slightly different. Nip and Wang [18] consider the *Two-Phase Knapsack Problem (2-PKP)* and obtain a $\frac{1}{4}$-approximation algorithm for this problem. in 2-PKP, items are assigned to knapsacks and the full knapsack needs to be packed in a cluster. Note that MKPC and 2-PKP are different because in MPKC only the costs assigned to a knapsack are restricted by the cluster capacity, whereas in 2-PKP each knapsack contributes with its maximum budget to the cluster. Dudzinski and Walukiewucz [6] study the *Nested Knapsack Problem (NKP)*, where there are multiple subsets of items and each subset has a capacity. That is, in NKP there are no predefined knapsacks to which the items can be assigned, but from each set we have to select the most profitable items. If these sets are disjoint, the problem is called the *Decomposed Knapsack Problem.* The authors present an exact branch-and-bound method, based on the Lagrangean relaxation. Xavier and Miyazawa [22] consider the *Class Constraint Shelf Knapsack Problem (CCSKP).* Here, there is one knapsack with a certain capacity and the items assigned to the knapsack should also be assigned to a shelf. The shelf has a maximum capacity, and between every two shelves a divisor needs to be placed. In the CCSKP, each item belongs to a class and each shelf must only have items of the same class. The authors derive a PTAS for CCSKP in [22].

2 Preliminaries

The *Maximum Coverage Problem with Cluster Constraints (MCPC)* considered in this paper is defined as follows: We are given a collection of m subsets $\mathcal{S} = \{S_1, \ldots, S_m\}$ over a ground set of items $\mathcal{I} = [n]$.[2] Each subset $S_j \subseteq \mathcal{I}$, $j \in [m]$, is associated with a cost $c_j > 0$ and each item $i \in \mathcal{I}$ has a profit $p_i > 0$. For notational simplicity, we identify $\mathcal{S} = [m]$, and for every subset $\mathcal{S}' \subseteq \mathcal{S}$, define $\cup\mathcal{S}' = \cup_{j \in \mathcal{S}'} S_j \subseteq \mathcal{I}$ as the set of items covered by $\mathcal{S}'$. Further, we use $\mathcal{S}(i) \subseteq \mathcal{S}$ to refer to the subsets in $\mathcal{S}$ that contain item $i \in \mathcal{I}$, i.e., $\mathcal{S}(i) = \{j \in \mathcal{S} \mid i \in S_j\}$. In addition, we are given a set of knapsacks $\mathcal{K} = [p]$ and each knapsack $k \in \mathcal{K}$ has a capacity $B_k > 0$. These knapsacks are partitioned into a set of q clusters $\mathcal{C} = [q]$ and each cluster $l \in \mathcal{C}$ has a separate capacity $U_l > 0$. We denote by $\mathcal{K}(l) \subseteq \mathcal{K}$ the subset of knapsacks that are contained in cluster l. The number of knapsacks in the set $\mathcal{K}(l)$ is given by $q(l) := |\mathcal{K}(l)|$. Also, we use $\mathcal{C}(k) \in \mathcal{C}$ to refer to the cluster containing knapsack $k \in \mathcal{K}$.

Our goal is to determine a feasible assignment $\sigma : \mathcal{S} \to \mathcal{K} \cup \{0\}$ of subsets to knapsacks such that the total profit of all covered items is maximized. Each subset $j \in \mathcal{S}$ can be assigned to at most one knapsack in $\mathcal{K}$ and we define $\sigma(j) = 0$ if j remains unassigned. We say that an assignment σ is *feasible* if (i) for every knapsack $k \in \mathcal{K}$, the total cost of all subsets assigned to k is at most B_k, and (ii) for every cluster $l \in \mathcal{C}$, the total cost of all subsets assigned to the knapsacks in $\mathcal{K}(l)$ of cluster l is at most U_l. Given an assignment σ, let $\mathcal{S}_\sigma(k) \subseteq \mathcal{S}$ be the set of subsets assigned to knapsack $k \in \mathcal{K}$ under σ, and let $\mathcal{S}_\sigma = \cup_{k \in \mathcal{K}} \mathcal{S}_\sigma(k) \subseteq \mathcal{S}$ be the set of all assigned subsets. Using the notation above, the MCPC problem is formally defined as follows:

$$\max_{\sigma} \left\{ \sum_{i \in \cup\mathcal{S}_\sigma} p_i \;\middle|\; \sum_{j \in \mathcal{S}_\sigma(k)} c_j \leq B_k \;\; \forall k \in \mathcal{K}, \; \sum_{k \in \mathcal{K}(l)} \sum_{j \in \mathcal{S}_\sigma(k)} c_j \leq U_l \;\; \forall l \in \mathcal{C} \right\}.$$

We make the following assumptions throughout the paper: We assume that for every cluster $l \in \mathcal{C}$: (i) for every knapsack $k \in \mathcal{K}(l)$, $B_k \leq U_l$, and (ii) $\sum_{k \in \mathcal{K}(l)} B_k > U_l$.

Note that these assumptions are without loss of generality: If the first condition is violated by some $k \in \mathcal{K}(l)$ then we can simply redefine $B_k = U_l$. If the second condition is not satisfied then the capacity of cluster l is redundant because each knapsack in cluster l can then contain its maximum capacity.

The special case of MCPC in which all cluster capacities are redundant is called the *Maximum Coverage Problem with Knapsack Constraints (MCPK)*. Another special case of MCPK is the classical *Multiple Knapsack Problem (MKP)* which we obtain if $\mathcal{S} = [n]$ and $S_i = \{i\}$ for all $i \in \mathcal{S}$. We refer to the generalization of MKP with (non-redundant) cluster capacities as the *Multiple Knapsack Problem with Cluster Constraints (MKPC)*.

[2] Throughout the paper, given an integer $n \geq 1$ we use $[n]$ to refer to the set $\{1, \ldots, n\}$.

3 Maximum Coverage with Knapsack Constraints

We start by deriving an approximation algorithm for the Maximum Coverage Problem with Knapsack Constraints (MCPK). The following is a natural integer linear programming (ILP) formulation:

$$
\begin{aligned}
\max \quad & L(x) = \textstyle\sum_{i \in \mathcal{I}} p_i y_i && && \text{(1a)}\\
\text{subject to} \quad & \textstyle\sum_{j \in \mathcal{S}} c_j x_{jk} \leq B_k && \forall k \in \mathcal{K} && \text{(1b)}\\
& \textstyle\sum_{k \in \mathcal{K}} x_{jk} \leq 1 && \forall j \in \mathcal{S} && \text{(1c)}\\
& \textstyle\sum_{j \in \mathcal{S}(i)} \sum_{k \in \mathcal{K}} x_{jk} \geq y_i && \forall i \in \mathcal{I} && \text{(1d)}\\
& x_{jk} \in \{0,1\} && \forall j \in \mathcal{S},\ \forall k \in \mathcal{K} && \text{(1e)}\\
& y_i \in \{0,1\} && \forall i \in \mathcal{I} && \text{(1f)}
\end{aligned}
$$

The decision variable x_{jk} indicates whether set $j \in \mathcal{S}$ is assigned to knapsack $k \in \mathcal{K}$. Constraint (1b) ensures that the total cost assigned to each knapsack is at most its budget and constraint (1c) makes sure that each set is assigned to at most one knapsack. In addition, the decision variable y_i indicates whether item $i \in \mathcal{I}$ is covered. Constraint (1d) ensures that an item can only be covered if at least one of the sets containing it is assigned to a knapsack. We refer to above ILP (1a)–(1f) as (IP) and to its corresponding linear programming relaxation as (LP). It is important to realize that for any feasible fixing of the x_{jk}-variables (satisfying (1b) and (1c)), the optimal y_i-variables are easily determined by setting $y_i = \min\{1, \sum_{j \in \mathcal{S}(i)} \sum_{k \in \mathcal{K}} x_{jk}\}$. As a consequence, an optimal solution (x, y) of the above program is fully specified by the corresponding x-part.[3]

We crucially exploit that we can find a solution to the LP-relaxation which satisfies the *bounded split property*. Let x be a fractional solution of (LP). We define the *support graph* H_x of x as the bipartite graph $H_x = (\mathcal{S} \cup \mathcal{K}, E_x)$ with node sets $\mathcal{S}$ and $\mathcal{K}$ on each side of the bipartition, respectively, and the edge set E_x contains all edges $\{j, k\}$ for which x_{jk} is non-integral, i.e., $E_x = \{\{j, k\} \in \mathcal{S} \times \mathcal{K} \mid 0 < x_{jk} < 1\}$. Let $M \subseteq E_x$ be a matching of the support graph H_x.[4] M *saturates* all nodes in $\mathcal{S}$ if for every (non-isolated) node $j \in \mathcal{S}$ there is a matching edge $\{j, k\} \in M$; we also say that M is an *$\mathcal{S}$-saturating matching*

[3] We comment on a subtle but important point here: Note that a set $j \in \mathcal{S}$ cannot be assigned to a knapsack $k \in \mathcal{K}$ whenever $c_j > B_k$. While these restrictions are taken care of implicitly in the ILP formulation (IP), they are not in the LP-relaxation (LP). In fact, we would have to add these restrictions explicitly by defining variables x_{jk} only for all $j \in \mathcal{S}$ and $k \in \mathcal{K}(j)$, where $\mathcal{K}(j) \subseteq \mathcal{K}$ is the set of knapsacks whose capacity is at least c_j, and adapt the constraints accordingly. However, these adaptations are straight-forward, and for notational convenience, we do not state them explicitly. In the remainder, whenever we refer to (LP) our understanding is that we refer to the corresponding LP-relaxation with these assignment restrictions incorporated. No confusion shall arise.

[4] Recall that a matching $M \subseteq E_x$ is a subset of the edges such that no two edges in M have a node in common.

of H_x. A feasible solution x of (LP) satisfies the *bounded split property* if the support graph H_x of x has an $\mathcal{S}$-saturating matching.

For MKP, it is known that there exists an optimal solution x^* of the LP-relaxation for which the bounded split property holds (see, e.g., [3,19]). Deriving a $\frac{1}{2}$-approximation algorithm for the problem is then easy: The idea is to decompose x^* into an integral and a fractional part. The integral part naturally corresponds to a feasible integral assignment and, exploiting the bounded split property, the corresponding $\mathcal{S}$-saturating matching gives rise to another feasible integral assignment. Thus, by taking the better of the two assignments, we recover at least half of the optimal LP-solution. Unfortunately, for our more general MCPK problem the optimal solution of (LP) does not necessarily satisfy the bounded split property.

Instead, below we show that there always exists a solution to the LP-relaxation of MCPK which satisfies the bounded split property and is only a factor $1 - \frac{1}{e}$ away from the optimal solution. We use the technique of *pipage rounding* [1]. Define a new program (CP) as follows:

$$\begin{aligned} \max \quad & F(x) = \textstyle\sum_{i \in \mathcal{I}} p_i \Big(1 - \prod_{j \in \mathcal{S}(i)} \big(1 - \sum_{k \in \mathcal{K}} x_{jk}\big)\Big) && && \text{(2a)} \\ \text{subject to} \quad & \textstyle\sum_{j \in \mathcal{S}} c_j x_{jk} \leq B_k && \forall k \in \mathcal{K} && \text{(2b)} \\ & \textstyle\sum_{k \in \mathcal{K}} x_{jk} \leq 1 && \forall j \in \mathcal{S} && \text{(2c)} \\ & x_{jk} \in [0,1] && \forall j \in \mathcal{S},\ \forall k \in \mathcal{K} && \text{(2d)} \end{aligned}$$

Obviously, a feasible solution x for the problem (LP) is also a feasible solution for (CP). In addition, the objective function values of (LP) and (CP) are the same for every integral solution, i.e., $L(x) = F(x)$ for every integer solution x.[5] Moreover, for fractional solutions x the value $F(x)$ is lower bounded by $L(x)$ as we show in the next lemma:

Lemma 1. *For every feasible solution x of (LP), we have that $F(x) \geq \left(1 - \frac{1}{e}\right) L(x)$.*

Proof. Let x be a feasible solution for (LP). Fix an item $i \in \mathcal{I}$ and let $s = |\mathcal{S}(i)|$ be the number of sets containing i. It can be shown that

$$1 - \prod_{j \in \mathcal{S}(i)} \Big(1 - \sum_{k \in \mathcal{K}} x_{jk}\Big) \geq 1 - \Big(\frac{1}{s} \sum_{j \in \mathcal{S}(i)} \Big(1 - \sum_{k \in \mathcal{K}} x_{jk}\Big)\Big)^s \geq \Big(1 - \frac{1}{e}\Big) y_i,$$

where the first inequality holds because for any n non-negative numbers $(a_i)_{i \in [n]}$, holds that $\prod_{i=1}^n a_i \leq (\frac{1}{n} \sum_{i=1}^n a_i)^n$ (see, e.g., [11]) and the second inequality follows from some standard arguments. The claim now follows from the above relation and the definitions of $F(x)$ and $L(x)$. □

In Theorem 1 below, we show that we can transform an optimal LP-solution x^*_{LP} into a solution x that satisfies the bounded split property without decreasing the objective value of (CP).

[5] Our formulation (CP) is even slightly stronger than standard pipage formulations in the sense that $L(x) = F(x)$ if for all $j \in \mathcal{S}$ it holds that $\sum_{k \in \mathcal{K}} x_{jk} \in \{0, 1\}$.

Theorem 1. *There exists a feasible solution x of (LP) that satisfies the bounded split property and for which $F(x) \geq F(x^*_{LP})$.*

Let x be a feasible solution to (LP) and let $H_x = (\mathcal{S} \cup \mathcal{K}, E_x)$ be the support graph of x. Consider a (maximal) path $P = \langle u_1, \ldots, u_t \rangle$ in H_x that starts and ends with a node of degree one. We call P an *$\mathcal{S}$-$\mathcal{S}$-path* if $u_1, u_t \in \mathcal{S}$, an *$\mathcal{S}$-$\mathcal{K}$-path* if $u_1 \in \mathcal{S}$ and $u_t \in \mathcal{K}$, and a *$\mathcal{K}$-$\mathcal{K}$-path* if $u_1, u_t \in \mathcal{K}$.

An outline of the proof of Theorem 1 is as follows: First, we show that there exists an optimal solution $x^* = x^*_{\text{LP}}$ for (LP) such that the support graph H_{x^*} is acyclic (Lemma 2). Second, we prove that from x^* we can derive a solution x' whose support graph $H_{x'}$ does not contain any $\mathcal{S}$-$\mathcal{S}$-paths such that the objective function value of (CP) does not decrease (Lemma 3). Finally, we show that this solution x' satisfies the bounded split property (Lemma 4). Combining these lemmas proves Theorem 1.

Lemma 2. *There is an optimal solution $x^* = x^*_{LP}$ of (LP) whose support graph H_{x^*} is acyclic.*

Proof. Let x be an optimal solution for (LP) and suppose $C = \langle u_1, \ldots, u_t, u_1 \rangle$ is a cycle in H_x. C has even length because H_x is bipartite and can thus be decomposed into two matchings M_1 and M_2 with $|M_1| = |M_2|$. Define

$$\varepsilon := \min\Big\{ \min_{\{j,k\} \in M_1} \{c_j x_{jk}\}, \min_{\{j,k\} \in M_2} \{c_j(1 - x_{jk})\} \Big\}.$$

We call each edge on C for which the minimum is attained a *critical edge*. We use ε to define a new solution $x(\varepsilon)$ as follows: $x_{jk}(\varepsilon) = x_{jk} + \varepsilon / c_j$ if $\{j,k\} \in M_1$, $x_{jk}(\varepsilon) = x_{jk} - \varepsilon / c_j$ if $\{j,k\} \in M_2$, and $x_{jk}(\varepsilon) = x_{jk}$ otherwise. In the way ε is defined, the value of each critical edge $\hat{e}$ on C with respect to $x(\varepsilon)$ is integral and thus $\hat{e}$ is not part of the support graph $H_{x(\varepsilon)}$. Hence, $H_{x(\varepsilon)}$ is a subgraph of H_x that has at least one cycle less. It can be verified that $x(\varepsilon)$ is a feasible solution to (LP) and has the same objective function value as x (see appendix). By repeating this procedure, we eventually obtain a feasible solution x^* of (LP) such that H_{x^*} is a subgraph of H_x that does not contain any cycles and $L(x^*) = L(x)$, i.e., x^* is an optimal solution. □

Lemma 3. *There exists a feasible solution x' of (LP) whose support graph $H_{x'}$ is acyclic and does not contain any $\mathcal{S}$-$\mathcal{S}$-paths, and which satisfies $F(x') \geq F(x^*_{LP})$.*

Proof. By Lemma 2, there exists an optimal solution x for (LP) whose support graph H_x is acyclic. Let $P = \langle u_1, \ldots, u_t \rangle$ be a $\mathcal{S}$-$\mathcal{S}$-path in H_x. Recall that nodes u_1 and u_t have degree one. P has even length because H_x is bipartite and can thus be decomposed into two matchings M_1 and M_2 with $|M_1| = |M_2|$. Let $x(\varepsilon)$ be defined as in the proof of Lemma 2 for every $\varepsilon \in [-\varepsilon_1, \varepsilon_2]$, where

$$\varepsilon_1 := \min\Big\{ \min_{\{j,k\} \in M_1} \{c_j x_{jk}\}, \min_{\{j,k\} \in M_2} \{c_j(1 - x_{jk})\} \Big\}$$

$$\varepsilon_2 := \min\Big\{ \min_{\{j,k\} \in M_1} \{c_j(1 - x_{jk})\}, \min_{\{j,k\} \in M_2} \{c_j x_{jk}\} \Big\}.$$

It is not hard to verify that $x(\varepsilon)$ is a feasible solution for (CP) for every $\varepsilon \in [-\varepsilon_1, \varepsilon_2]$ (see appendix).

Next, we show that $F(x(-\varepsilon_1))$ or $F(x(\varepsilon_2))$ is at least as large as $F(x)$ by proving that $F(x(\varepsilon))$ as a function of ε is convex; in fact, we show convexity for each item $i \in \mathcal{I}$ separately. Observe that the first and the last edge of P are in different matchings, say $\{u_1, u_2\} \in M_1$ and $\{u_{t-1}, u_t\} \in M_2$. The contribution of i to the objective function $F(x(\varepsilon))$ can be written as $p_i f_i(\varepsilon)$, where

$$f_i(\varepsilon) = \Big(1 - \prod_{j \in \mathcal{S}(i) \cap \{u_1, u_t\}} \Big(1 - \sum_{k \in \mathcal{K}} x_{jk}(\varepsilon)\Big) \prod_{j \in \mathcal{S}(i) \setminus \{u_1, u_t\}} \Big(1 - \sum_{k \in \mathcal{K}} x_{jk}\Big)\Big).$$

(We adopt the convention that the empty product is defined to be 1.) Note that the latter product $\prod_{j \in \mathcal{S}(i) \setminus \{u_1, u_t\}} (1 - \sum_{k \in \mathcal{K}} x_{jk})$ is independent of ε and has a value between 0 and 1.

Case 1: $|\{u_1, u_t\} \cap \mathcal{S}(i)| = 2$. In this case, we have

$$\prod_{j \in \mathcal{S}(i) \cap \{u_1, u_t\}} \Big(1 - \sum_{k \in \mathcal{K}} x_{jk}(\varepsilon)\Big) = \Big(1 - (x_{u_1 u_2} - \frac{\varepsilon}{c_{u_1}})\Big)\Big(1 - (x_{u_t u_{t-1}} + \frac{\varepsilon}{c_{u_t}})\Big)$$

That is, $f_i(\varepsilon)$ is a quadratic function of ε and the coefficient of the quadratic term is $1/(c_{u_1} c_{u_t})$, which is positive. Thus, $f_i(\varepsilon)$ is convex.

Case 2: $|\{u_1, u_t\} \cap \mathcal{S}(i)| = 1$. In this case, $f_i(\varepsilon)$ is a linear function in ε.

Case 3: $|\{u_1, u_t\} \cap \mathcal{S}(i)| = 0$. In this case, $f_i(\varepsilon)$ is independent of ε.

We conclude that $F(x(\varepsilon))$ is convex in ε and its maximum over $[-\varepsilon_1, \varepsilon_2]$ is thus attained at one of the endpoints, i.e., $\max\{F(x(-\varepsilon_1)), F(x(\varepsilon_2))\} = \max_{\varepsilon \in [-\varepsilon_1, \varepsilon_2]}\{F(x(\varepsilon))\} \geq F(x(0)) = F(x)$.

As a result, we can find a feasible solution $x' \in \{x(-\varepsilon_1), x(\varepsilon_2)\}$ with the property that $F(x') \geq F(x)$. Further, x' has at least one fractional variable on P less than x. Thus, $H_{x'}$ is a subgraph of H_x with at least one edge of P removed. By repeating this procedure, we eventually obtain a feasible solution x' of (CP) whose support graph does not contain any $\mathcal{S}$-$\mathcal{S}$-paths, and for which $F(x') \geq F(x)$. □

Lemma 4. *Let x' be a feasible solution of (LP) whose support graph $H_{x'}$ is acyclic and does not contain any $\mathcal{S}$-$\mathcal{S}$-paths. Then x' satisfies the bounded split property.*

Proof. We show that there exists an $\mathcal{S}$-saturating matching M in $H_{x'}$. By assumption, $H_{x'}$ is acyclic and thus a forest. We construct a matching M_T for each tree T of the forest $H_{x'}$. Our final matching M is then simply the union of all these matchings, i.e., $M = \cup_T M_T$.

Consider a tree T of $H_{x'}$ and root it at an arbitrary node $r \in \mathcal{K}$. By assumption, T has at most one leaf in $\mathcal{S}$ because otherwise there would exist an $\mathcal{S}$-$\mathcal{S}$-path in $H_{x'}$. We can construct a matching M_T that matches all the nodes in $T \cap \mathcal{S}$ as follows:[6] If there is a (unique) $\mathcal{S}$-leaf, say $j \in \mathcal{S}$, then we match each $\mathcal{S}$-node on

[6] We slightly abuse notation here by letting T refer to the tree and the set of nodes it spans.

Algorithm 1: $\frac{1}{2}(1-\frac{1}{e})$-approximation algorithm for MCPK.

1 Compute an optimal solution x^* to (LP). ;
2 Derive a solution x from x^* satisfying the bounded split property (Theorem 1). ;
3 Let M be the corresponding $\mathcal{S}$-saturating matching. ;
4 Decompose the fractional solution x into x^1, x^2 as follows:

$$x^1_{jk} = \begin{cases} x^1_{jk} = 1 & \text{if } x_{jk} = 1 \\ x^1_{jk} = 0 & \text{otherwise} \end{cases} \quad \text{and} \quad x^2_{jk} = \begin{cases} x^2_{jk} = 1 & \text{if } x_{jk} \in (0,1),\ \{j,k\} \in M \\ x^2_{jk} = 0 & \text{otherwise.} \end{cases}$$

5 Output $x_{\text{ALG}} \in \arg\max\{L(x^1), L(x^2)\}$;
;

the path from j to the root r to its unique parent in T. Each remaining $\mathcal{S}$-node in T is matched to one of its children (chosen arbitrarily); there always is at least one child as j is the only $\mathcal{S}$-leaf in T. Note that this defines a matching M_T that matches all nodes in $T \cap \mathcal{S}$ as desired. □

In light of Theorem 1, our approximation algorithm (Algorithm 1) simply chooses the better of the integral and the rounded solution obtained from an optimal LP-solution.

Theorem 2. *Algorithm 1 is a* $\frac{1}{2}\left(1-\frac{1}{e}\right)$*-approximation algorithm for MCPK.*

Proof. We bound the approximation ratio here, we have

$$\begin{aligned} L(x_{\text{ALG}}) &= \max\{L(x^1), L(x^2)\} \geq \frac{1}{2}(L(x^1) + L(x^2)) = \frac{1}{2}(F(x^1) + F(x^2)) \\ &\geq \frac{1}{2}F(x) \geq \frac{1}{2}F(x^*) \geq \frac{1}{2}\left(1-\frac{1}{e}\right)L(x^*) \geq \frac{1}{2}\left(1-\frac{1}{e}\right)\text{OPT}. \end{aligned}$$

Here, the second equality follows from the fact that the objective function values of (LP) and (CP) are the same for integer solutions (as observed above). To see that the second inequality holds, note that by the definitions of x^1 and x^2 we have

$$\sum_{j\in\mathcal{S}}\sum_{k\in\mathcal{K}} x_{jk} \leq \sum_{j\in\mathcal{S}}\sum_{k\in\mathcal{K}} (x^1_{jk} + x^2_{jk})$$

and the function $F(x)$ is non-decreasing in x. The third inequality follows from Lemma 3 and the fourth inequality holds because of Lemma 1. The final inequality holds because (LP) is a relaxation of the integer program (IP) of MCPK. □

4 Maximum Coverage with Cluster Constraints

We derive an approximation algorithm for the Maximum Coverage Problem with Cluster Constraints (MCPC). Our algorithm exploits the existence of an LP-based approximation algorithm for the problem without cluster constraints. In particular, we use our algorithm for MCPK derived in the previous section as a subroutine to obtain a $\frac{1}{3}(1-\frac{1}{e})$-approximation algorithm for MCPC.

A key element of our approach is integrating the cluster capacities into the IP formulation of MCPK by introducing a variable z_{kl} for every cluster $l \in \mathcal{C}$ and every knapsack $k \in \mathcal{K}(l)$, which specifies the fraction of the cluster capacity U_l that is assigned to knapsack k. The MIP formulation for MCPCis as follows:

$$
\begin{aligned}
\max \quad & L(x,z) = \sum_{i \in \mathcal{I}} p_i y_i && && \text{(3a)}\\
\text{subject to} \quad & (1b)-(1f) && && \text{(3b)}\\
& \textstyle\sum_{j \in \mathcal{S}} c_j x_{jk} \le U_l z_{kl} && \forall l \in \mathcal{C},\ \forall k \in \mathcal{K}(l) && \text{(3c)}\\
& \textstyle\sum_{k \in \mathcal{K}(l)} z_{kl} \le 1 && \forall l \in \mathcal{C} && \text{(3d)}\\
& z_{kl} \ge 0 && \forall l \in \mathcal{C},\ \forall k \in \mathcal{K}(l) && \text{(3e)}
\end{aligned}
$$

Note that constraints (3c)–(3d) ensure that the capacity of cluster $l \in \mathcal{C}$ is not exceeded. The advantage of this formulation is that, once we know the optimal values of the z_{kl}-variables, the remaining problem basically reduces to an instance of MCPK (though a subtle point remains, as explained below). We use (IP) and (LP) to refer to the integer formulation above and its relaxation, respectively. Let OPT refer to the objective function value of an optimal solution to MCPC. Note that also for this formulation, the remarks that are given in Footnote 3 apply. Similar to MCPK, for every feasible fixing of the x_{jk}, z_{kl}-variables, the optimal y_i-variables can be determined as before. As a consequence, an optimal solution (x, y, z) of the above program is fully specified by (x, z).

It will be convenient to assume that the knapsacks $\mathcal{K}(l) = \{1, \dots, q(l)\}$ of each cluster $l \in \mathcal{C}$ are ordered by non-increasing capacities (breaking ties arbitrarily), i.e., if $k, k' \in \mathcal{K}(l)$ with $k < k'$ then $B_k \ge B_{k'}$. The following notion will be crucial: The knapsack $\kappa(l) \in \mathcal{K}(l)$ which satisfies $\sum_{k=1}^{\kappa(l)-1} B_k \le U_l < \sum_{k=1}^{\kappa(l)} B_k$ is called the *critical knapsack* of cluster l.[7]

We first show that the z-variables of an optimal LP-solution admit a specific structure. Intuitively, the lemma states that the cluster capacity U_l of each cluster l is shared maximally among the first $\kappa(l) - 1$ knapsacks in $\mathcal{K}(l)$ and the remaining capacity is assigned to the critical knapsack $\kappa(l)$.

Lemma 5. *There is an optimal solution (x^*, z^*) of (LP) such that for every cluster $l \in \mathcal{C}$, $z^*_{kl} = B_k/U_l$ for $k < \kappa(l)$, $z^*_{kl} = 1 - \sum_{t=1}^{\kappa(l)-1} z^*_{tl}$ for $k = \kappa(l)$ and $z^*_{kl} = 0$ otherwise.*

An approach that comes to one's mind is as follows: Fix the z^*-values as in Lemma 5 and let LP(z^*) be the respective LP-relaxation. Note that LP(z^*) is basically the same as the LP-relaxation of MCPK, where each knapsack $k \in \mathcal{K}(l)$, $l \in \mathcal{C}$, has a *reduced capacity* of $\min\{B_k, U_l z^*_{kl}\}$. So we could

[7] Note that there always exists such a critical knapsack by the assumption that $\sum_{k \in \mathcal{K}(l)} B_k > U_l$.

Algorithm 2: $\frac{1}{3}(1-\frac{1}{e})$-approximation algorithm for the MCPC.

1 Fix z^* as in Lemma 5 and compute an optimal solution x^* to (LP(z^*)). ;
2 Derive a solution x from x^* that satisfies the bounded split property (Theorem 1). ;
3 Let M be the corresponding $\mathcal{S}$-saturating matching. ;
4 Decompose the fractional solution x into x^1, x^2, x^3 as follows:

$$x^1_{jk} = \begin{cases} x^1_{jk} = 1 & \text{if } x_{jk} = 1 \\ x^1_{jk} = 0 & \text{otherwise} \end{cases}$$

$$x^2_{jk} = \begin{cases} x^2_{jk} = 1 & \text{if } x_{jk} \in (0,1),\ k \text{ not critical, } \{j,k\} \in M \\ x^2_{jk} = 0 & \text{otherwise} \end{cases}$$

$$x^3_{jk} = \begin{cases} x^3_{jk} = 1 & \text{if } x_{jk} \in (0,1),\ k \text{ critical, } \{j,k\} \in M \\ x^3_{jk} = 0 & \text{otherwise} \end{cases}$$

5 Output $x_{\text{ALG}} \in \arg\max\{L(x^1), L(x^2), L(x^3)\}$. ;

compute an optimal solution x^* to LP(z^*) and use our LP-based approximation algorithm (Algorithm 1) to derive an integral solution x_{ALG} satisfying $L(x_{\text{ALG}}) \geq \frac{1}{2}(1-\frac{1}{e})L(x^*, z^*) \geq \frac{1}{2}(1-\frac{1}{e})\text{OPT}$. Unfortunately, however, this approach fails because of the following subtle point: If a set $j \in \mathcal{S}$ is fractionally assigned to some critical knapsack $k \in \mathcal{K}$ in the optimal LP-solution x^* of LP(z^*), then it might be infeasible to assign j to k integrally. We could exclude these infeasible assignments beforehand (i.e., by setting $x_{jk} = 0$ whenever $c_j > U_l z^*_{kl}$ for a critical knapsack k), but then an optimal LP-solution might not recover a sufficiently large fraction of OPT.

Instead, we can fix this problem by using a slightly more refined algorithm: We decompose the fractionally assigned sets into two solutions, one using non-critical knapsacks and one using critical knapsacks only, and then choose the better of those and the integral solution. This way, we obtain a $\frac{1}{3}(1-\frac{1}{e})$-approximation algorithm for MCPC.

Theorem 3. *Algorithm 2 is a $\frac{1}{3}(1-\frac{1}{e})$-approximation algorithm for MCPC.*

Proof. Note that by Theorem 1 the fractional solution x derived in Step 2 of the algorithm is a feasible solution to (LP(z^*)). Clearly, x^1 is a feasible (integral) solution. Further, the $\mathcal{S}$-saturating matching M ensures that $c_j \leq B_k$ for every $\{j,k\} \in M$. In particular, this implies that the solution x^2 is feasible because for every (non-critical) knapsack $k \in \mathcal{K}(l)$, $l \in \mathcal{C}$, we have $B_k = U_l z^*_{kl}$ by Lemma 5, and thus $\sum_{j \in \mathcal{S}} c_j x^2_{jk} \leq B_k = U_l z^*_{kl}$.

It remains to argue that x^3 is feasible. But this holds because for every cluster $l \in \mathcal{C}$ there is at most one set $j \in \mathcal{S}$ assigned to this cluster, namely the one (if any) assigned to the critical knapsack $\kappa(l) \in \mathcal{K}(l)$ with $\{j,k\} \in M$. In particular, for $k = \kappa(l)$ we have $\sum_{j \in \mathcal{S}} c_j x^3_{jk} \leq B_k \leq U_l$.

It remains to bound the approximation factor of the algorithm. Using the same arguments as in the proof of Theorem 2, we obtain

$$L(x_{\text{ALG}}) \geq \frac{1}{3}(L(x^1) + L(x^2) + L(x^3)) = \frac{1}{3}(F(x^1) + F(x^2) + F(x^3)) \geq \frac{1}{3}F(x)$$
$$\geq \frac{1}{3}F(x^*) \geq \frac{1}{3}\left(1 - \frac{1}{e}\right)L(x^*) = \frac{1}{3}\left(1 - \frac{1}{e}\right)L(x^*, z^*) \geq \frac{1}{3}\left(1 - \frac{1}{e}\right)\text{OPT}.$$

Note that the last equality holds because x^* is an optimal solution to (LP(z^*)).

5 Multiple Knapsack with Cluster Constraints

Our technique introduced in the previous section can also be applied to other cluster capacitated problems. One such example is the Multiple Knapsack Problem with Cluster Constraints (MKPC). Here we only sketch the ideas to derive a $\frac{1}{3}$-approximation algorithm for MKPC and a more sophisticated iterative rounding scheme that provides a $\frac{1}{2}$-approximation algorithm for certain special cases of MKPC (details will be given in the full version of the paper).

Note that for MKPC the notions of sets and items coincide and we simply refer to them as items; in particular, each item $j \in \mathcal{S}$ now has a profit p_j and a cost c_j. Thus, we can also drop the y-variables in the ILP formulation of the problem. Throughout this section, we assume that the knapsacks in $\mathcal{K} = [p]$ are ordered by non-increasing capacities, i.e., if $k, k' \in \mathcal{K}$ with $k < k'$ then $B_k \geq B_{k'}$.

5.1 $\frac{1}{3}$-Approximation for MKPC with General Clusters

MKPC is a generalization of the classical multiple knapsack problem (MKP). For MKP, the optimal solution of the LP-relaxation satisfies the bounded split property (see, e.g., [19]) and there is a natural greedy algorithm to find an optimal solution of the LP-relaxation (see, e.g., [13]). Here we exploit some ideas of the previous section to derive a greedy algorithm (called GREEDY subsequently) for MKPC and prove that it computes an optimal solution to the LP-relaxation; this algorithm is also used at the core of our iterative rounding scheme in the next section. Further, we show that the constructed solution satisfies the bounded split property. Exploiting this, we can then easily obtain a $\frac{1}{3}$-approximation algorithm for MKPC.

Our algorithm first fixes optimal z^*-variables as defined in Lemma 5 and then runs an adapted version of the greedy algorithm in [13] on the instance with the reduced capacities.[8] GREEDY assigns items in non-increasing order of their *efficiency ratios* (breaking ties arbitrarily), where the efficiency ratio of item $j \in \mathcal{S}$ is defined as p_j/c_j. When item j is considered, it is assigned to the knapsack with the smallest capacity that can hold the item and has some residual capacity. More formally, a knapsack $k \in \mathcal{K}$ *can hold* item $j \in \mathcal{S}$ if $B_k \geq c_j$. Further, we say that a knapsack k has *residual capacity* with respect to (x^*, z^*) if $res_k(x^*, z^*) := U_{\mathcal{C}(k)} z^*_{k\mathcal{C}(k)} - \sum_{j \in \mathcal{S}} c_j x^*_{jk} > 0$. (Recall that $\mathcal{C}(k)$

[8] The greedy algorithm for MKP described in [13] operates on a per-knapsack basis, while our algorithm proceeds on a per-item basis.

denotes the cluster to which knapsack k belongs.) We continue this way until either item j is assigned completely (possibly split over several knapsacks) or all knapsacks that can hold j have zero residual capacity. We then continue with the next item in the order.

We show that GREEDY computes an optimal fractional solution which satisfies the bounded split property. We say that an item $j \in \mathcal{S}$ is a *split item* if it is fractionally assigned to one or multiple knapsacks. Let SS refer to the set of all split items. An item $j \in SS$ is a *split item of knapsack* $k \in \mathcal{K}$ if k is the knapsack with the smallest capacity to which j is assigned in GREEDY.

Lemma 6. GREEDY *computes an optimal solution* (x^*, z^*) *to the LP-relaxation of MKPC which satisfies the bounded split property.*

In the analysis of Algorithm 2 for the MCPC, a factor $(1-\frac{1}{e})$ is lost in the approximation guarantee by transforming the optimal solution of the LP-relaxation to a solution that satisfies the bounded split property. By Lemma 6, we can avoid this loss here.

Theorem 4. *Algorithm 2 is a* $\frac{1}{3}$*-approximation algorithm for MKPC.*

5.2 $\frac{1}{2}$-Approximation for MKPC with Isolation Property

We derive an iterative rounding $\frac{1}{2}$-approximation algorithm for instances of MKPC that satisfy a certain *isolation property*. A practical situation in which the isolation property hold is when standardized containers that all have the same capacity are used.

Let (x^*, z^*) be the optimal solution to the LP-relaxation of MKPC computed by GREEDY. We say that an item $j \in \mathcal{S}$ is an *unsplit item* if it is integrally assigned to some knapsack, i.e., $x^*_{jk} = 1$ for some $k \in \mathcal{K}$. Let US and SS refer to the sets of unsplit and split items, respectively. As argued in the proof of Lemma 6, the split item $j \in SS$ of knapsack $k \in \mathcal{K}$ is unique; we use $\varsigma_k = j$ to refer to the split item of knapsack k. Further, we use $US(k)$ to denote the set of all unsplit items assigned to knapsack k. For each cluster $l \in \mathcal{C}$, we denote by $US(l) = \cup_{k \in \mathcal{K}(l)} US(k)$ the set of all unsplit items and by $SS(l) = \cup_{k \in \mathcal{K}(l)} \varsigma_k$ the set of all split items assigned to l.

A cluster $\iota \in \mathcal{C}$ is said to be *isolated* if for every item $j \in US(l) \cup SS(l)$ assigned to some cluster $l \neq \iota$ it holds that $x^*_{jk} = 0$ for all $k \in \mathcal{K}(\iota)$. We say that a class of instances of MKPC satisfies the *isolation property* if after the removal of an arbitrary set of clusters there always exists an isolated cluster. For example, the isolation property holds true for instances whose clusters can be *disentangled* in the sense that we can impose an order on the set of clusters $\mathcal{C} = [q]$ such that for any two clusters $l, l' \in \mathcal{C}$ with $l < l'$ it holds that $\min_{k \in \mathcal{K}(l)}\{B_k\} \geq \max_{k \in \mathcal{K}(l')}\{B_k\}$.

The following Rounding Lemma provides a crucial building block of our iterative rounding scheme. It shows that we can always find a feasible assignment σ which recovers at least half of the fractional profit of an isolated cluster ι. Recall that $\mathcal{S}_\sigma$ refers to the set of items assigned under σ.

Algorithm 3: Iterative rounding algorithm for MKPC with isolation property.

1 Let $\text{LP}^{(1)}$ be the original LP-relaxation of MKPC. ;
2 **for** $i = 1, \ldots, q$ **do**
3   Compute an optimal solution $(x^{(i)}, z^{(i)})$ to $\text{LP}^{(i)}$ using GREEDY. ;
4   Identify an isolated cluster $\iota^{(i)}$ with respect to $(x^{(i)}, z^{(i)})$. ;
5   Obtain a feasible assignment $\sigma^{(i)}$ for cluster $\iota^{(i)}$ using the Rounding Lemma. ;
6   Add the constraints (5a)–(5b) fixing the assignment $\sigma^{(i)}$ for $\iota^{(i)}$ to obtain $\text{LP}^{(i+1)}$.

Lemma 7 (Rounding Lemma). *Let ι be an isolated cluster. Then there exists a feasible assignment $\sigma : US(\iota) \cup SS(\iota) \to \mathcal{K}(\iota)$ such that*

$$\sum_{j \in \mathcal{S}_\sigma} p_j \geq \sum_{j \in \mathcal{S}_\sigma} \sum_{k \in \mathcal{K}(\iota)} p_j x^*_{jk} \geq \frac{1}{2} \Big(\sum_{j \in \mathcal{S}} \sum_{k \in \mathcal{K}(\iota)} p_j x^*_{jk} \Big). \tag{4}$$

We next describe our iterative rounding scheme (see Algorithm 3). We first compute an optimal solution (x^*, y^*) to the LP relaxation of MKPC using GREEDY. Because of the isolation property, there exists an isolated cluster ι. We then apply the Rounding Lemma to obtain an assignment σ for cluster ι. After that, we fix the corresponding variables in the LP-relaxation accordingly and repeat. By iterating this procedure, we obtain a sequence of isolated clusters $\iota^{(1)}, \ldots, \iota^{(q)}$ and assignments $\sigma^{(1)}, \ldots, \sigma^{(q)}$, where $\sigma^{(l)}$ is the assignment obtained by applying the Rounding Lemma to cluster $\iota^{(l)}$.

Let $\sigma(i) = \langle \sigma^{(1)}, \ldots, \sigma^{(i)} \rangle$ be the (combined) assignment for the first i clusters $\iota^{(1)}, \ldots, \iota^{(i)}$ that we obtain at the end of iteration i. The LP-relaxation $\text{LP}^{(i+1)}$ that we solve in iteration $i+1$ is then defined as the LP (3a)–(3e) with the following additional constraints:

$$x_{jk} = 1 \quad \forall l \in \{\iota^{(1)}, \ldots, \iota^{(i)}\} \quad \forall k \in \mathcal{K}(l) \quad \forall j \in \mathcal{S}_{\sigma(i)}(k), \tag{5a}$$

$$x_{jk} = 0 \quad \forall l \in \{\iota^{(1)}, \ldots, \iota^{(i)}\} \quad \forall k \in \mathcal{K}(l) \quad \forall j \notin \mathcal{S}_{\sigma(i)}(k). \tag{5b}$$

Let $(x^{(i+1)}, z^{(i+1)})$ be the optimal solution of $\text{LP}^{(i+1)}$ computed by GREEDY in iteration $i+1$. The next lemma establishes that the final assignment recovers at least half of the optimal solution.

Lemma 8. *Fix some* $1 \leq i \leq q$ *and let* $\sigma(i)$ *be the assignment at the end of iteration* i*. Further, let* (x^*, z^*) *be the optimal solution to the original LP-relaxation constructed by* GREEDY*. Then*

$$\sum_{j \in \mathcal{S}_{\sigma(i)}} p_j \geq \frac{1}{2} \sum_{l \in \{\iota^{(1)}, \ldots, \iota^{(i)}\}} \sum_{k \in \mathcal{K}(l)} \sum_{j \in \mathcal{S}} p_j x^*_{jk}. \tag{6}$$

We summarize the result in the following theorem.

Theorem 5. *Algorithm 3 is a* $\frac{1}{2}$*-approximation algorithm for instances of MKPC satisfying the isolation property.*

Proof. By Lemma 8, the final assignment $\sigma = \sigma(q)$ returned by the algorithm has profit at least

$$\sum_{j \in S_\sigma} p_j \geq \frac{1}{2} \sum_{l \in C} \sum_{k \in \mathcal{K}(l)} \sum_{j \in \mathcal{S}} p_j x^*_{jk} \geq \frac{1}{2} \text{OPT}.$$

References

1. Ageev, A., Sviridenko, M.: Pipage rounding: a new method of constructing algorithms with proven performance guarantee. J. Comb. Optim. **8**(3), 307–328 (2004). https://doi.org/10.1023/B:JOCO.0000038913.96607.c2
2. Badanidiyuru, A., Vondrák, J.: Fast algorithms for maximizing submodular functions. In: Proceedings of the 2014 Annual ACM-SIAM Symposium on Discrete Algorithms, pp. 1497–1514 (2014)
3. Chekuri, C., Khanna, S.: A polynomial time approximation scheme for the multiple knapsack problem. SIAM J. Comput. **35**, 719–728 (2005)
4. Chekuri, C., Vondrák, J., Zenklusen, R.: Submodular function maximization via the multilinear relaxation and contention resolution schemes. SIAM J. Comput. **43**(6), 1831–1879 (2014)
5. Chekuri, C., Kumar, A.: Maximum coverage problem with group budget constraints and applications. In: Jansen, K., Khanna, S., Rolim, J.D.P., Ron, D. (eds.) APPROX/RANDOM -2004. LNCS, vol. 3122, pp. 72–83. Springer, Heidelberg (2004). https://doi.org/10.1007/978-3-540-27821-4_7
6. Dudzinski, K., Walukiewicz, S.: Exact methods for the knapsack problem and its generalizations. Eur. J. Oper. Res. **28**, 3–21 (1987)
7. Ene, A., Nguyen, H.L.: A nearly-linear time algorithm for submodular maximization with a knapsack constraint. In: Baier, C., Chatzigiannakis, I., Flocchini, P., Leonardi, S. (eds.) 46th International Colloquium on Automata, Languages, and Programming (ICALP 2019). Leibniz International Proceedings in Informatics (LIPIcs), vol. 132, pp. 1–12 (2019). https://doi.org/10.4230/LIPIcs.ICALP.2019.53, http://drops.dagstuhl.de/opus/volltexte/2019/10629
8. Fairstein, Y., Kulik, A., Naor, J., Raz, D., Shachnai, H.: A $(1 - e^{-1} - \varepsilon)$-approximation for the monotone submodular multiple knapsack problem. arXiv:2004.12224 (2020)
9. Farbstein, B., Levin, A.: Maximum coverage problem with group budget constraints. J. Comb. Optim. **34**, 725–735 (2017)

10. Feige, U.: A threshold of ln n for approximating set cover. J. ACM **45**(4), 634–652 (1998)
11. Goemans, M.X., Williamson, D.P.: New $\frac{3}{4}$-approximation algorithms for the maximum satisfiability problem. SIAM J. Discrete Math. **7**(4), 656–666 (1994). https://search-proquest-com.vu-nl.idm.oclc.org/docview/925617906?accountid=10978
12. Guo, L., Li, M., Xu, D.: Efficient approximation algorithms for maximum coverage with group budget constraints. Theoret. Comput. Sci. **788**, 53–65 (2019)
13. Kellerer, H., Pferschy, U., Pisinger, D.: Knapsack Problems. Springer, Heidelberg (2004). https://doi.org/10.1007/978-3-540-24777-7
14. Khuller, S., Moss, A., Naor, J.: The budgeted maximum coverage problem. Inf. Process. Lett. **70**, 39–45 (1999)
15. Kulik, A., Shachnai, H., Tamir, T.: Approximations for monotone and nonmonotone submodular maximization with knapsack constraints. Math. Oper. Res. **38**(4), 729–739 (2013)
16. Lee, J., Mirrokni, V., Nagarajan, V., Sviridenko, M.: Maximizing nonmonotone submodular functions under matroid or knapsack constraints. SIAM J. Discrete Math. **23**(4), 2053–2078 (2010)
17. Nemhauser, G., Wolsey, L.: An analysis of approximations for maximizing submodular set functions-I. Math. Program. **14**, 265–294 (1978)
18. Nip, K., Wang, Z.: Approximation algorithms for a two-phase knapsack problem. In: Wang, L., Zhu, D. (eds.) COCOON 2018. LNCS, vol. 10976, pp. 63–75. Springer, Cham (2018). https://doi.org/10.1007/978-3-319-94776-1_6
19. Shmoys, D., Tardos, E.: An approximation algorithm for the generalized assignment problem. Math. Program. A **62**, 461–474 (1993)
20. Sun, X., Zhang, J., Zhang, Z.: Deterministic algorithms for the submodular multiple knapsack problem. arXiv:2003.11450 (2020)
21. Sviridenko, M.: A note on maximizing a submodular set function subject to a knapsack constraint. Oper. Res. Lett. **32**, 41–43 (2004)
22. Xavier, E., Miyazawa, F.: Approximation schemes for knapsack problems with shelf divisions. Theoret. Comput. Sci. **352**, 71–84 (2006)

A Constant-Factor Approximation Algorithm for Vertex Guarding a WV-Polygon

Stav Ashur[1], Omrit Filtser[2], and Matthew J. Katz[3](✉)

[1] Ben-Gurion University of the Negev, Beersheba, Israel
stavshe@post.bgu.ac.il
[2] Stony Brook University, Stony Brook, USA
[3] Ben-Gurion University of the Negev, Beersheba, Israel
matya@cs.bgu.ac.il

Abstract. The problem of vertex guarding a simple polygon was first studied by Subir K. Ghosh (1987), who presented a polynomial-time $O(\log n)$-approximation algorithm for placing as few guards as possible at vertices of a simple n-gon P, such that every point in P is visible to at least one of the guards. Ghosh also conjectured that this problem admits a polynomial-time algorithm with constant approximation ratio. Due to the centrality of guarding problems in the field of computational geometry, much effort has been invested throughout the years in trying to resolve this conjecture. Despite some progress (surveyed below), the conjecture remains unresolved to date. In this paper, we confirm the conjecture for the important case of weakly visible polygons, by presenting a $(2+\varepsilon)$-approximation algorithm for guarding such a polygon using vertex guards. A simple polygon P is *weakly visible* if it has an edge e, such that every point in P is visible from some point on e. We also present a $(2+\varepsilon)$-approximation algorithm for guarding a weakly visible polygon P, where guards may be placed anywhere on P's boundary (except in the interior of the edge e). Finally, we present an $O(1)$-approximation algorithm for vertex guarding a polygon P that is weakly visible from a chord.

Our algorithms are based on an in-depth analysis of the geometric properties of the regions that remain unguarded after placing guards at the vertices to guard the polygon's boundary. Finally, our algorithms may become useful as part of the grand attempt of Bhattacharya et al. to prove the original conjecture, as their approach is based on partitioning the underlying simple polygon into a hierarchy of weakly visible polygons.

Keywords: Geometric optimization · Approximation algorithms · Visibility · Art gallery problems

O. Filtser was supported by the Eric and Wendy Schmidt Fund for Strategic Innovation, by the Council for Higher Education of Israel, and by Ben-Gurion University of the Negev. M. Katz was supported by grant 1884/16 from the Israel Science Foundation.

C. Kaklamanis and A. Levin (Eds.): WAOA 2020, LNCS 12806, pp. 81–96, 2021.
https://doi.org/10.1007/978-3-030-80879-2_6

1 Introduction

The Art Gallery Problem is a classical problem in computational geometry, posed by Victor Klee in 1973: Place a minimum number of points (representing guards) in a given simple polygon P (representing an art gallery), so that every point in P is seen by at least one of the placed points. We say that a point $p \in P$ *sees* (or *guards*) a point $q \in P$ if the line segment $\overline{pq}$ is contained in P. We say that a subset $G \subseteq P$ *guards* P, if every point in P is seen by at least one point in G.

There are numerous variants of the art gallery problem, which are also referred to as the art gallery problem. These variants differ from one another in (i) the underlying domain, e.g., simple polygon, polygon with holes, orthogonal polygon, or terrain, (ii) which parts of the domain must be guarded, e.g., only its vertices, only its boundary, or the entire domain, (iii) the type of guards, e.g., static, mobile, or with various restrictions on their coverage area such as limited range, (iv) the restrictions on the location of the guards, e.g., only at vertices (vertex-guards), only on the boundary (boundary-guards), or anywhere (point-guards), and (v) the underlying notion of visibility, e.g., line of sight, rectangle visibility, or staircase visibility. It is impossible to survey here the vast literature on the art gallery problem, ranging from combinatorial and optimization results to hardness of computation results, so we only mention the book by O'Rourke [26] and a small sample of recent papers [1,8].

In this paper, we deal with the version of the art gallery problem, where the guards are confined to the boundary of the underlying polygon, and in particular to its vertices. Such guards are referred to as *boundary* guards or *vertex* guards, respectively. The first to present results for this version was Ghosh [14,15], who gave a polynomial-time $O(\log n)$-approximation algorithm for guarding either a simple polygon or a polygon with holes using vertex (or edge) guards. In the related work paragraph below we survey many of the subsequent results for this version.

We consider an important family of simple polygons, namely, the family of *weakly visible* polygons. A simple polygon P is *weakly visible* if it has an edge e, such that every point in P is visible from some point on e, or, in other words, a guard patrolling along e can see the entire polygon. We also consider polygons that are weakly visible from a chord, rather than an edge, where a *chord* in a polygon P is a line segment whose endpoints are on the boundary of P and whose interior is contained in the interior of P.

The problem of guarding a weakly visible polygon (WV-polygon) P by vertex guards was studied by Bhattacharya et al. [6]. They first present a 4-approximation algorithm for vertex guarding *only* the vertices of P. Next, they claim that this algorithm places the guards at vertices in such a way that each of the remaining unguarded regions of P has a boundary edge which is contained in an edge of P. Based on this claim, they devise a 6-approximation algorithm for vertex guarding P's boundary (by adding vertex guards to their set of vertex guards that guards P's vertices), and present it as a 6-approximation algorithm for guarding P (boundary plus interior). Unfortunately, this claim is false, i.e., the interior of P might still contain unguarded regions; counterexamples were

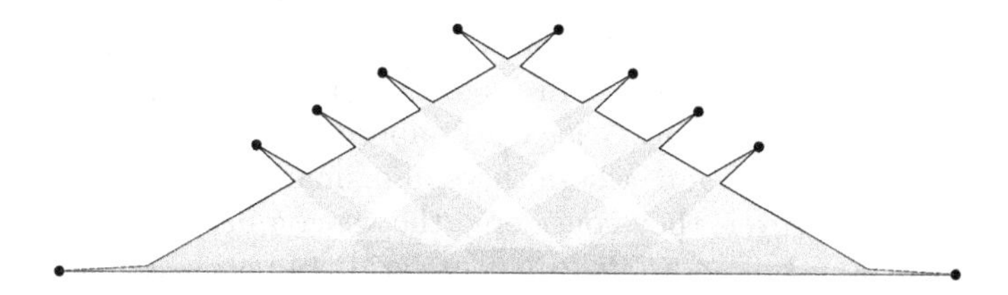

Fig. 1. A WV-polygon P and a subset of its vertices that guards P's boundary but not its interior.

constructed and approved by the authors (who are now attempting to fix their algorithm, so as to obtain an algorithm for vertex guarding P entirely) [7]. Thus, the challenge of obtaining a constant-factor approximation algorithm for guarding a WV-polygon with vertex guards or boundary guards is still on. Figure 1 depicts a WV-polygon P and a subset of its vertices (which is not necessarily the one returned by the algorithm of Bhattacharya et al. [6]) that guards P's boundary but not its interior.

The main result of this paper is such an algorithm. Specifically, denote by OPT the size of a minimum-cardinality subset of the vertices of P that guards P. We present a polynomial-time algorithm that finds a subset I of the vertices of P, such that I guards P and $|I| \leq (2+\varepsilon)\text{OPT}$, for any constant $\varepsilon > 0$.

Already in 1987, Ghosh conjectured that there exists a constant-factor approximation algorithm for vertex guarding a simple polygon. Recently, Bhattacharya et al. [5] managed to devise such an algorithm for vertex guarding the *vertices* or the *boundary* of a simple polygon P, by first partitioning P into a hierarchy of WV-polygons according to the link distance from a starting vertex. They also present such an algorithm for vertex guarding P (boundary plus interior), however, this algorithm is erroneous, since it relies on the false statement mentioned above. Thus, our result may prove useful towards resolving Ghosh's conjecture.

We note that our algorithm for vertex guarding a WV-polygon is more general than that of Bhattacharya et al. [6] (assuming it can be fixed), since unlike theirs, it is not based on a *specific* procedure for placing guards at vertices so as to see all the polygon's vertices. This we believe makes our algorithm a more suitable tool for dealing with the general problem (i.e., for simple polygons), and will enable a better approximation ratio, if successful.

Prior to our result, the only (non-trivial) family of polygons for which a constant-factor approximation algorithm for guarding a member of the family was known, is the family of monotone polygons. Specifically, Krohn and Nilsson [22] showed that vertex guarding a monotone polygon is NP-hard, and presented a constant-factor approximation algorithm for *point* guarding such a polygon (as well as an $O(\text{OPT}^2)$-approximation algorithm for point guarding an orthogonal polygon).

Related Work. Several improvements to Ghosh's $O(\log n)$-approximation algorithm were obtained over the years. In 2006, Efrat and Har-Peled [11] described a randomized polynomial-time $O(\log \text{OPT})$-approximation algorithm for vertex guarding a simple polygon, where the approximation factor is correct with high probability. Moreover, they considered the version in which the guards are restricted to the points of an arbitrarily dense grid, and presented a randomized algorithm which returns an $O(\log \text{OPT})$-approximation with high probability, where OPT is the size of an optimal solution to the modified problem and the running time depends on the ratio between the diameter of the polygon and the grid size. Combining ideas from the latter algorithm and from the work of Deshpande et al. [10], Bonnet and Miltzow [8] presented a randomized polynomial-time $O(\log \text{OPT})$-approximation algorithm for point guards, assuming integer coordinates and a general position assumption on the vertices of the underlying simple polygon.

In 2011, King and Kirkpatrick [21] observed that by applying methods that were developed for the Set Cover problem after the publication of Ghosh's algorithm, one can obtain an $O(\log \text{OPT})$ approximation factor for vertex guarding a simple polygon (and an $O(\log h \log \text{OPT})$ factor for vertex guarding a polygon with h holes). Moreover, they improved the approximation factor to $O(\log\log \text{OPT})$ for guarding a simple polygon either with vertex guards or boundary guards, where in the former case the running time is polynomial in n and in the latter case it is polynomial in n and the spread of the vertices.

Most of the variants of the Art Gallery Problem are NP-hard. O'Rourke and Supowit [27] proved this for polygons with holes and point guards, Lee and Lin [24] proved this for simple polygons and vertex guards, and Aggarwal [2] generalized the latter proof to simple polygons and point guards. Eidenbenz et al. [12] presented a collection of hardness results. In particular, they proved that it is unlikely that a PTAS exists for vertex guarding or point guarding a simple polygon. Recently, Abrahamsen et al. [1] proved that the Art Gallery Problem is $\exists\mathbb{R}$-complete, for simple polygons and point guards.

WV-polygons were defined and studied in the context of mobile guards [26]. An *edge guard* is a guard that traverses an edge of the polygon. Thus, a simple polygon is weakly visible if and only if it can be guarded by a single edge guard, and the problem of vertex guarding a WV-polygon is equivalent to the problem of replacing the single edge guard by as few vertex guards as possible. Avis and Toussaint [4] presented a linear-time algorithm for detecting whether a polygon is weakly visible from a given edge. Subsequently, Sack and Suri [28] and Das et al. [9] devised linear-time algorithms which output all the edges (if any) from which the polygon is weakly visible. Algorithms for finding either an edge or a chord from which the polygon is weakly visible were given by Ke [20] and by Ghosh et al. [16]. Finally, Bhattacharya et al. [6] proved that the problem of point guarding a WV-polygon is NP-hard, and that there does not exist a polynomial-time algorithm for vertex guarding a WV-polygon with holes with approximation factor better than $((1-\varepsilon)/12)\ln n$, unless $\text{P} = \text{NP}$.

Our algorithm for vertex guarding a WV-polygon uses a solution to the problem of guarding the boundary of a WV-polygon using vertex guards. This problem admits a local-search-based PTAS (see [3, 19]), which is similar to the local-search-based PTAS of Gibson et al. [17] for vertex guarding the vertices of a 1.5D-terrain. The proof of both these PTASs is based on the proof scheme of Mustafa and Ray [25].

Results. Our algorithm for vertex guarding a WV-polygon P (presented in Sect. 3) consists of two main parts. In the first part, it computes a subset G of the vertices of P that guards P's boundary. This is done by applying a known algorithm for this task. In the second part, it computes a subset G' of the vertices of P of size at most $|G|$, such that $G \cup G'$ guards P (boundary plus interior). Thus, if we apply the algorithm of [3] for computing G, then the approximation ratio of our algorithm is $2 + \varepsilon$, since the former algorithm guarantees that $|G| \leq (1 + \varepsilon/2)\mathrm{OPT}_\partial$, where OPT_∂ is the size of a minimum-cardinality subset of the vertices of P that guards P's boundary, and clearly $\mathrm{OPT}_\partial \leq \mathrm{OPT}$.

Let x be a vertex in G and let $Vis(x)$ be the visibility polygon of x (i.e., $Vis(x)$ is the set of all points of P that are visible from x), then $P \setminus Vis(x)$ is a set of connected regions, which we refer to as *pockets*. Moreover, a connected subset H of P is a *hole* in P w.r.t. G if (i) there is no point in H that is visible from G, and (ii) H is maximal in the sense that any connected subset of P that strictly contains H has a point that is visible from G. The second part of our algorithm (and its proof) are based on a deep structural analysis and characterization of the pockets and holes in P (presented in Sect. 2).

The requirement that G is a subset of the vertices of P is actually not necessary; the second part of our algorithm only needs a set of boundary points that guards P's boundary. This observation enables us to use a smaller number of guards, assuming that boundary guards are allowed.

Finally, in Sect. 4, we consider the more general family of polygons, those that are weakly visible from a chord. Notice that a chord $\overline{uv}$ in P slices P into two polygons, such that each of them is weakly visible w.r.t. the edge $\overline{uv}$. After updating two of the geometric claims presented in Sect. 2, we show how to apply our algorithm to a polygon that is weakly visible from a chord. The approximation ratio in this case is $3|G|$ (rather than $2|G|$). However, the best known algorithm for computing G in this case is that of Bhattacharya et al. [5] (for simple polygons), which computes a c-approximation, for some relatively big constant c. Therefore, the approximation ratio of our algorithm in this case is $3c$.

2 Structural Analysis

For two points $x, y \in \mathbb{R}^2$, we denote by $\overline{xy}$ the line segment whose endpoints are x and y, and by ℓ_{xy} the line through x and y. We denote by ℓ_x the horizontal line through x.

Let P be a polygon whose set of vertices is $V = \{u = v_1, v_2, \ldots, v_n = v\}$, and which is weakly visible from its edge $e = \overline{uv}$. We denote the boundary of P

by ∂P. The edges of P are the segments $\overline{v_1v_2}, \overline{v_2v_3}, \ldots, \overline{v_{n-1}v_n}$ and $\overline{v_1v_n} = \overline{uv}$. We assume w.l.o.g. that $\overline{uv}$ is contained in the x-axis, and u is to the left of v. Furthermore, we assume that P is contained in the (closed) halfplane above the x-axis; in particular, the angles at u and v are convex. This assumption can be easily removed, as we show towards the end of Sect. 3.

2.1 Visibility Polygons

For a point $p \in P$, let $Vis(p) = \{q \mid \overline{pq} \subseteq P\}$ be the ***visibility polygon*** of p. In other words, $Vis(p)$ is the set of all points of P that are visible from p. By definition $Vis(p)$ is a star-shaped polygon, and thus clearly also a simple polygon, contained in P (see Fig. 2).

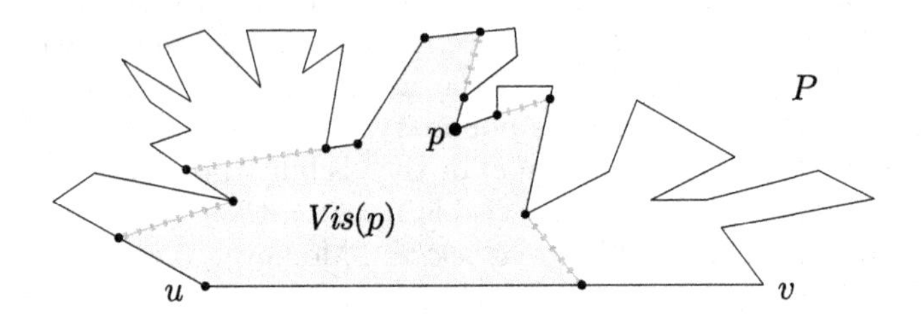

Fig. 2. The visibility polygon of a point $p \in P$.

Any vertex of P that belongs to $Vis(p)$ is also considered a vertex of $Vis(p)$. Consider the set of edges of $Vis(p)$. Some of these edges are fully contained in ∂P. The edges that are not contained in ∂P are ***constructed edges***: these are edges whose endpoints are on ∂P and whose interior is contained in the interior of P (the gray dotted edges in Fig. 2).

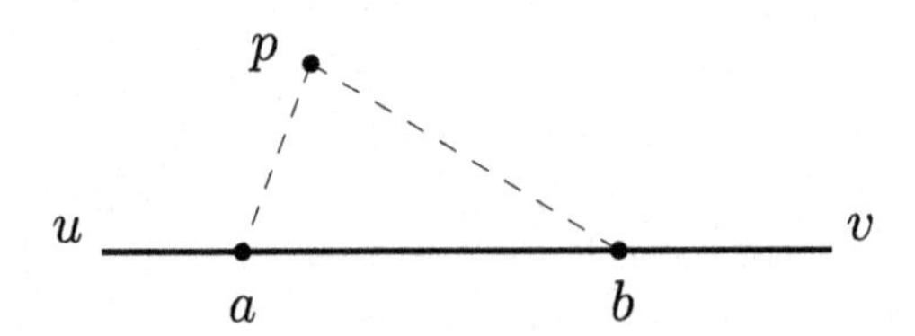

Fig. 3. If p sees both a and b, then p sees the entire segment $\overline{ab}$.

Claim 1. *For any $p \in P$, there exists a single edge of $Vis(p)$ that is contained in $\overline{uv}$.*

Proof. Since p is visible from $\overline{uv}$, $\overline{uv} \cap Vis(p) \neq \emptyset$. Let a (resp. b) be the leftmost (resp. rightmost) point on $\overline{uv}$ that belongs to $Vis(p)$ (see Fig. 3). The triangle $\triangle pab$ cannot contain points of ∂P in its interior, because ∂P cannot cross the segments $\overline{pa}$ and $\overline{bp}$. Therefore, p sees every point in $\triangle pab$, and in particular it sees every point between a and b on $\overline{uv}$.

2.2 Pockets

Consider $P \setminus Vis(p)$ (see Fig. 2). This is a set of connected regions, which we refer to as *pockets*. Since P is a simple polygon, each pocket C is adjacent to a single constructed edge, and therefore the intersection of ∂P and C is connected. We refer to $\partial P \cap C$ as the boundary of the pocket C, and denote it by ∂C. (Thus the constructed edge itself is not part of ∂C.)

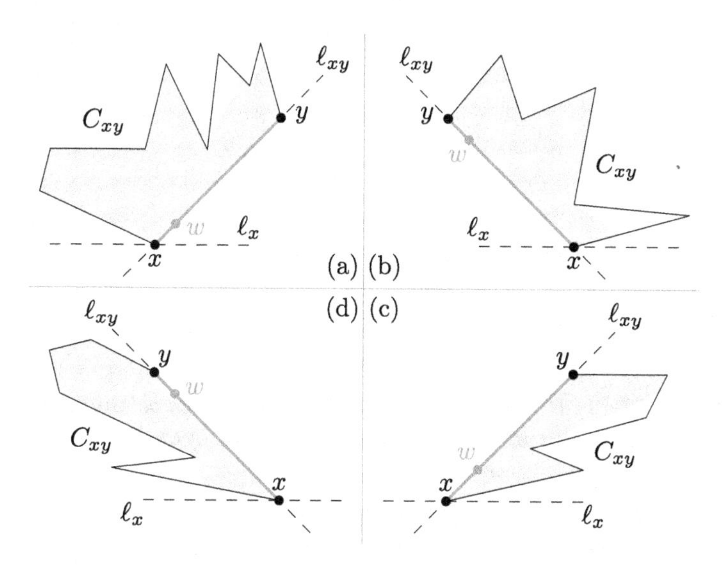

Fig. 4. $\overline{xy}$ is a constructed edge, and the gray area is the set of all points of $Vis(w)$ that lie below/above ℓ_{xy}. In (a) and (b) the pocket C_{xy} lies above ℓ_{xy}, and thus $\overline{xy}$ is a lower edge. In (c) and (d) the pocket C_{xy} lies below ℓ_{xy}, and $\overline{xy}$ is an upper edge. In (a) and (c) the slope of $\overline{xy}$ is positive, and in (b) and (d) it is negative.

Let $\overline{xy}$ be a constructed edge such that x is below y (w.r.t. the y-axis), and denote by C_{xy} the pocket adjacent to $\overline{xy}$. We say that C_{xy} lies above (resp. below) ℓ_{xy}, if for any point w in the interior of $\overline{xy}$, all the points that w sees (points of $Vis(w)$) that lie above (resp. below) ℓ_{xy}, belong to C_{xy} (see Fig. 4). Notice that since $\overline{xy}$ divides P into two parts, C_{xy} and $P \setminus C_{xy}$, 'stepping off' $\overline{xy}$ to one side, places us in the interior of C_{xy} (and 'stepping off' $\overline{xy}$ to the other side places us in the interior of $P \setminus C_{xy}$).

Note that if C_{xy} lies above (resp. below) ℓ_{xy}, it does not necessarily mean that all the points of C_{xy} lie above (resp. below) ℓ_{xy}. Indeed, when $x \in \overline{uv}$, C_{xy} may have points on both sides of ℓ_{xy} (see Fig. 5).

We say that $\overline{xy}$ is an *upper edge* (resp. *lower edge*), if C_{xy} lies below (resp. above) ℓ_{xy}. Notice that by Claim 1, $Vis(p)$ has at most two constructed edges with an endpoint in $\overline{uv}$. Moreover, at most one of these edges is an upper edge.

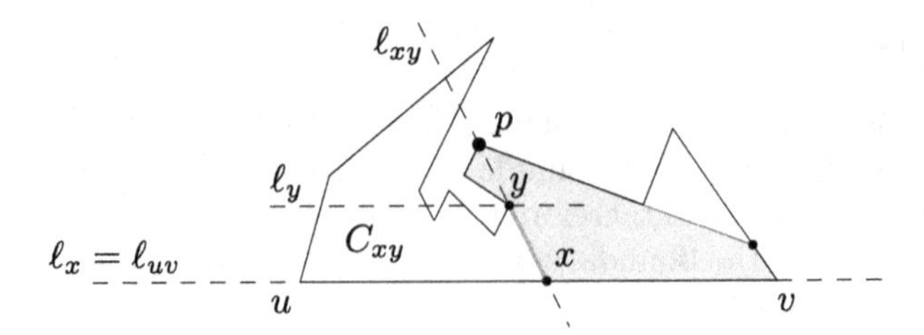

Fig. 5. The gray area is $Vis(p)$. $\overline{xy}$ is an upper edge (i.e., C_{xy} lies below ℓ_{xy}) and x is on $\overline{uv}$, but some points of C_{xy} lie above ℓ_{xy}.

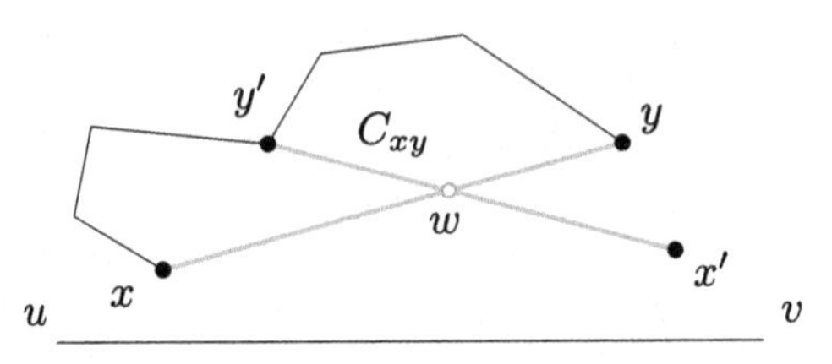

Fig. 6. Two constructed edges that cross each other.

Claim 2. *Let $\overline{xy}$ and $\overline{x'y'}$ be two constructed edges (which belong to two different visibility polygons), such that $\overline{xy}$ crosses $\overline{x'y'}$, and y' is on the same side of ℓ_{xy} as C_{xy}. Then, $y' \in \partial C_{xy}$ (see Fig. 6).*

Proof. Let w be the crossing point of $\overline{xy}$ and $\overline{x'y'}$. Then, w is a point in the interior of $\overline{xy}$ that sees y', and y' is on the same side of ℓ_{xy} as C_{xy}, so $y' \in C_{xy}$. Since $\overline{x'y'}$ is a constructed edge, $y' \in \partial P$ and thus $y' \in \partial C_{xy}$.

Claim 3. *Let $\overline{xy}$ be a constructed edge of $Vis(p)$ such that x is below y, then C_{xy} lies above ℓ_x (in the weak sense), i.e., every point in C_{xy} is either above or on ℓ_x. Moreover, if $x \notin \overline{uv}$, then C_{xy} lies strictly above ℓ_x, i.e., every point in C_{xy} is above ℓ_x.*

Proof. If $x \in \overline{uv}$, then since P is above $\ell_{uv} = \ell_x$, we get that C_{xy} is above ℓ_x.

Observe that if $\partial C_{xy} \cap \overline{uv} \neq \emptyset$, then $x \in \overline{uv}$. Otherwise, it follows that $\overline{uv} \subseteq \partial C_{xy}$, and since the entire pocket C_{xy} is not visible from p, we get that p is not visible from $\overline{uv}$, which contradicts the fact that P is a WV-polygon.

If $x \notin \overline{uv}$, then x is above ℓ_{uv} (see Fig. 4). By the observation above, $\partial C_{xy} \cap \overline{uv} = \emptyset$ (and thus of course $C_{xy} \cap \overline{uv} = \emptyset$). Let z be any point in C_{xy}. Since P is a WV-polygon, there exists a point z' on $\overline{uv}$ such that $\overline{zz'} \subseteq P$. The segment $\overline{zz'}$ has to cross $\overline{xy}$, because $z \in C_{xy}$ and $z' \notin C_{xy}$. Since x is below y, this crossing point is above x, which in turn is above z', so we get that z is above x. We conclude that C_{xy} lies strictly above ℓ_x.

2.3 Holes

Let $G \subseteq P$ be a set of points that guards ∂P. A connected subset H of P is a *hole* in P w.r.t. G if (i) there is no point in H that is visible from G, and (ii) H

is maximal in the sense that any connected subset of P that strictly contains H has a point that is visible from G.

Let H be a hole in P w.r.t. G, then clearly H is a simple polygon. Each edge of H lies on some constructed edge e, and we say that H (and this edge of H) *lean* on the edge e. Notice that H is fully contained in the pocket adjacent to e. Moreover, since $H \cap \partial P = \emptyset$, we can view H as an intersection of halfplanes (defined by the lines containing the constructed edges on which the edges of H lean), and thus obtain the following observation.

Observation 4. *Any hole H in P w.r.t. G is a convex polygon.*

Another immediate but useful observation is the following.

Observation 5. *Any hole H in P w.r.t. G leans on at least one upper edge and at least one lower edge.*

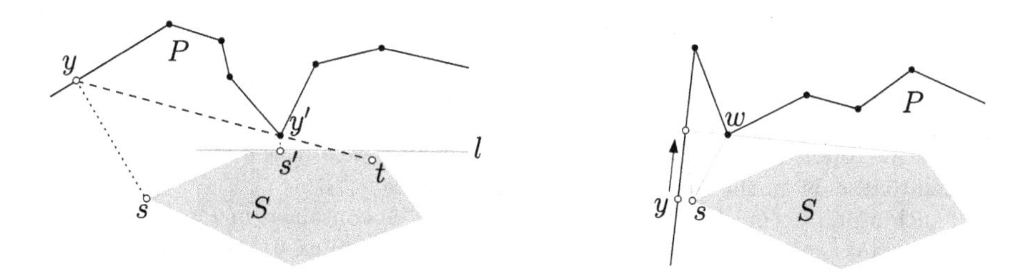

Fig. 7. Proof of Lemma 6.

Next, we show that any hole can be guarded by a single vertex, and that such a vertex can be found in polynomial time. We actually prove a slightly more general claim.

Lemma 6. *Let S be a simple convex polygonal region contained in P. Then, there exists a vertex w of P, such that S is visible from w, i.e., every point in S is visible from w. Moreover, w can be found in polynomial time.*

Proof. If P and S share a vertex, then we are done. Otherwise, let y be a point on ∂P that is closest to S, where the distance between a point p and S is $dist(p, S) = \min\{||p - s|| \mid s \in S\}$, and let s be a point in S closest to y. We first prove that S is visible from y. If y is also on ∂S, then y clearly sees S, so assume that y is not on ∂S and that there exists a point $t \in S$ that is not visible from y. Then, there exists a point y' on ∂P that lies in the interior of the segment $\overline{yt}$. Let s' be a point in S closest to y', and let l be the line through s' and perpendicular to $\overline{y's'}$. Since S is convex, it is contained in one of the two halfplanes supported by l. Assume w.l.o.g. that l is horizontal and that S is contained in the bottom halfplane supported by l (see Fig. 7, left). Then y' is above l and its projection onto l is s'. Now, it is impossible that y is not above

y' (in terms of y-coordinate), since this would imply that t is above l. So y must be above y', but then $||y - s|| > ||y' - s'||$ and we have reached a contradiction.

Now, if y is a vertex of P, then we are done. Otherwise, we slide y along ∂P, in any one of the directions, until the convex hull of $S \cup y$ meets a vertex w of P (see Fig. 7, right). The vertex w is either an endpoint of the edge of P on which we slide y, or it is another vertex of P that lies on one of the tangents to S through y. In both cases, w clearly sees S.

3 The Algorithm

We show that given a set G of vertices that guards ∂P, one can find a set of vertices G' of size at most $|G|$ such that $G \cup G'$ guards P (boundary plus interior). Let E be the set of the constructed edges of the visibility polygons of the vertices in G.

Algorithm 1. Guarding the interior of P

For each upper edge $e = \overline{xy}$ in E with $x \in \overline{uv}$:

1. Find the topmost intersection point p of e with an edge $e' = \overline{x'y'}$ in E with $x' \in \overline{uv}$ such that x' is on the same side of ℓ_{xy} as the pocket C_{xy}.
2. If such a point p exists (then the triangle $\triangle xpx'$ is contained in P), find a vertex that guards $\triangle xpx'$ (see description in the proof of Lemma 6) and add it to G'.

Now our goal is to show that for any hole H in P w.r.t. G there exists a vertex in G' that guards it. More precisely, we show that H is contained in one of the triangles considered by the algorithm.

Let H be a hole with vertices $h_1, \ldots, h_k$. Using the following lemma, we first show that H leans on an upper edge with an endpoint in $\overline{uv}$. By Observation 5, H leans on at least one upper edge.

Lemma 7. *Assume that h_1h_2 leans on an upper edge $\overline{xy}$, such that x is below y, $x \notin \overline{uv}$, and h_1 is below h_2. Then h_2h_3 also leans on an upper edge and h_2 is below h_3.*

Proof. Since $\overline{xy}$ is an upper edge, H lies below ℓ_{xy}, and by Claim 3, H lies strictly above ℓ_x (see Fig. 8).

Let $\overline{x'y'}$ be the constructed edge on which h_2h_3 is leaning, and assume that y' is on the same side of ℓ_{xy} as h_3. (h_2 is the crossing point of $\overline{xy}$ and $\overline{x'y'}$, so x' and y' are on different sides of ℓ_{xy}.) Since h_3 is a vertex of the hole H, both h_3 and y' are on the same side of ℓ_{xy} as C_{xy}. Moreover, by Claim 2 we get that $y' \in \partial C_{xy}$, and therefore y' is strictly above x.

Next we show that h_2 is below h_3, by showing that x' is below y'. Indeed, if x' is above y', then since h_1 is a vertex of the hole H, both h_1 and x are on

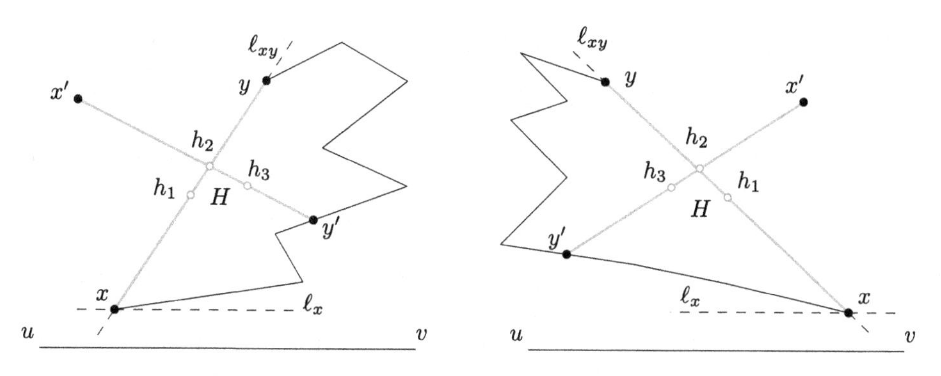

Fig. 8. Two edges of the hole H: h_1h_2 and h_2h_3. The edge h_1h_2 leans on an upper edge and h_1 is below h_2, thus h_2h_3 has to lean on an upper edge such that h_2 is below h_3 (and not as drawn in the figure).

the same side of $\ell_{x'y'}$ as $C_{x'y'}$. Again by Claim 2 we get that $x \in \partial C_{x'y'}$, and therefore x is strictly above y', a contradiction.

Finally, $\overline{x'y'}$ is clearly an upper edge, since h_1 (which is in $C_{x'y'}$) is below $\overline{x'y'}$.

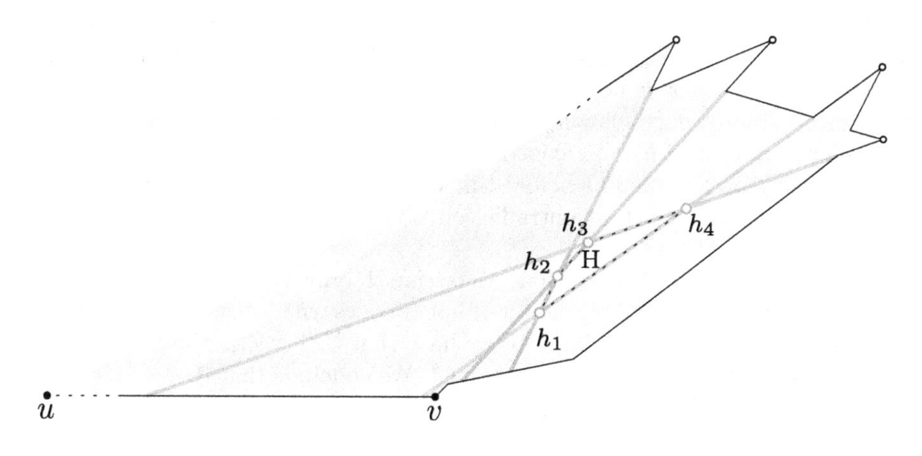

Fig. 9. The edges of H lean on a sequence of upper edges.

This implies that every hole H leans on an upper edge with an endpoint in $\overline{uv}$: H has at least one edge h_1h_2 that leans on an upper edge $\overline{x_1y_1}$, and h_1 is below h_2. If x_1 or y_1 are on $\overline{uv}$ then we are done. Otherwise, by Lemma 7, the edge h_2h_3 also leans on an upper edge $\overline{x_2y_2}$, and h_2 is below h_3. Again, if one of x_2, y_2 is on $\overline{uv}$ then we are done, otherwise by Lemma 7 h_3h_4 leans on an upper edge and h_3 is below h_4. The process must end before it reaches $h_{k-1}h_k$, since, if $h_{k-1}h_k$ leans on an upper edge $\overline{x_{k-1}y_{k-1}}$ such that none of x_{k-1}, y_{k-1} is on

$\overline{uv}$, then by applying Lemma 7 once again we get that h_k is below h_1, which is impossible (see Fig. 9). But the only way for the process to end is when it reaches an upper edge $\overline{x_t y_t}$ such that x_t or y_t are on $\overline{uv}$.

Claim 8. *Assume that h_1h_2 leans on an upper edge $\overline{xy}$ such that x is below y and $x \in \overline{uv}$, and that h_1 is below h_2. Then h_2h_3 leans on a constructed edge $\overline{x'y'}$ (where x' is below y') such that $x' \in \overline{uv}$.*

Proof. Let $\overline{x'y'}$ be the constructed edge on which h_2h_3 is leaning such that x' is below y'. The constructed edges $\overline{xy}$ and $\overline{x'y'}$ cross each other at the point h_2, so by Claim 2 we get that $x \in \partial C_{x'y'}$, because h_1 and x are on the same side of $\ell_{x'y'}$. Now, by Claim 3, x lies (weakly) above $\ell_{x'}$, but $x \in \overline{uv}$, so $x' \in \ell_{uv}$.

We are now ready to prove the correctness of our algorithm.

Claim 9. *Any hole H in P w.r.t. G is contained in one of the triangles considered by Algorithm 1.*

Proof. By the argument immediately following Lemma 7, H has an edge which leans on an upper edge with an endpoint in $\overline{uv}$. Let h_1h_2, where h_1 is below h_2, be such an edge of H of minimum (absolute) slope, and let $\overline{xy}$, where $x \in \overline{uv}$, be the upper edge on which h_1h_2 is leaning. By Claim 8, the edge h_2h_3 leans on a constructed edge $\overline{x'y'}$ such that $x' \in \overline{uv}$.

Assume first that the algorithm found a point p on $\overline{xy}$. If p is not below h_2, then the triangle corresponding to p contains H and we are done. If p is below h_1 (or $p = h_1$), then x' is not on the same side of $\overline{xy}$ as C_{xy} (because otherwise, p would be a point not below h_2). But, if so, then the (absolute) slope of h_2h_3 is smaller than that of h_1h_2, a contradiction. Now, if the algorithm did not find such a point p on $\overline{xy}$, then as before this means that x' is not on the same side of $\overline{xy}$ as C_{xy} and we reach a contradiction.

Approximation Ratio. Notice that Algorithm 1 only uses a subset E' of the constructed edges, namely, the subset of upper edges with an endpoint in $\overline{uv}$, and, as already observed (using Claim 1), we have that $|E'| \leq |G|$. Moreover, in each iteration, at most one vertex is added to G'. We conclude that $|G'| \leq |E'| \leq |G|$. Now, since $|G| \leq (1 + \varepsilon/2)\text{OPT}$ (assuming G is computed by the algorithm of [3]), the approximation ratio of Algorithm 1 is $2 + \varepsilon$, for any $\varepsilon > 0$.

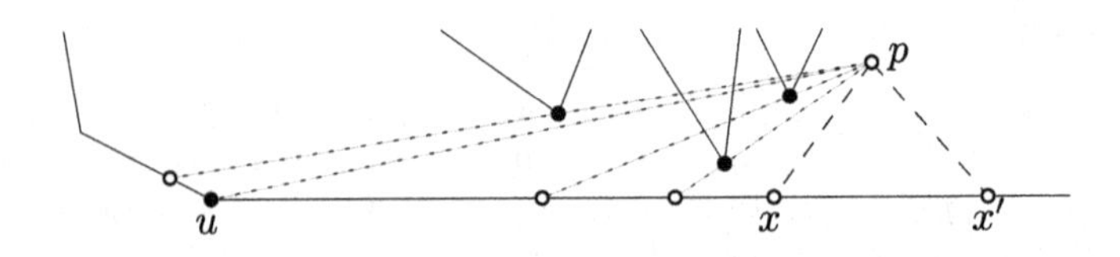

Fig. 10. Finding a vertex of P that guards the triangle $\Delta xpx'$.

Running Time. The running time of the algorithm of [3] for finding a set of vertices G of size $(1+\varepsilon/2)$OPT that guards ∂P is $O(n^{O(1/\varepsilon^2)})$, for any $\varepsilon > 0$. The set of constructed edges E can be computed in $O(|G|n)$ time [18,23]. Since $|E'| \leq |G|$, the total number of intersection points (Step 1 of Algorithm 1) is $O(|G|^2)$. Moreover, a vertex that guards the triangle $\triangle xpx'$ (Step 2 of Algorithm 1) can be found in linear time by the following simple algorithm. (We could use the algorithm described in the proof of Lemma 6, but in this special case it is not necessary.) Assume, w.l.o.g., that x is to the left of x'. Then, for each vertex w of P such that w is below p, compute the crossing point (if it exists) between $\overline{uv}$ and the ray emanating from p and passing through w. Now, among the vertices whose corresponding crossing point is between u and x, pick the one closest to x (see Fig. 10). Thus, the total running time of Algorithm 1 (given the set G) is $O(|G|n) = O(n^2)$. Note that for any $\varepsilon < \frac{1}{\sqrt{2}}$, the total running time is dominated by $O(n^{O(1/\varepsilon^2)})$.

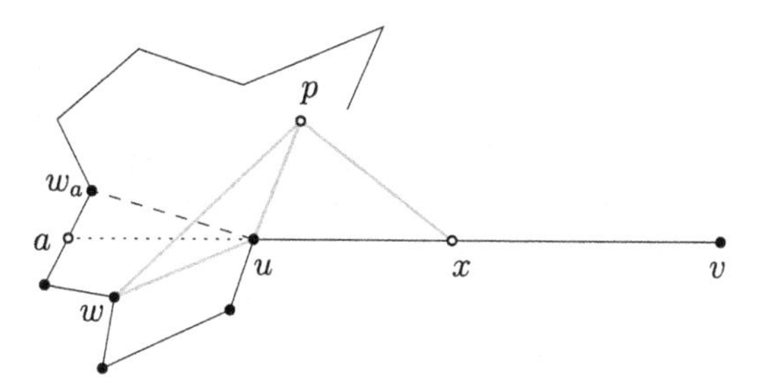

Fig. 11. A polygon with a concave angle at u.

Removing the Convexity Assumption. Up to now, we have assumed that the angles at u and at v are convex. As in [3], this assumption can be easily removed. Assume, e.g., that the angle at u is concave, and let a be the first point on P's boundary (moving clockwise from u) that lies on the x-axis (see Fig. 11). Then, every point in the open portion of the boundary between u and a is visible from u and is not visible from any other point on $\overline{uv}$. Moreover, for any vertex w in this portion of P's boundary, if w sees some point p in P above the x-axis, then so does u. Indeed, since P is weakly-visible from $\overline{uv}$, there exists a point $x \in \overline{uv}$ that sees p. In other words, $\overline{xp}$ is contained in P, as well as $\overline{uw}$ and $\overline{wp}$. Thus, the quadrilateral $uwpx$ does not contain points of ∂P in its interior, and since $\overline{up}$ is contained in it, we conclude that u sees p. Therefore, we may assume that an optimal guarding set does not include a vertex from this portion. Moreover, we may assume that the size of an optimal guarding set is greater than some appropriate constant, since otherwise we can find such a set in polynomial time. Now, let w_a be the first vertex following a. We place a guard at u and replace the portion of P's boundary between u and w_a by the edge $\overline{uw_a}$. Notice that every point in the removed region is visible from u. Similarly, if the angle at v is concave, we define the point b and the vertex w_b (by moving counterclockwise from v), place a guard at v, and replace the portion of P's boundary between v and w_b by the edge $\overline{vw_b}$. Finally, we apply the algorithm of [3] to the resulting polygon, after adjusting its parameters so that together with u and v we still get a $(1+\varepsilon/2)$-approximation of an optimal guarding set for ∂P.

Theorem 10. *Given a WV-polygon P with n vertices and $\varepsilon > 0$, one can find in $O(n^{\max\{2,O(1/\varepsilon^2)\}})$ time a subset G of the vertices of P, such that G guards P (boundary plus interior) and G is of size at most $(2+\varepsilon)OPT$, where OPT is the size of a minimum-cardinality such set.*

Boundary Guards. Let OPT^B be the size of a minimum-cardinality set of points on P's boundary (except the interior of the edge $\overline{uv}$) that guards P (boundary plus interior), and let OPT^B_∂ be the size of a minimum-cardinality such set that guards P's boundary; clearly, $\mathrm{OPT}^B_\partial \leq \mathrm{OPT}^B \leq \mathrm{OPT}$. A PTAS for finding a set G^B of points on $(\partial P \setminus \overline{uv}) \cup \{u,v\}$ that guards ∂P is described in [3], that is, $|G^B| \leq (1+\varepsilon)\mathrm{OPT}^B_\partial$, for any $\varepsilon > 0$. Its running time is $O(n^{O(1/\varepsilon^2)})$, and it is similar to the corresponding PTAS of Friedrichs et al. [13] for the case of 1.5D-terrains. Given the set G^B as input, we can apply our algorithm as is and obtain a set G' of boundary points of size at most $|G^B|$ such that $G^B \cup G'$ guards P. Thus, we have $|G^B| + |G'| \leq 2|G^B| \leq (2+\varepsilon)\mathrm{OPT}^B$.

Corollary 11. *Given a WV-polygon P (w.r.t to edge $\overline{uv}$) with n vertices and $\varepsilon > 0$, one can find in $O(n^{\max\{2,O(1/\varepsilon^2)\}})$ time a set G of points on $(\partial P \setminus \overline{uv}) \cup \{u,v\}$, such that G guards P (boundary plus interior) and G is of size at most $(2+\varepsilon)OPT^B$, where OPT^B is the size of a minimum-cardinality such set.*

4 Polygons Weakly Visible from a Chord

A *chord* in a simple polygon P is a line segment whose endpoints are on the boundary of P and whose interior is contained in the interior of P. In particular, any diagonal of P is a chord in P.

In this section, we show that our method can be extended to the case where P is weakly visible from a chord $\overline{uv}$ (in P), i.e., every point in P is visible from some point on $\overline{uv}$. More precisely, we show that given a set G of vertices that guards the boundary of such a polygon P, one can find a set G' of size at most $2|G|$ such that $I = G \cup G'$ guards P (boundary plus interior). Thus, given a c-approximation algorithm for vertex guarding the boundary of a polygon P weakly visible from a chord, we provide a $3c$-approximation algorithm for guarding P. The only known such c-approximation algorithm is that of Bhattacharya et al. [5] (for simple polygons), and there $c = 18$.

The sequence of claims leading to the following theorem can be found in the full version of this paper.

Theorem 12. *Given a polygon P with n vertices that is weakly visible from a chord, and a set G of vertices of P (or a set G of points on ∂P) that guards ∂P,* *one can find in polynomial time a set G' of size at most $2|G|$ such that $G \cup G'$ guards P (boundary plus interior). If G is a c-approximation for guarding ∂P, then $G \cup G'$ is a 3c-approximation for guarding the entire polygon P.*

References

1. Abrahamsen, M., Adamaszek, A., Miltzow, T.: The art gallery problem is $\exists\,\mathbb{R}$-complete. In: Proceedings of the 50th ACM SIGACT Sympos. on Theory of Computing, STOC, pp. 65–73 (2018)
2. Aggarwal, A.: The art gallery theorem: its variations, applications and algorithmic aspects. Ph.D. thesis, Johns Hopkins University (1984)
3. Ashur, S., Filtser, O., Katz, M.J., Saban, R.: Terrain-like graphs: PTASs for guarding weakly-visible polygons and terrains. In: Proceedings of the 17th Workshop on Approximation and Online Algorithms, WAOA, pp. 1–17 (2019)
4. Avis, D., Toussaint, G.T.: An optimal algorithm for determining the visibility of a polygon from an edge. IEEE Trans. Comput. **30**(12), 910–914 (1981)
5. Bhattacharya, P., Ghosh, S.K., Pal, S.: Constant approximation algorithms for guarding simple polygons using vertex guards. arXiv:1712.05492 (2017)
6. Bhattacharya, P., Ghosh, S.K., Roy, B.: Approximability of guarding weak visibility polygons. Discret. Appl. Math. **228**, 109–129 (2017)
7. Bhattacharya, P., Ghosh, S.K., Roy, B.: Personal communication (2018)
8. Bonnet, É., Miltzow, T.: An approximation algorithm for the art gallery problem. In: Proceedings of the 33rd International Symposium on Computational Geometry, SoCG, pp. 20:1–20:15 (2017)
9. Das, G., Heffernan, P.J., Narasimhan, G.: Finding all weakly-visible chords of a polygon in linear time. Nord. J. Comput. **1**(4), 433–457 (1994)
10. Deshpande, A., Kim, T., Demaine, E.D., Sarma, S.E.: A pseudopolynomial time $O(\log n)$-approximation algorithm for art gallery problems. In: Algorithms and Data Structures, 10th International Workshop, WADS, pp. 163–174 (2007)
11. Efrat, A., Har-Peled, S.: Guarding galleries and terrains. Inf. Process. Lett. **100**(6), 238–245 (2006)
12. Eidenbenz, S., Stamm, C., Widmayer, P.: Inapproximability results for guarding polygons and terrains. Algorithmica **31**(1), 79–113 (2001)
13. Friedrichs, S., Hemmer, M., King, J., Schmidt, C.: The continuous 1.5D terrain guarding problem: Discretization, Optimal solutions, and PTAS. JoCG 7(1), 256–284 (2016)
14. Ghosh, S.K.: Approximation algorithms for art gallery problems, pp. 429–434 (1997)
15. Ghosh, S.K.: Approximation algorithms for art gallery problems in polygons. Discret. Appl. Math. **158**(6), 718–722 (2010)
16. Ghosh, S.K., Maheshwari, A., Pal, S.P., Saluja, S., Veni Madhavan, C.E.: Characterizing and recognizing weak visibility polygons. Comput. Geom. **3**, 213–233 (1993)
17. Gibson, M., Kanade, G., Krohn, E., Varadarajan, K.: Guarding terrains via local search. J. Comput. Geometry **5**(1), 168–178 (2014)
18. Joe, B., Simpson, R.B.: Corrections to Lee's visibility polygon algorithm. BIT **27**(4), 458–473 (1987)
19. Katz, M.J.: A PTAS for vertex guarding weakly-visible polygons - an extended abstract. arXiv:1803.02160 (2018)

20. Ke, Y.: Detecting the weak visibility of a simple polygon and related problems. Technical report, Johns Hopkins University (1987)
21. King, J., Kirkpatrick, D.G.: Improved approximation for guarding simple galleries from the perimeter. Discrete Comput. Geom. **46**(2), 252–269 (2011)
22. Krohn, E., Nilsson, B.: Approximate guarding of monotone and rectilinear polygons. Algorithmica **66**, 564–594 (2013)
23. Lee, D.T.: Visibility of a simple polygon. Comput. Vis. Graph. Image Process. **22**(2), 207–221 (1983)
24. Lee, D.T., Lin, A.K.: Computational complexity of art gallery problems. IEEE Trans. Inf. Theory **32**, 276–282 (1986)
25. Mustafa, N.H., Ray, S.: PTAS for geometric hitting set problems via local search. In: Proceedings 25th Symposium on Computational Geometry, SoCG, pp. 17–22 (2009)
26. O'Rourke, J.: Art gallery theorems and algorithms. Oxford Univ. Press, Oxford (1987)
27. O'Rourke, J., Supowit, K.J.: Some NP-hard polygon decomposition problems. IEEE Trans. Info. Theory **29**(2), 181–189 (1983)
28. Sack, J.-R., Suri, S.: An optimal algorithm for detecting weak visibility of a polygon (preliminary version). In: Proceedings of the 5th Symposium on Theoretical Aspects of Computer Science, STACS, pp. 312–321 (1988)

Lasserre Integrality Gaps for Graph Spanners and Related Problems

Michael Dinitz[1(✉)], Yasamin Nazari[1], and Zeyu Zhang[2]

[1] Johns Hopkins University, Baltimore, MD, USA
{mdinitz,ynazari}@jhu.edu
[2] Facebook, Inc., Menlo Park, CA, USA

Abstract. There has been significant recent progress on algorithms for approximating graph spanners, i.e., algorithms which approximate the best spanner for a given input graph. Essentially all of these algorithms use the same basic LP relaxation, so a variety of papers have studied the limitations of this approach and proved integrality gaps for this LP. We extend these results by showing that even the strongest lift-and-project methods cannot help significantly, by proving polynomial integrality gaps even for $n^{\Omega(\varepsilon)}$ levels of the Lasserre hierarchy, for both the directed and undirected spanner problems. We also extend these integrality gaps to related problems, notably DIRECTED STEINER NETWORK and SHALLOW-LIGHT STEINER NETWORK.

1 Introduction

A spanner is a subgraph which approximately preserves distances. Formally, a t-spanner of a graph G is a subgraph H such that $d_H(u,v) \leq t \cdot d_G(u,v)$ for all $u, v \in V$ (where d_H and d_G denote shortest-path distances in H and G respectively). Since H is a subgraph it is also the case that $d_G(u,v) \leq d_H(u,v)$, and thus a t-spanner preserves all distances up to a multiplicative factor of t, which is known as the *stretch*. Graph spanners originally appeared in the context of distributed computing [26,27], but have since been used as fundamental building blocks in applications ranging from routing in computer networks [30] to property testing of functions [5] to parallel algorithms [20].

Most work on graph spanners has focused on tradeoffs between various parameters, particularly the size (number of edges) and the stretch. Most notably, a seminal result of Althöfer et al. [1] is that every graph admits a $(2k-1)$-spanner with at most $n^{1+1/k}$ edges, for every integer $k \geq 1$. This tradeoff is also known to be tight, assuming the Erdős girth conjecture [19], but extensions to this fundamental result have resulted in an enormous literature on graph spanners.

Alongside this work on tradeoffs, there has been a line of work on *optimizing* spanners. In this line of work, we are usually given a graph G and a value t,

Supported in part by NSF award CCF-1909111.

C. Kaklamanis and A. Levin (Eds.): WAOA 2020, LNCS 12806, pp. 97–112, 2021.
https://doi.org/10.1007/978-3-030-80879-2_7

and are asked to find the t-spanner of G with the fewest number of edges. If G is undirected then this is known as Basic t-Spanner, while if G is directed then this is known as Directed t-Spanner. The best known approximation for Directed t-Spanner is an $O(n^{1/2})$-approximation [4], while for Basic t-Spanner the best known approximations are $O(n^{1/3})$ when $t = 3$ [4] and when $t = 4$ [17], and $O\left(n^{\frac{1}{\lfloor (t+1)/2 \rfloor}}\right)$ when $t > 4$. Note that this approximation for $t > 4$ is directly from the result of [1] by using the trivial fact that the optimal solution is always at least $n-1$ (in a connected graph). So this result is in a sense "generic", as both the upper bound and the lower bound are universal rather than applying to the particular input graph.

One feature of the algorithms of [4,17], as well as earlier work [14] and extensions to related settings (such as approximating fault-tolerant spanners [15,17] and minimizing the maximum or norm of the degrees [8,9]), is that they all use some variant of the same basic LP: a flow-based relaxation originally introduced for spanners by [14]. The result of [4] uses a slightly different LP (based on cuts rather than flows), but it is easy to show that the LP of [4] is no stronger than the LP of [14].

The fact that for Basic t-Spanner we cannot do better than the "generic" bound when $t > 4$, as well as the common use of a standard LP relaxation, naturally gives rise to a few questions. Is it possible to do better than the generic bound when $t > 4$? Can this be achieved with the basic LP? Can we improve our analysis of this LP to get improvements for Directed t-Spanner? In other words: what is the power of convex relaxations for spanner problems? It seems particularly promising to use lift-and-project methods to try for stronger LP relaxations, since one of the very few spanner approximations that uses a different LP relaxation was the use of the Sherali-Adams hierarchy to give an approximation algorithm for the Lowest Degree 2-Spanner problem [12].

It has been known since [18,21] that Directed t-Spanner does not admit an approximation better than $2^{\log^{1-\varepsilon} n}$ for any constant $\varepsilon > 0$ (under standard complexity assumptions), and it was more recently shown [13] that Basic t-Spanner cannot be approximated any better than $2^{(\log^{1-\varepsilon} n)/t}$ for any constant $\varepsilon > 0$. Thus no convex relaxation can do better than these bounds. But it is possible to prove stronger integrality gaps: it was shown in [14] that the integrality gap of the basic LP for Directed t-Spanner is at least $\tilde{\Omega}(n^{\frac{1}{3}-\varepsilon})$, while in [17] it was shown that the basic LP for Basic t-Spanner has an integrality gap of at least $\Omega(n^{\frac{2}{(1+\varepsilon)(t+1)+4}})$, nearly matching the generic upper bound (particularly for large t). But this left open a tantalizing prospect: perhaps there are stronger relaxations which could be used to get improved approximation bounds. Of course, the hardness of approximation results prove a limit to this. But even with the known hardness results and integrality gaps, it is possible that there is, say, an $O(n^{1/1000})$-approximation for Directed t-Spanner and an $O(n^{1/(1000t)})$-approximation for Basic t-Spanner that uses more advanced relaxations. It is also possible that a more complex relaxation, which perhaps cannot be solved in polynomial time, could lead to better approximations (albeit not in polynomial time).

1.1 Our Results and Techniques

This is the problem which we investigate: can we design stronger relaxations for spanners and related problems? While we cannot rule out all possible relaxations, we show that an extremely powerful lift-and-project technique, the Lasserre hierarchy [22] applied to the basic LP, does not give relaxations which are significantly better than the basic LP. This is true despite the fact that Lasserre is an SDP hierarchy rather than an LP hierarchy, and despite the fact that we allow a polynomial number of levels in the hierarchy (even though only a constant number of levels can be solved in polynomial time). And since the Lasserre hierarchy is at least as strong as other hierarchies such as the Sherali-Adams hierarchy [29] and the Lovasz-Schrijver hierarchy [24], our results also imply integrality gaps for these hierarchies.

In other words: *we show that even super-polynomial time algorithms which are based on any of the most well-known lift-and-project methods cannot approximate spanners much better than the current approximations based on the basic LP.*

Slightly more formally, we first rewrite the basic LP in a way that is similar to [4] but is equivalent to the stronger original formulation [14]. This makes the Lasserre lifts of the LP easier to reason about, thanks to the new structure of this formulation. We then consider the Lasserre hierarchy applied to this LP, and prove the following theorems.

Theorem 1. *For every constant* $0 < \varepsilon < 1$ *and sufficiently large* n*, the integrality gap of the* $n^{\Omega(\varepsilon)}$*-th level Lasserre SDP for* Directed $(2k-1)$-Spanner *is at least* $\left(\frac{n}{k}\right)^{\frac{1}{18}-\Theta(\varepsilon)}$.

Theorem 2. *For every constant* $0 < \varepsilon < 1$ *and sufficiently large* n*, the integrality gap of the* $n^{\Omega(\varepsilon)}$*-th level Lasserre SDP for* Basic $(2k-1)$-Spanner *is at least* $\frac{1}{k} \cdot \left(\frac{n}{k}\right)^{\min\left\{\frac{1}{18}, \frac{5}{32k-6}\right\}-\Theta(\varepsilon)} = n^{\Theta\left(\frac{1}{k}-\varepsilon\right)}$.

While the constant in the exponent is different, Theorem 2 is similar to [17] in that it shows that the integrality gap "tracks" the trivial approximation from [1] as a function of k. Thus for undirected spanners, even using the Lasserre hierarchy cannot give too substantial an improvement over the trivial greedy algorithm.

At a very high level, we follow the approach to building spanner integrality gaps of [14,17]. They started with random instances of the Unique Games problem, which are known to not admit any good solutions [6]. They then used these Unique Games instances to build spanner instances with the property that every spanner had to be large (or else the Unique Games instance would have had a good solution), but by "splitting flow" the LP could be very small.

In order to apply this framework to the Lasserre hierarchy, we need to make a number of changes. First, since Unique Games can be solved reasonably well by Lasserre [3,7], starting with a random instance of Unique Games will not work. Instead, we start with a more complicated problem known as PROJECTION GAMES (the special case of LABEL COVER in which all the edge relations are functions). Hardness reductions for spanners have long used PROJECTION GAMES [13,18,21], but previous integrality gaps [14,17] have used UNIQUE GAMES since it was sufficient and was easier to work with. But we will need to overcome the complications that come from PROJECTION GAMES.

Fortunately, an integrality gap for the Lasserre hierarchy for PROJECTION GAMES was recently given by [10,25] (based on an integrality gap for CSPs from [31]), so we can use this as our starting point and try to plug it into the integrality gap framework of [14,17] to get an instance of either directed or undirected spanners. Unfortunately, the parameters and structure that we get from this are different enough from the parameters used in the integrality gap of the basic LP that we cannot use [14,17] as a black box. We need to reanalyze the instance using different techniques, even for the "easy" direction of showing that there are no good integral solutions. In order to do this, we also need some additional properties of the gap instance for PROJECTION GAMES from [10] which were not stated in their original analysis. So we cannot even use [10] as a black box.

The main technical difficulty, though, is verifying that there is a "low-cost" fractional solution to the resulting SDP. For the basic LP this is straightforward, but for Lasserre we need to show that the associated slack moment matrices are all PSD. This turns out to be surprisingly tricky, because the slack moment matrices for the spanner SDP are much more complicated than the constraints of the PROJECTION GAME SDP. But by decomposing these matrices carefully we can show that each matrix in the decomposition is PSD. At a high level, we decompose the slack moment matrices as a summation of several matrices in a way that allows us to use the consistency properties of the feasible solution to the PROJECTION GAMES instance in [10] to show that the overall sum is PSD.

Doing this requires us to use some properties of the feasible fractional solution provided by [10], some of which we need to prove as they were not relevant in the original setting. In particular, one important property which makes our task much easier is that their fractional solution actually satisfies *all* of the edges in the PROJECTION GAMES instance. That is, their integrality gap is in a particular "place": the fractional solution has value 1 while every integral solution has much smaller value. This is enormously useful to us since spanners and the other network design problems we consider are minimization problems rather than maximization problems like PROJECTION GAMES, as it essentially allows us to use essentially "the same" fractional solution (as it will also be feasible for the minimization version since it satisfies all edges). Technically, we end up combining this fact about the fractional solution of [10] with several properties of the Lasserre hierarchy to infer some more refined structural properties of the derived fractional solution for spanners, allowing us to argue that they are feasible for the Lasserre lifts.

Extensions. A number of other network design problems exhibit behavior that is similar to spanners, and we can extend our integrality gaps to these problems. In particular, we give a new integrality gap for Lasserre for Directed Steiner Network (DSN) (also called Directed Steiner Forest) and Shallow-Light Steiner Network (SLSN) [2]. In DSN we are given a directed graph $G = (V, E)$ (possibly with weights) and a collection of pairs $\{(s_i, t_i)\}_{i \in [p]}$, and are asked to find the cheapest subgraph such that there is an s_i to t_i path for all $i \in [p]$. In SLSN the graph is undirected, but each s_i and t_i is required to be connected within a global distance bound L. The best known approximation for DSN is an $O(n^{3/5+\varepsilon})$-approximation for arbitrarily small constant $\varepsilon > 0$ [11], which uses a standard flow-based LP relaxation. We can use the ideas we developed for spanners to also give integrality gaps for the Lasserre lifts of these problems. We provide the theorems here; details and proofs can be found in the full version [16].

Theorem 3. *For every constant $0 < \varepsilon < 1$ and sufficiently large n, the integrality gap of the $n^{\Omega(\varepsilon)}$-th level Lasserre SDP for* Directed Steiner Network *is at least $n^{\frac{1}{16}-\Theta(\varepsilon)}$.*

Theorem 4. *For every constant $0 < \varepsilon < 1$ and sufficiently large n, the integrality gap of the $n^{\Omega(\varepsilon)}$-th level Lasserre SDP for* Shallow-Light Steiner Network *is at least $n^{\frac{1}{16}-\Theta(\varepsilon)}$.*

2 Preliminaries: Lasserre Hierarchy

The Lasserre hierarchy lifts a polytope to a higher dimensional space, and then optionally projects this lift back to the original space in order to get tighter relaxations. The standard characterization for Lasserre (when the base polytope includes the hypercube constraints that every variable is in $[0, 1]$) is as follows [22,23,28]:

Definition 1 (Lasserre Hierarchy). *Let $A \in \mathbb{R}^{m \times n}$ and $b \in \mathbb{R}^m$, and define the polytope $K = \{x \in \mathbb{R}^n : Ax \geq b\}$. The r-th level of the Lasserre hierarchy $L_r(K)$ consists of the set of vectors $y \in [0, 1]^{\mathscr{P}([n])}$ (where $\mathscr{P}$ denotes the power set) that satisfy the following constraints:*

$$y_\varnothing = 1, \qquad M_{r+1}(y) := (y_{I \cup J})_{|I|,|J| \leq r+1} \succeq 0,$$

$$\forall \ell \in [m] : \quad M_r^\ell(y) := \left(\sum_{i=1}^{n} A_{\ell i} y_{I \cup J \cup \{i\}} - b_\ell y_{I \cup J} \right)_{|I|,|J| \leq r} \succeq 0$$

The matrix M_{r+1} is called the base moment matrix, and the matrices M_r^ℓ are called slack moment matrices.

Let us review (see, e.g., [28]) multiple helpful properties that we will use later. We include proofs in the full version [16] for completeness.

Claim. If $M_r(y) \succcurlyeq 0$, $|I| \le r$, and $y_I = 1$, then $y_{I \cup J} = y_J$ for all $|J| \le r$.

Claim. If $M_r(y) \succcurlyeq 0$, $|I| \le r$, and $y_I = 0$, then $y_{I \cup J} = 0$ for all $|J| \le r$.

Lemma 1. *If $M_{r+1}(y) \succeq 0$ then $M^{i,1}(y) = (y_{I \cup J \cup \{i\}})_{|I|,|J| \le r} \succeq 0$ and $M^{i,0}(y) = (y_{I \cup J} - y_{I \cup J \cup \{i\}})_{|I|,|J| \le r} \succeq 0$ for all $i \in [n]$.*

3 PROJECTION GAMES: Background and Previous Work

In this section we discuss the PROJECTION GAMES problem, its Lasserre relaxation, and the integrality gap that was recently developed for it [10] which form the basis of our integrality gaps for spanners and related problems. We begin with the problem definition.

Definition 2 (Projection Games). *Given a bipartite graph (L, R, E), a (label) alphabet set Σ, and projections (functions) $\pi_e : \Sigma \to \Sigma$ for each $e \in E$, the objective is to find a label assignment $\alpha : L \cup R \to \Sigma$ that maximizes the number of edges $e = (v_L, v_R) \in E$ where $\pi_e(\alpha(v_L)) = \alpha(v_R)$. We refer to these as satisfied edges.*

We sometimes use relation notation for the projection π_e, e.g., we use $(\sigma_1, \sigma_2) \in \pi_e$. PROJECTION GAMES is the standard LABEL COVER problem, but where every relation is required to be a projection (and hence we inherit the relation notation when useful). Similarly, if we further restrict every function π_e to be a bijection, then we have the UNIQUE GAMES problem. So PROJECTION GAMES lies between UNIQUE GAMES and LABEL COVER.

The basis of our integrality gaps is the integrality gap instance recently shown by [10] for Lasserre relaxations of PROJECTION GAMES. We first formally define this SDP. For every $\Psi \subseteq (L \cup R) \times \Sigma$ we will have a variable y_Ψ. Then the r-th level Lasserre SDP for PROJECTION GAMES is the following:

$$\mathrm{SDP}^r_{Proj} : \quad \begin{aligned} &\max \sum_{(v_L, v_R) \in E, (\sigma_L, \sigma_r) \in \pi_{(u,v)}} y_{(v_L, \sigma_L), (v_R, \sigma_R)} \\ &s.t.\ y_\varnothing = 1 \\ &\quad M_r(y) := \left(y_{\Psi_1 \cup \Psi_2}\right)_{|\Psi_1|, |\Psi_2| \le r} \succcurlyeq 0 \\ &\quad M_r^v(y) := \left(\sum_{\sigma \in \Sigma} y_{\Psi_1 \cup \Psi_2 \cup \{(v,\sigma)\}} - y_{\Psi_1 \cup \Psi_2}\right)_{|\Psi_1|, |\Psi_2| \le r} = \mathbf{0}\ \forall v \in V \end{aligned}$$

It is worth noting that for simplicity we are not using the original presentation of this SDP given by [10]: they used a vector inner product representation. But it can be shown (and we do so in the full version [16]) that these representations are equivalent. To prove their integrality gap, [10] gave a PROJECTION GAMES instance with the following properties. One of the properties is not proven in their paper, but is essentially trivial. We give a proof of this property, as well as a discussion of how the other properties follow from their construction, in the full version [16].

Lemma 2. *For any constant $0 < \varepsilon < 1$, there exists an instance of* PROJECTION GAMES $(L, R, E_{Proj}, \Sigma, (\pi_e)_{e \in E_{Proj}})$ *with the following properties:*

1. $\Sigma = \left[n^{\frac{3-3\varepsilon}{5}}\right]$, $R = \{x_1, \ldots, x_n\}$, $L = \{c_1, \ldots, c_m\}$, *where* $m = n^{1+\varepsilon}$.
2. *There exists a feasible solution* $\mathbf{y}^*$ *for the* $r = n^{\Omega(\varepsilon)}$*-th level* SDP^r_{Proj}*, such that for all* $\{c_i, x_j\} \in E_{Proj}$, $\sum_{(\sigma_L, \sigma_R) \in \pi_{(c_i, x_j)}} y_{(c_i, \sigma_L), (x_j, \sigma_R)} = 1$.
3. *At most* $O\left(\frac{n^{1+\varepsilon} \ln n}{\varepsilon}\right)$ *edges can be satisfied.*
4. *The degree of vertices in L is* $K = n^{\frac{1-\varepsilon}{5}} - 1$*, and the degree of vertices in R is at most* $2Kn^{\varepsilon}$.

4 Lasserre Integrality Gap for DIRECTED $(2k-1)$-SPANNER

In this section we prove our main result for the DIRECTED $(2k-1)$-SPANNER problem: a polynomial integrality gap for polynomial levels of the Lasserre hierarchy.

4.1 Spanner LPs and Their Lasserre Lifts

The standard flow-based LP for spanners (including both the directed and basic $(2k-1)$-spanner problems) was introduced by [14], and has subsequently been used in many other spanner problems [4,8,15]. However, it is an awkward fit with Lasserre, most notably since it has an exponential number of variables (and hence so too do all the Lasserre lifts). We instead will work with an equivalent LP which appeared implicitly in [14] and is slightly stronger than the "antispanner" LP of [4].

First we review the standard LP, as stated in [14], and then describe the new LP. Let $\mathcal{P}_{u,v}$ denote the set of all stretch-$(2k-1)$ paths from u to v.

$$\begin{aligned}
\mathrm{LP}^{Flow}_{Spanner}: \quad & \min \sum_{e \in E} x_e \\
s.t. \quad & \sum_{P \in \mathcal{P}_{u,v} : e \in P} f_P \le x_e && \forall (u,v) \in E, \forall e \in E \\
& \sum_{P \in \mathcal{P}_{u,v}} f_P \ge 1 && \forall (u,v) \in E \\
& x_e \ge 0 && \forall e \in E \\
& f_P \ge 0 && \forall (u,v) \in E, P \in \mathcal{P}_{u,v}
\end{aligned}$$

Since the f_P variables do not appear in the objective function, we can project the polytope defined by $\mathrm{LP}^{Flow}_{Spanner}$ onto the x_e variables and use the same objective function to get an equivalent LP but with only the x_e variables. To define this LP more formally, let $\mathcal{Z}^{u,v} = \{\mathbf{z} \in [0,1]^{|E|} : \sum_{e \in P} z_e \ge 1 \ \forall P \in \mathcal{P}_{u,v}\}$ be the polytope bounded by $0 \le z_e \le 1$ for all $e \in E$ and $\sum_{e \in P} z_e \ge 1$ for all $P \in \mathcal{P}_{u,v}$. Then we will use the following LP relaxation as our starting point, which informally says that every "stretch-$(2k-1)$ fractional cut" must be covered, and hence in an integral solution there is a stretch-$(2k-1)$ path for all edges (and so corresponds to a $(2k-1)$-spanner).

$$\begin{aligned} \text{LP}_{Spanner}: \quad & \min \sum_{e \in E} x_e \\ & s.t. \sum_{e \in E} z_e x_e \geq 1 \; \forall (u,v) \in E, \text{ and } \mathbf{z} \in \mathcal{Z}^{u,v} \\ & x_e \geq 0 \qquad \forall e \in E \end{aligned}$$

While as written there are an infinite number of constraints, it is easy to see by convexity that we need to only include the (exponentially many) constraints corresponding to vectors $\mathbf{z}$ that are vertices in the polytope $\mathcal{Z}^{u,v}$, for each $(u,v) \in E$. Thus there are only an exponential number of constraints, but for simplicity we will analyze this LP as if there were constraints for all possible $\mathbf{z}$. We show in the full version [16] that $\text{LP}_{Spanner}$ and $\text{LP}^{Flow}_{Spanner}$ are equivalent.

Using Definition 1, we can write the r-th level Lasserre SDP of $\text{LP}_{Spanner}$ (which is the SDP we consider in our integrality gap results) as follows. We let $\text{SDP}^r_{Spanner}$ denote this SDP.

$$\begin{aligned} & \min \textstyle\sum_{e \in E} y_e \\ & s.t. \; y_{\varnothing} = 1 \\ & \quad M_{r+1}(y) := (y_{I \cup J})_{|I|,|J| \leq r+1} \succcurlyeq 0 \\ & \quad M^{\mathbf{z}}_r(y) := \left(\sum_{e \in E} z_e y_{I \cup J \cup \{e\}} - y_{I \cup J} \right)_{|I|,|J| \leq r} \succcurlyeq 0 \; \forall (u,v) \in E, \mathbf{z} \in \mathcal{Z}^{u,v} \end{aligned}$$

4.2 Spanner Instance

We now formally define the instance of Directed $(2k-1)$-Spanner that we will analyze to prove the integrality gap. We follow the framework of [14], who showed how to use the hardness framework of [18,21] to prove integrality gaps for the basic flow LP. We start with a different instance (integrality gaps instances for Projection Games rather than random instances of Unique Games), and also change the reduction in order to obtain a better dependency on k.

Roughly speaking, given a Projection Games instance, we start with the "label-extended" graph. For each original vertex in the projection game, we create a group of vertices in the spanner instance of size $|\Sigma|$, so each vertex in the group can be thought as a label assignment for the Projection Games vertex. We add paths between these groups corresponding to each function π_e (we add a path if the associated assignment satisfies the Projection Games edges). We add many copies of the Projection Games graph itself as the "outer graph", and then connect each Projection Games vertex to the group associated with it. The key point is to prove that any integral solution must contain either many outer edges or many "connection edges" (in order to span the outer edges), while the fractional solution can buy connection and inner edges fractionally and simultaneously span all of the outer edges. A schematic version of this graph is presented in Fig. 1.

More formally, given the Projection Games instance $(L = \cup_{i \in [m]} \{c_i\}, R = \cup_{i \in [n]} \{x_i\}, E_{Proj}, \Sigma = [n^{\frac{3-3\varepsilon}{5}}], (\pi_e)_{e \in E_{Proj}})$ from Lemma 2, we create a directed graph $G = (V, E)$ as follows (note that K is the degree of the vertices in L).

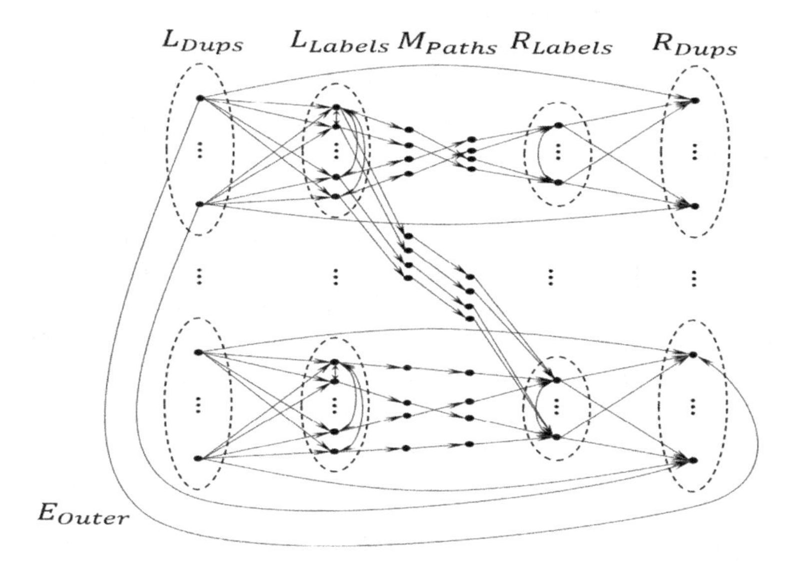

Fig. 1. DIRECTED $(2k-1)$-SPANNER instance

For every $c_i \in L$, we create $|\Sigma| + kK|\Sigma|$ vertices: $c_{i,\sigma}$ for all $\sigma \in \Sigma$ and c_i^l for all $l \in [kK|\Sigma|]$. We also create edges $(c_i^l, c_{i,\sigma})$ for each $\sigma \in \Sigma$ and $l \in [kK|\Sigma|]$. We call this edge set E_L. For every $x_i \in R$, we create $|\Sigma|+kK|\Sigma|$ vertices: $x_{i,\sigma}$ for $\sigma \in \Sigma$ and x_i^l for $l \in [kK|\Sigma|]$. We also create an edge $(x_i^l, x_{i,\sigma})$ for each $\sigma \in \Sigma$ and $l \in [kK|\Sigma|]$. We call this edge set E_R. For every $e = \{c_i, x_j\} \in E_{Proj}$, we create edges (c_i^l, x_j^l) for each $l \in [kK|\Sigma|]$. We call this edge set E_{Outer}. For each $e = \{c_i, x_j\} \in E_{Proj}$ and $(\sigma_L, \sigma_R) \in \pi_{i,j}$, we create vertices $w_{i,j,\sigma_L,\sigma_R,t}$ for $t \in [2k-4]$ and edges $(c_{i,\sigma_L}, w_{i,j,\sigma_L,\sigma_R,1}), (w_{i,j,\sigma_L,\sigma_R,1}, w_{i,j,\sigma_L,\sigma_R,2}), \ldots, (w_{i,j,\sigma_L,\sigma_R,2k-4}, x_{j,\sigma_R})$. We call this edge set E_M. Finally, for technical reasons we need some other edges E_{LStars} and E_{RStars} inside each of the groups groups.

More formally, $V = L_{Dups} \cup L_{Labels} \cup M_{Paths} \cup R_{Labels} \cup R_{Dups}$ and $E = E_L \cup E_{LStars} \cup E_M \cup E_{RStars} \cup E_R \cup E_{Outer}$, such that:

$$
\begin{aligned}
&L_{Labels} = \{c_{i,\sigma} \mid i \in |L|, \sigma \in \Sigma\}, \qquad L_{Dups} = \{c_i^l \mid i \in |L|, l \in [kK|\Sigma|]\},\\
&R_{Labels} = \{x_{i,\sigma} \mid i \in |R|, \sigma \in \Sigma\}, \qquad R_{Dups} = \{x_i^l \mid i \in |R|, l \in [kK|\Sigma|]\},\\
&M_{Paths} = \{w_{i,j,\sigma_L,\sigma_R,t} \mid \{c_i, x_j\} \in E_{Proj}, (\sigma_L, \sigma_R) \in \pi_{i,j}, t \in [2k-4]\}\\
&E_L^{i,l} = \{(c_i^l, c_{i,\sigma}) \mid \sigma \in \Sigma\}, \qquad E_L^l = \cup_{i \in |L|} E_L^{i,l},\ E_L = \cup_{l \in [kK|\Sigma|]} E_L^l,\\
&E_R^{i,l} = \{(x_i^l, x_{i,\sigma}) \mid \sigma \in \Sigma\}, \qquad E_R^l = \cup_{i \in |R|} E_R^{i,l},\ E_R = \cup_{l \in [kK|\Sigma|]} E_R^l,\\
&E_M^{i,j,\sigma_L,\sigma_R} = \{(c_{i,\sigma_L}, w_{i,j,\sigma_L,\sigma_R,1}), (w_{i,j,\sigma_L,\sigma_R,1}, w_{i,j,\sigma_L,\sigma_R,2}),\\
&\qquad\qquad (w_{i,j,\sigma_L,\sigma_R,2}, w_{i,j,\sigma_L,\sigma_R,3}), \ldots, (w_{i,j,\sigma_L,\sigma_R,2k-4}, x_{j,\sigma_R})\}\\
&E_M^{i,j} = \cup_{(\sigma_L,\sigma_R) \in \pi_{i,j}} E_M^{i,j,\sigma_L,\sigma_R} \qquad E_M = \cup_{i,j:\{c_i,x_j\} \in E_{Proj}} E_M^{i,j}\\
&E_{Outer} = \{(c_i^l, x_j^l) \mid \{c_i, x_j\} \in E_{Proj}, l \in [kK|\Sigma|]\}\\
&E_{LStars} = \{(c_{i,1}, c_{i,\sigma}), (c_{i,\sigma}, c_{i,1}) \mid i \in |L|, \sigma \in \Sigma \setminus \{1\}\},\\
&E_{RStars} = \{(x_{i,1}, x_{i,\sigma}), (x_{i,\sigma}, x_{i,1}) \mid i \in |R|, \sigma \in \Sigma \setminus \{1\}\},
\end{aligned}
$$

Note that if $k < 3$ then then there is no vertex set M_{Paths}, but only E_M which directly connects c_{i,σ_L} and x_{j,σ_R} for each $\{c_i, x_j\} \in E_{Proj}$ and $(\sigma_L, \sigma_R) \in \pi_{i,j}$.

4.3 Fractional Solution

We now provide a low-cost feasible vector solution for the r-th level Lasserre lift of the spanner instance described above. Slightly more formally, we define values $\{y'_S : S \subseteq E\}$ and show that they form a feasible solution for the r-th level Lasserre lift $\mathrm{SDP}^r_{Spanner}$, and show that the objective value is $O(|V|)$. We start with a feasible solution $\{y^*_\Psi : \Psi \subseteq (L \cup R) \times \Sigma\}$ to the $(r+2)$-th level Lasserre lift $\mathrm{SDP}^{r+2}_{Proj}$ for the Projection Games instance (based on Lemma 2) that was used to construct our directed spanner instance, and adapt it to a solution for the spanner problem. We first define a function $\Phi : E \setminus E_{Outer} \to \mathscr{P}((L \cup R) \times \Sigma)$ (where $\mathscr{P}$ indicates the power set) as follows:

$$\Phi(e) = \begin{cases} \varnothing, & \text{if } e \in E_{LStars} \cup E_M \cup E_{RStars} \\ \{(c_i, \sigma)\}, & \text{if } e \in E_L \text{ and } e \text{ has an endpoint } c_{i,\sigma} \in L_{labels} \\ \{(x_i, \sigma)\}, & \text{if } e \in E_R \text{ and } e \text{ has an endpoint } x_{i,\sigma} \in R_{labels} \end{cases}$$

We then extend the definition of Φ to $\mathscr{P}(E \setminus E_{Outer}) \to \mathscr{P}((L \cup R) \times \Sigma)$ by setting $\Phi(S) = \cup_{e \in S} \Phi(e)$. Next, we define the solution $\{y'_S \mid S \subseteq E\}$: For any set S containing any edge in E_{Outer}, we define $y'_S = 0$, otherwise, let $y'_S = y^*_{\Phi(S)}$. Note that based on how we defined the function Φ, for all edges in $E_{LStars} \cup E_M \cup E_{RStars}$ we have $y'_S = y^*_\varnothing = 1$. In other words, these edges will be picked integrally in our feasible solution. At a very high level, what we are doing is fractionally buying edges in E_L, E_R and integrally buying edges in E_M in order to span edges in E_{Outer}. The edges in E_{LStars} and E_{RStars} are used to span edges in E_L and E_R. For the cost, a straightforward calculation (which appears in the full version [16]) implies the following.

Lemma 3. *The objective value of* $\mathbf{y}'$ *in* $SDP^r_{Spanner}$ *is* $O(|V|)$.

Feasibility. We now show that the described vector solution $\mathbf{y}'$ is feasible for the r-th level of Lasserre, i.e., that all the moment matrices defined in $SDP^r_{Spanner}$ are PSD. This is the most technically complex part of the analysis, particularly for the slack moment matrices for edges in E_{outer}.

We first prove that the base moment matrix in $SDP^r_{Spanner}$ is PSD for our solution $\mathbf{y}'$ by using the fact that the base moment matrix in SDP^{r+2}_{Proj} is PSD for the Projection Games solution $\mathbf{y}^*$ (the proof be can found in the full version [16]).

Theorem 5. *The base moment matrix* $M_{r+1}(\mathbf{y}') = (y'_{I \cup J})_{|I|,|J| \le r+1}$ *is PSD.*

Showing that the slack moment matrices of our spanner solution are all PSD is more subtle and requires a case by case analysis. We divide this argument into three cases (given by the next three theorems), from simplest to most complex, depending on the edge in the underlying linear constraint.

Theorem 6. *Slack moment matrix* $M_r^{\mathbf{z}}(\mathbf{y}') = \left(\sum_{e\in E} z_e y'_{I\cup J\cup\{e\}} - y'_{I\cup J}\right)_{|I|,|J|\le r}$ *is PSD for all* $(u,v) \in E_{LStars} \cup E_M \cup E_{RStars}$ *and* $\mathbf{z} \in \mathcal{Z}^{u,v}$.

This first case corresponds to pairs (u,v) for which we assigned $y'_{(u,v)} = 1$, so is comparatively simple to analyze via basic properties of the Lasserre hierarchy. The proof is included in the full version [16].

Theorem 7. *Slack moment matrix* $M_r^{\mathbf{z}}(\mathbf{y}') = \left(\sum_{e\in E} z_e y'_{I\cup J\cup\{e\}} - y'_{I\cup J}\right)_{|I|,|J|\le r}$ *is PSD for every* $(u,v) \in E_L \cup E_R$ *and* $\mathbf{z} \in \mathcal{Z}^{u,v}$.

This second case corresponds to the edges in E_L and E_R, and is a bit more complex since these edges are only bought fractionally in our solution. At a high-level, we first partition the terms of the summation $\sum_{e\in E} z_e(y'_{I\cup J\cup\{e\}})_{|I|,|J|\le r}$ depending on which middle edge e is included in the path that spans (u,v), and then show that each part in the partition is PSD even after subtracting an auxiliary matrix. Finally, we show that the sum of these auxiliary matrices is PSD after subtracting $(y'_{I\cup J})_{|I|,|J|\le r}$. Each step can be derived from the fact that $\mathbf{y}^*$ is a feasible solution of SDP^{r+2}_{Proj}. The complete proof is included in the full version [16].

Now we move to the main technical component of our integrality gap analysis: proving that the slack moment matrices corresponding to outer edges are PSD.

Theorem 8. *Slack moment matrix* $M_r^{\mathbf{z}}(\mathbf{y}') = \left(\sum_{e\in E} z_e y'_{I\cup J\cup\{e\}} - y'_{I\cup J}\right)_{|I|,|J|\le r}$ *is PSD for every* $(u,v) \in E_{outer}$ *and* $\mathbf{z} \in \mathcal{Z}^{u,v}$.

Proof. In this case the edge (u,v) is not included in the solution, and is thus only spanned using other (inner) edges. The main difficulty here is that the spanning path is longer, and each hop of the path contributes to the solution differently. To prove this theorem, we have to again partition the summation $\sum_{e\in E} z_e(y'_{I\cup J\cup\{e\}})_{|I|,|J|\le r}$, while subtracting and adding a more carefully designed set of auxiliary matrices. Note that in this argument we crucially use the fact that the SDP solution of [10] satisfies all of the demands (property 2. in Lemma 2).

We first show how to decompose $M_r^{\mathbf{z}}(\mathbf{y}')$ as the sum of several simpler matrices. This will let us reason about each matrix differently based on the assigned values and their connection to the Projection Games constraints. We will then explain why each of these matrices is PSD. Observe that for each $(u,v) = (c_i^l, x_j^l) \in E_{outer}$, the set of stretch-$(2k-1)$ paths consist of the outer edge, or one of the paths that go through some labels (σ_L, σ_R). It is not hard to see that any other path connecting such pairs has length larger than $(2k-1)$. More formally:

Claim. For every pair $(c_i^l, x_j^l) \in E_{Outer}$, the length $(2k-1)$ paths from c_i^l to x_j^l are: the path consisting of only the edge (c_i^l, x_j^l), or the paths consisting of edges $\{(c_i^l, c_{i,\sigma_L})\} \cup E_M^{i,j,\sigma_L,\sigma_R} \cup \{(x_{j,\sigma_R}, x_j^l)\}$ for some $(\sigma_L, \sigma_R) \in \pi_{i,j}$.

We use this observation, and the fact that $y_e = 0$ for all $e \in E_{Outer}$ to break $M_r^{\mathbf{z}}(\mathbf{y}')$ into several pieces, and argue that each piece is PSD.

$$\left(\sum_{e\in E} z_e y'_{I\cup J\cup\{e\}} - y'_{I\cup J}\right)_{|I|,|J|\le r} \tag{1}$$

$$= \sum_{e\in E} z_e \left(y^*_{\Phi(I\cup J)\cup\Phi(e)}\right)_{|I|,|J|\le r} - \left(y^*_{\Phi(I\cup J)}\right)_{|I|,|J|\le r}$$

$$= \sum_{\sigma_L\in\Sigma} z_{(c_i^l, c_i, \sigma_L)} \left(y^*_{\Phi(I\cup J)\cup\{(c_i,\sigma_L)\}}\right)_{|I|,|J|\le r}$$

$$+ \sum_{(\sigma_L,\sigma_R)\in\pi_{i,j}} \sum_{e\in E_M^{i,j,\sigma_L,\sigma_R}} z_e \left(y^*_{\Phi(I\cup J)\cup\varnothing}\right)_{|I|,|J|\le r}$$

$$+ \sum_{\sigma_R\in\Sigma} z_{(x_j,\sigma_R,x_j^l)} \left(y^*_{\Phi(I\cup J)\cup\{(x_j,\sigma_R)\}}\right)_{|I|,|J|\le r}$$

$$+ \sum_{e\in E\setminus(E_L^{i,l}\cup E_M^{i,j}\cup E_R^{j,l})} z_e \left(y'_{I\cup J\cup\{e\}}\right)_{|I|,|J|\le r} - \left(y^*_{\Phi(I\cup J)}\right)_{|I|,|J|\le r}$$

$$= \sum_{\sigma_L\in\Sigma} z_{(c_i^l, c_i, \sigma_L)} \left(y^*_{\Phi(I\cup J)\cup\{(c_i,\sigma_L)\}} - \sum_{\sigma_R:(\sigma_L,\sigma_R)\in\pi_{i,j}} y^*_{\Phi(I\cup J)\cup\{(c_i,\sigma_L),(x_j,\sigma_R)\}}\right)_{|I|,|J|\le r} \tag{2}$$

$$+ \sum_{(\sigma_L,\sigma_R)\in\pi_{i,j}} \sum_{e\in E_M^{i,j,\sigma_L,\sigma_R}} z_e \left(y^*_{\Phi(I\cup J)} - y^*_{\Phi(I\cup J)\cup\{(c_i,\sigma_L),(x_j,\sigma_R)\}}\right)_{|I|,|J|\le r} \tag{3}$$

$$+ \sum_{\sigma_R\in\Sigma} z_{(x_j,\sigma_R,x_j^l)} \left(y^*_{\Phi(I\cup J)\cup\{(x_j,\sigma_R)\}} - \sum_{\sigma_L:(\sigma_L,\sigma_R)\in\pi_{i,j}} y^*_{\Phi(I\cup J)\cup\{(c_i,\sigma_L),(x_j,\sigma_R)\}}\right)_{|I|,|J|\le r} \tag{4}$$

$$+ \sum_{e\in E\setminus(E_L^{i,l}\cup E_M^{i,j}\cup E_R^{j,l})} z_e \left(y'_{I\cup J\cup\{e\}}\right)_{|I|,|J|\le r} \tag{5}$$

$$- \left(y^*_{\Phi(I\cup J)} - \sum_{(\sigma_L,\sigma_R)\in\pi_{i,j}} y^*_{\Phi(I\cup J)\cup\{(c_i,\sigma_L),(x_j,\sigma_R)\}}\right)_{|I|,|J|\le r} \tag{6}$$

$$+ \sum_{(\sigma_L,\sigma_R)\in\pi_{i,j}} \left(z_{(c_i^l,c_i,\sigma_L)} + \sum_{e\in E_M^{i,j,\sigma_L,\sigma_R}} z_e + z_{(x_j,\sigma_R,x_j^l)} - 1 \right) \tag{7}$$

$$\times \left(y^*_{\Phi(I\cup J)\cup\{(c_i,\sigma_L),(x_j,\sigma_R)\}}\right)_{|I|,|J|\le r} \tag{8}$$

To prove the third equality, we have subtracted and added the sum over $(\sigma_L, \sigma_R) \in \pi_{i,j}$, and then partitioned it over the other sums.

The matrix in (3) is PSD by basic properties of Lasserre. The matrix in (5) is PSD because it is a principal submatrix of $M_{r+1}(\mathbf{y}')$, which is PSD by Theorem 5. The matrix in (8) is equal to $\left(y'_{I\cup J\cup\{(c_i^l,c_{i,\sigma_L}),(x_{j,\sigma_R},x_j^l)\}}\right)_{|I|,|J|\le r}$, and so it is also a principal submatrix of $M_{r+1}(\mathbf{y}')$ and is hence PSD. The coefficient in (7) is non-negative because $\sum_{e\in P} z_e \ge 1$ for all path $P \in \mathcal{P}_{c_i^l,x_j^l}$ by definition of $\mathbf{z}$.

We now argue that the matrices in (2), (4), and (6) are all-zero matrices, which will complete the proof. In order to show this, we need the following claim, which uses the fact that the fractional solution $\mathbf{y}^*$ to the Projection Games instance satisfies all of the demands.

Lemma 4. *$y^*_{\Psi\cup\{(c_i,\sigma_L),(x_j,\sigma_R)\}} = 0$ for all $\{c_i, x_j\} \in E_{Proj}$, $(\sigma_L, \sigma_R) \notin \pi_{i,j}$, $|\Psi| \leq 2r$.*

Proof. By Property 2. of Lemma 2, we know that $\sum_{(\sigma_L,\sigma_R)\in\pi_{i,j}} y^*_{(c_i,\sigma_L),(x_j,\sigma_R)} = 1$. By the slack moment matrix constraints in SDP^r_{Proj}, we know that $M^{c_i}_r(\mathbf{y}^*)$ and $M^{x_j}_r(\mathbf{y}^*)$ are both all-zero matrices. Hence,

$$\sum_{\sigma_L\in\Sigma}\sum_{\sigma_R\in\Sigma} y^*_{(c_i,\sigma_L),(x_j,\sigma_R)} = \sum_{\sigma_R\in\Sigma} y^*_{(x_j,\sigma_R)} = 1 = \sum_{(\sigma_L,\sigma_R)\in\pi_{i,j}} y^*_{(c_i,\sigma_L),(x_j,\sigma_R)}$$

In other words, since $y^*_{(c_i,\sigma_L),(x_j,\sigma_R)} \geq 0$ for all $\sigma_L \in \Sigma$ and $\sigma_R \in \Sigma$, it follows for all $(\sigma_L, \sigma_R) \notin \pi_{i,j}$ that $y^*_{(c_i,\sigma_L),(x_j,\sigma_R)} = 0$. Then basic properties of Lasserre imply $y^*_{\Psi\cup\{(c_i,\sigma_L),(x_j,\sigma_R)\}} = 0$.

Next, we use this lemma to complete the proof. We argue that all entries of the matrix in (2) are zero. This is because for any entry with index I and J we have:

$$\begin{aligned} y^*_{\Phi(I\cup J)\cup\{(c_i\sigma_L)\}} &= \sum_{\sigma_R\in\Sigma} y^*_{\Phi(I\cup J)\cup\{(c_i,\sigma_L),(x_j,\sigma_R)\}} \\ &= \sum_{\sigma_R:(\sigma_L,\sigma_R)\in\pi_{i,j}} y^*_{\Phi(I\cup J)\cup\{(c_i,\sigma_L),(x_j,\sigma_R)\}}, \end{aligned}$$

where the last equality follows from above Lemma 4. A similar argument implies that the matrix in (4) is also all-zero.

To prove the matrix in (6) is also all-zero, consider an entry with index I and J. Then

$$\begin{aligned} y^*_{\Phi(I\cup J)} &= \sum_{\sigma_L\in\Sigma} y_{\Phi(I\cup J)\cup\{(c_i,\sigma_L)\}} = \sum_{\sigma_L\in\Sigma}\sum_{\sigma_R\in\Sigma} y^*_{\Phi(I\cup J)\cup\{(c_i,\sigma_L),(x_j,\sigma_R)\}} \\ &= \sum_{(\sigma_L,\sigma_R)\in\pi_{i,j}} y^*_{\Phi(I\cup J)\cup\{(c_i,\sigma_L),(x_j,\sigma_R)\}}. \end{aligned}$$

Again, we have used Lemma 4 in the last equality.

Therefore, (3), (5), (8) are PSD, (2), (4), and (6) are all-zero, and (7) is non-negative, which proves the theorem.

4.4 Proof of Theorem 1

In order to prove Theorem 1, we now just need to analyze the integral optimum solution. In the full version [16] we prove the following.

Lemma 5. *The optimal $(2k-1)$-spanner of G has at least $nkK|\Sigma|\sqrt{K}$ edges.*

The proof of Theorem 1 then follows (after plugging in all parameters) from Lemma 5, the feasibility of the fractional solutions (Theorems 5, 6, 7, and 8), and the objective value of the fractional solution (Lemma 3).

To see this, remember from Lemma 2 that $m = n^{1+\varepsilon}$, $K = n^{\frac{1-\varepsilon}{5}} - 1$, $|\Sigma| = n^{\frac{3-3\varepsilon}{5}}$. Also

$$\begin{aligned} |V| &= |L_{Dups}| + |L_{Labels}| + |M_{Paths}| + |R_{Labels}| + |R_{Dups}| \\ &= mkK|\Sigma| + m|\Sigma| + (2k-4)mK|\Sigma| + n|\Sigma| + nkK|\Sigma| \\ &= O(mkK|\Sigma|) = O\left(kn^{\frac{9+\varepsilon}{5}}\right) \end{aligned}$$

Therefore, $\frac{nkK|\Sigma|\sqrt{K}}{O(|V|)} = \Omega\left(\frac{\sqrt{K}}{n^{\varepsilon}}\right) = \Omega\left(n^{\frac{1-11\varepsilon}{10}}\right) = \Omega\left(\left(\frac{|V|}{k}\right)^{\frac{1}{18}-\Theta(\varepsilon)}\right)$, proving the theorem.

5 Undirected $(2k-1)$-Spanner

In order to extend these techniques to the undirected case (Theorem 2), we need to make a number of changes. This is the most technically complex result in this paper, so due to space constraints we defer all details to the full version [16]. First, for technical reasons we need to replace the middle paths E_M with edges, and instead add "outside" paths (as was done in [17]). More importantly, though, we have the same fundamental problem that always arises when moving from directed to undirected spanners: without directions on edges, there can be many more short cycles (and thus ways of spanning edges) in the resulting graph. In particular, if we directly change every edge on the integrality gap instance of Directed $(2k-1)$-Spanner in the previous section to be undirected, then there are more paths available to span the outer edges (more formally, the analog of Claim 4.3 is false), so the integral optimum might no longer be large.

This difficulty is fundamentally caused by the fact that the graph of the Projection Games instance from [10] that we use as our starting point might have short cycles. These turn into short spanning paths of outer edges in our spanner instance. In order to get around this, we first carefully subsample and prune to remove a selected subset of edges in E_{Proj}, causing the remaining graph to have large girth (at least $2k+2$) but without losing too much of its density or any of the other properties that we need. This is similar to what was done by [17] to prove an integrality gap for the base LP, but here we are forced to start with the instance of [10], which has a far more complicated structure than the random Unique Games instances used by [17]. This is the main technical difficulty, but once we overcome it we can use the same ideas as in Sect. 4 to prove Theorem 2.

Acknowledgements. The authors would like to thank Amitabh Basu for many helpful discussions.

References

1. Althöfer, I., Das, G., Dobkin, D., Joseph, D., Soares, J.: On sparse spanners of weighted graphs. Discrete Comput. Geom. **9**(1), 81–100 (1993)

2. Babay, A., Dinitz, M., Zhang, Z.: Characterizing demand graphs for (fixed-parameter) shallow-light steiner network. arXiv preprint arXiv:1802.10566 (2018)
3. Barak, B., Raghavendra, P., Steurer, D.: Rounding semidefinite programming hierarchies via global correlation. In: Proceedings of the 2011 IEEE 52Nd Annual Symposium on Foundations of Computer Science, pp. 472–481. FOCS 2011 (2011)
4. Berman, P., Bhattacharyya, A., Makarychev, K., Raskhodnikova, S., Yaroslavtsev, G.: Approximation algorithms for spanner problems and directed Steiner forest. Inf. Comput. **222**, 93–107 (2013)
5. Bhattacharyya, A., Grigorescu, E., Jung, K., Raskhodnikova, S., Woodruff, D.P.: Transitive-closure spanners. SIAM J. Comput. **41**(6), 1380–1425 (2012)
6. Charikar, M., Hajiaghayi, M., Karloff, H.J.: Improved approximation algorithms for label cover problems. Algorithmica **61**(1), 190–206 (2011)
7. Charikar, M., Makarychev, K., Makarychev, Y.: Near-optimal algorithms for unique games. In: Proceedings of the Thirty-eighth Annual ACM Symposium on Theory of Computing, pp. 205–214. STOC 2006 (2006)
8. Chlamtáč, E., Dinitz, M.: Lowest-degree k-spanner: approximation and hardness. Theor. Comput. **12**(15), 1–29 (2016)
9. Chlamtác, E., Dinitz, M., Robinson, T.: Approximating the Norms of Graph Spanners. In: Approximation, Randomization, and Combinatorial Optimization. Algorithms and Techniques (APPROX/RANDOM 2019). Leibniz International Proceedings in Informatics (LIPIcs), vol. 145, pp. 11:1–11:22 (2019)
10. Chlamtác, E., Manurangsi, P., Moshkovitz, D., Vijayaraghavan, A.: Approximation algorithms for label cover and the log-density threshold. In: Proceedings of the Twenty-Eighth Annual ACM-SIAM Symposium on Discrete Algorithms. SODA 2017 (2017)
11. Chlamtáč, E., Dinitz, M., Kortsarz, G., Laekhanukit, B.: Approximating spanners and directed Steiner forest: upper and lower bounds. In: Proceedings of the 28th Annual ACM-SIAM Symposium on Discrete Algorithms. SODA 2017 (2017)
12. Chlamtáč, E., Dinitz, M., Krauthgamer, R.: Everywhere-sparse spanners via dense subgraphs. In: Proceedings of the 53rd IEEE Annual Symposium on Foundations of Computer Science, pp. 758–767. FOCS 2012 (2012)
13. Dinitz, M., Kortsarz, G., Raz, R.: Label cover instances with large girth and the hardness of approximating basic k-spanner. ACM Trans. Algorithms **12**(2), 25:1–25:16 (2015)
14. Dinitz, M., Krauthgamer, R.: Directed spanners via flow-based linear programs. In: Proceedings of the Forty-third Annual ACM Symposium on Theory of Computing, pp. 323–332. STOC 2011 (2011)
15. Dinitz, M., Krauthgamer, R.: Fault-tolerant spanners: better and simpler. In: Proceedings of the 30th Annual ACM SIGACT-SIGOPS Symposium on Principles of Distributed Computing, pp. 169–178. PODC 2011 (2011)
16. Dinitz, M., Nazari, Y., Zhang, Z.: Lasserre integrality gaps for graph spanners and related problems. CoRR abs/1905.07468 (2019). http://arxiv.org/abs/1905.07468
17. Dinitz, M., Zhang, Z.: Approximating low-stretch spanners. In: Proceedings of the 27th Annual ACM-SIAM Symposium on Discrete Algorithms. SODA (2016)
18. Elkin, M., Peleg, D.: The hardness of approximating spanner problems. Theor. Comput. Syst. **41**(4), 691–729 (2007)
19. Erdős, P.: Extremal problems in graph theory. In: Theory Graphs and its Applications: Proceedings of the Symposium held in Smolenice 1963, pp. 29–36 (1964)
20. Friedrichs, S., Lenzen, C.: Parallel metric tree embedding based on an algebraic view on Moore-Bellman-Ford. J. ACM (JACM) **65**(6), 43 (2018)

21. Kortsarz, G.: On the hardness of approximating spanners. Algorithmica **30**(3), 432–450 (2001)
22. Lasserre, J.B.: An explicit exact SDP relaxation for nonlinear 0-1 programs. In: Aardal, K., Gerards, B. (eds.) IPCO 2001. LNCS, vol. 2081, pp. 293–303. Springer, Heidelberg (2001). https://doi.org/10.1007/3-540-45535-3_23
23. Laurent, M.: A comparison of the sherali-adams, lovász-schrijver, and lasserre relaxations for 0–1 programming. Math. Oper. Res. **28**(3), 470–496 (2003)
24. Lovász, L., Schrijver, A.: Cones of matrices and set-functions and 0–1 optimization. SIAM J. Optim. **1**(2), 166–190 (1991)
25. Manurangsi, P.: On approximating projection games. Master's thesis, Massachusetts Institute of Technology (2015)
26. Peleg, D., Schäffer, A.A.: Graph spanners. J. Graph Theor. **13**(1), 99–116 (1989)
27. Peleg, D., Ullman, J.D.: An optimal synchronizer for the hypercube. SIAM J. Comput. **18**(4), 740–747 (1989)
28. Rothvoß, T.: The Lasserre hierarchy in approximation algorithms. Lecture Notes for the MAPSP, pp. 1–25 (2013)
29. Sherali, H.D., Adams, W.P.: A hierarchy of relaxations between the continuous and convex hull representations for zero-one programming problems. SIAM J. Discret. Math. **3**(3), 411–430 (1990)
30. Thorup, M., Zwick, U.: Compact routing schemes. In: Proceedings of the Thirteenth Annual ACM Symposium on Parallel Algorithms and Architectures, pp. 1–10. ACM (2001)
31. Tulsiani, M.: CSP gaps and reductions in the Lasserre hierarchy. In: Proceedings of the Forty-First Annual ACM Symposium on Theory of Computing, pp. 303–312. ACM (2009)

To Close Is Easier Than To Open: Dual Parameterization To k-Median

Jarosław Byrka[1], Szymon Dudycz[1], Pasin Manurangsi[2], Jan Marcinkowski[1], and Michał Włodarczyk[3](✉)

[1] University of Wrocław, Wrocław, Poland
{jby,szymon.dudycz,jan.marcinkowski}@cs.uni.wroc.pl
[2] Google Research, Mountain View, USA
pasin@google.com
[3] Eindhoven University of Technology, Eindhoven, The Netherlands
m.wlodarczyk@tue.nl

Abstract. The k-MEDIAN problem is one of the well-known optimization problems that formalize the task of data clustering. Here, we are given sets of facilities F and clients C, and the goal is to open k facilities from the set F, which provides the best division into clusters, that is, the sum of distances from each client to the closest open facility is minimized. In the CAPACITATED k-MEDIAN, the facilities are also assigned capacities specifying how many clients can be served by each facility.

Both problems have been extensively studied from the perspective of approximation algorithms. Recently, several surprising results have come from the area of parameterized complexity, which provided better approximation factors via algorithms with running times of the form $f(k) \cdot poly(n)$. In this work, we extend this line of research by studying a different choice of parameterization. We consider the parameter $\ell = |F| - k$, that is, the number of facilities that remain closed. It turns out that such a parameterization reveals yet another behavior of k-MEDIAN. We observe that the problem is W[1]-hard but it admits a parameterized approximation scheme. Namely, we present an algorithm with running time $2^{\mathcal{O}(\ell \log(\ell/\varepsilon))} \cdot poly(n)$ that achieves a $(1+\varepsilon)$-approximation. On the other hand, we show that under the assumption of Gap Exponential Time Hypothesis, one cannot extend this result to the capacitated version of the problem.

1 Introduction

Recent years have brought many surprising algorithmic results originating from the intersection of the areas of approximation algorithms and parameterized complexity. It turns out that the combination of techniques from these theories can be very fruitful and a new research area has emerged, devoted to studying *parameterized approximation algorithms*. The main goal in this area it to design

Part of this work was done while the third and the fifth author were visiting University of Wroclaw. The fifth author was supported by the Foundation for Polish Science (FNP).

C. Kaklamanis and A. Levin (Eds.): WAOA 2020, LNCS 12806, pp. 113–126, 2021.
https://doi.org/10.1007/978-3-030-80879-2_8

an algorithm processing an instance (I, k) in time $f(k) \cdot |I|^{\mathcal{O}(1)}$, where f is some computable function, and producing an approximate solution to the optimization problem in question. Such algorithms, called *FPT approximations*, are particularly interesting in the case of problems for which (1) we fail to make progress on improving the approximation factors in polynomial time, and (2) there are significant obstacles for obtaining exact parameterized algorithms. Some results of this kind are FPT approximations for k-Cut [22], Directed Odd Cycle Transversal [27], and Planar Steiner Network [10]. A good introduction to this area can be found in the survey [20].

One problem that has recently enjoyed a significant progress in this direction is the famous k-Median problem. Here, we are given a set F of facilities, a set C of clients, a metric d over $F \cup C$ and an upper bound k on the number of facilities we can open. A solution is a set $S \subseteq F$ of at most k open facilities and a connection assignment $\phi : C \to S$ of clients to the open facilities. The goal is to find a solution that minimizes the connection cost $\sum_{c \in C} d(c, \phi(c))$. The problem can be approximated in polynomial time up to a constant factor [3,9] with the currently best approximation factor being $(2.675 + \varepsilon)$ [5]. On the other hand, we cannot hope for a polynomial-time $(1 + 2/e - \varepsilon)$-approximation, since it would entail P = NP [21]. Therefore, there is a gap in our understanding of the optimal approximability of k-Median.

Surprisingly, the situation becomes simpler if we consider parameterized algorithms, with k as the natural choice of parameterization. Such a parameterized problem is W[2]-hard [1] so it is unlikely to admit an exact algorithm with running time of the form $f(k) \cdot n^{\mathcal{O}(1)}$, where n is the size of an instance. However, Cohen-Addad et al. [12] have obtained an algorithm with approximation factor $(1+2/e+\varepsilon)$ and running time[1] $2^{\mathcal{O}(k \log k)} \cdot n^{\mathcal{O}(1)}$. This result is essentially tight, as the existence of an FPT-approximation with factor $(1 + 2/e - \varepsilon)$ would contradict the *Gap Exponential Time Hypothesis*[2] (Gap-ETH) [12]. The mentioned hardness result has also ruled out running time of the form $f(k) \cdot n^{g(k)}$, where $g = k^{poly(1/\varepsilon)}$. This lower bound has been later strengthened: under Gap-ETH no algorithm with running time $f(k) \cdot n^{o(k)}$ can achieve approximation factor $(1 + 2/e - \varepsilon)$ [28].

The parameterized approach brought also a breakthrough to the understanding of Capacitated k-Median. In this setting, each facility f is associated with a capacity $u_f \in \mathbb{Z}_{\geqslant 0}$ and the connection assignment ϕ must satisfy $|\phi^{-1}(f)| \leqslant u_f$ for every facility $f \in S$. The best known polynomial-time approximation for Capacitated k-Median is burdened with a factor $\mathcal{O}(\log k)$ [1,8] and relies on the generic technique of metric tree embeddings with expected logarithmic distortion [19]. All the known constant-factor approximations violate either the number of facilities or the capacities. Li has provided such an algorithm by

[1] We omit the dependency on ε in the running time except for approximation schemes.

[2] The Gap Exponential Time Hypothesis [18,29] states that, for some constant $\gamma > 0$, there is no $2^{o(n)}$-time algorithm that can, given a 3SAT instance, distinguish between (1) the instance is fully satisfiable or (2) any assignment to the instance violates at least γ fraction of the clauses.

opening $(1+\varepsilon)\cdot k$ facilities [24,25]. Afterwards analogous results, but violating the capacities by a factor of $(1+\varepsilon)$ were also obtained [6,17]. This is in contrast with other capacitated clustering problems such as FACILITY LOCATION or k-CENTER, for which constant factor approximation algorithms have been constructed [15,23]. However, no superconstant lower bound for CAPACITATED k-MEDIAN is known.

When it comes to parameterized algorithms, Adamczyk et al. [1] have presented a $(7+\varepsilon)$-approximation algorithm with running time $2^{\mathcal{O}(k\log k)}\cdot n^{\mathcal{O}(1)}$ for CAPACITATED k-MEDIAN. Xu et al. [31] proposed a similar algorithm for the related CAPACITATED k-MEANS problem, where one minimizes the sum of squares of distances. These results have been improved by Cohen-Addad and Li [13], who obtained factor $(3+\varepsilon)$ for CAPACITATED k-MEDIAN and $(9+\varepsilon)$ for CAPACITATED k-MEANS, within the same running time.

Our Contribution. In this work, we study a different choice of parameterization for k-MEDIAN. Whereas k is the number of facilities to open, we consider the dual parameter $\ell=|F|-k$: the number of facilities to be closed. We refer to this problem as CO-ℓ-MEDIAN in order to avoid ambiguity. Note that even though this is the same task from the perspective of polynomial-time algorithms, it is a different problem when seen through the lens of parameterized complexity. First, we observe that CO-ℓ-MEDIAN is W[1]-hard (Theorem 3), which motivates the study of approximation algorithms also for this choice of parameterization. It turns out that switching to the dual parameterization changes the approximability status dramatically and we can obtain an arbitrarily good approximation factor. More precisely, we present an efficient parameterized approximation scheme (EPAS), i.e., $(1+\varepsilon)$-approximation with running time of the form $f(\ell,\varepsilon)\cdot n^{\mathcal{O}(1)}$. This constitutes our main result.

Theorem 1. *The* CO-ℓ-MEDIAN *problem admits a deterministic* $(1+\varepsilon)$-*approximation algorithm running in time* $2^{\mathcal{O}(\ell\log(\ell/\varepsilon))}\cdot n^{\mathcal{O}(1)}$ *for any constant* $\varepsilon>0$.

We obtain this result by combining the technique of color-coding from the FPT theory with a greedy approach common in the design of approximation algorithms. The running time becomes polynomial whenever we want to open all but $\mathcal{O}\left(\frac{\log n}{\log\log n}\right)$ facilities. To the best of our knowledge, this is the first non-trivial setting with general metric space which admits an approximation scheme.

A natural question arises about the behavior of the capacitated version of the problem in this setting, referred to as CAPACITATED CO-ℓ-MEDIAN. Both in polynomial-time regime or when parameterized by k, there is no evidence that the capacitated problem is any harder and the gap between the approximation factors might be just a result of our lack of understanding. Somehow surprisingly, for the dual parameterization ℓ we are able to show a clear separation between the capacitated and uncapacitated case. Namely, we present a reduction from the MAX k-COVERAGE problem which entails the same approximation lower bound as for the uncapacitated problem parameterized by k.

Theorem 2. *Assuming Gap-ETH, there is no* $f(\ell) \cdot n^{o(\ell)}$*-time algorithm that can approximate* CAPACITATED CO-ℓ-MEDIAN *to within a factor of* $(1+2/e-\epsilon)$ *for any function* f *and any constant* $\epsilon > 0$.

Related Work. A simple example of dual parameterization is given by k-INDEPENDENT SET and ℓ-VERTEX COVER. From the perspective of polynomial-time algorithms, these problems are equivalent (by setting $\ell = |V(G)| - k$), but they differ greatly when analyzed as parameterized problems: the first one is W[1]-hard while the latter is FPT and admits a polynomial kernel [14]. Another example originates from the early work on k-DOMINATING SET, which is a basic W[2]-complete problem. When parameterized by $\ell = |V(G)| - k$, the problem is known as ℓ-NONBLOCKER. This name can be interpreted as a task of choosing ℓ vertices so that none is blocked by the others, i.e., each chosen vertex has a neighbor which has not been chosen. Under this parameterization, the problem is FPT and admits a linear kernel [16]. The best known running time for ℓ-NONBLOCKER is $1.96^\ell \cdot n^{\mathcal{O}(1)}$ [30]. It is worth noting that ℓ-NONBLOCKER is a special case of CO-ℓ-MEDIAN with a graph metric and $F = C = V(G)$, however this analogy works only in a non-approximate setting.

The Gap Exponential Time Hypothesis was employed for proving parameterized inapproximability by Chalermsook et al. [7], who presented hardness results for k-CLIQUE, k-DOMINATING SET, and DENSEST k-SUBGRAPH. It was later used to obtain lower bounds for DIRECTED ODD CYCLE TRANSVERSAL [27], DIRECTED STEINER NETWORK [10], PLANAR STEINER ORIENTATION [11], and UNIQUE SET COVER [28], among others. Moreover, Gap-ETH turned out to be a sufficient assumption to rule out the existence of an FPT algorithm for k-EVEN SET [4].

2 Preliminaries

Parameterized Complexity and Reductions. A parameterized problem instance is created by associating an input instance with an integer parameter k. We say that a problem is *fixed parameter tractable* (FPT) if it admits an algorithm solving an instance (I, k) in time $f(k) \cdot |I|^{\mathcal{O}(1)}$, where f is some computable function. Such an algorithm we shall call an *FPT algorithm*.

To show that a problem is unlikely to be FPT, we use *parameterized reductions* analogous to those employed in the classical complexity theory (see [14]). Here, the concept of W-hardness replaces the one of NP-hardness, and we need not only to construct an equivalent instance in time $f(k) \cdot |I|^{\mathcal{O}(1)}$, but also to ensure that the value of the parameter in the new instance depends only on the value of the parameter in the original instance. In contrast to the NP-hardness theory, there is a hierarchy of classes FPT $=$ W[0] $\subseteq$ W[1] $\subseteq$ W[2] $\subseteq \ldots$ and these containments are believed to be strict. If there exists a parameterized reduction transforming a W[t]-hard problem to another problem Π, then the problem Π is W[t]-hard as well. If a parameterized reduction transforms parameter linearly, i.e., maps an instance (I_1, k) to $(I_2, \mathcal{O}(k))$, then it also preserves running time of the form $f(k) \cdot |I|^{o(k)}$.

In order to prove hardness of parameterized approximation, we use parameterized reductions between *promise problems*. Suppose we are given an instance (I_1, k_1) of a minimization problem with a promise that the answer is at most D_1 and we want to find a solution of value at most $\alpha \cdot D_1$. Then a reduction should map (I_1, k_1) to such an instance (I_2, k_2) so that the answer to it is at most D_2 and any solution to (I_2, k_2) of value at most $\alpha \cdot D_2$ can be transformed in time $f(k_1) \cdot |I_1|^{\mathcal{O}(1)}$ to a solution to (I_1, k_1) of value at most $\alpha \cdot D_1$. If an FPT α-approximation exists for the latter problem, then it exists also for the first one. Again, if we have $k_2 = \mathcal{O}(k_1)$, then this relation holds also for algorithms with running time of the form $f(k) \cdot |I|^{o(k)}$.

Problem Definitions. Below we formally introduce the main studied problem and the problems employed in reductions.

(CAPACITATED) CO-ℓ-MEDIAN **Parameter:** ℓ
Input: set of facilities F, set of clients C, metric d over $F \cup C$, sequence of capacities $u_f \in \mathbb{Z}_{\geqslant 0}$, integer ℓ
Task: find a set $S \subseteq F$ of at most $|F| - \ell$ facilities and a connection assignment $\phi : C \to F \setminus S$ that satisfies $|\phi^{-1}(f)| \leqslant u_f$ for all $f \in F \setminus S$, and minimizes $\sum_{c \in C} d(c, \phi(c))$

A metric $d : (F \cup C) \times (F \cup C) \to \mathbb{R}_{\geqslant 0}$ is a symmetric function that obeys the triangle inequality $d(x, y) + d(y, z) \geq d(x, z)$ and satisfies $d(x, x) = 0$. In the uncapacitated version we assume that all capacities are equal $|C|$, so any assignment $\phi : C \to F \setminus S$ is valid. In the approximate version of CAPACITATED CO-ℓ-MEDIAN we treat the capacity condition $|\phi^{-1}(f)| \leqslant u_f$ as a hard constraint and we allow only the connection cost $\sum_{c \in C} d(c, \phi(c))$ to be larger than the optimum.

k-INDEPENDENT SET **Parameter:** k
Input: graph $G = (V, E)$, integer k
Task: decide whether there exists a set $S \subseteq V(G)$ of size k such that for all pairs $u, v \in S$ we have $uv \notin E(G)$

MAX k-COVERAGE **Parameter:** k
Input: universe U, family of subsets $T_1, \ldots, T_n \subseteq U$, integer k
Task: find k subsets $T_{i_1}, \ldots, T_{i_k}$ that maximizes $|T_{i_1} \cup \cdots \cup T_{i_k}|$

3 Uncapacitated CO-ℓ-MEDIAN

We begin with a simple reduction, showing that the exact problem remains hard under the dual parameterization.

Theorem 3. *The* CO-ℓ-MEDIAN *problem is W[1]-hard.*

Proof. We reduce from ℓ-INDEPENDENT SET, which is W[1]-hard. We transform a given graph G into a CO-ℓ-MEDIAN instance by setting $F = V(G)$, and placing a client in the middle of each edge. The distance from a client to both endpoints of its edge is 1 and the shortest paths of such subdivided graph induce the metric d.

If we did not close any facilities, the cost of serving all clients would equal $|E(G)|$. The same holds if each client has an open facility within distance 1, so the set of closed facilities forms an independent set of vertices in G. On the other hand, if we close a set of facilities containing two endpoints of a single edge then the cost increases. Therefore the answer to the created instance is $|E(G)|$ if and only if G contains an independent set of size ℓ.

We move on to designing a parameterized approximation scheme for CO-ℓ-MEDIAN. We use notation $d(c, S)$ for the minimum distance between c and any element of the set S. In the uncapacitated setting the connection assignment ϕ_S is unique for a given set of closed facilities S: each client is assigned to the closest facility outside S. Whenever we consider a solution set $S \subseteq F$, we mean that this is the set of closed facilities and denote $cost(S) = \sum_{c \in C} d(c, F \setminus S)$. We define $V(f)$ to be the *Voronoi cell* of facility f, i.e., the set of clients for which f is the closest facility. We can break ties arbitrarily and for the sake of disambiguation we assume an ordering on F and whenever two distances are equal we choose the facility that comes first in the ordering.

Let $C(f)$ denote the cost of the cell $V(f)$, i.e., $\sum_{c \in V(f)} d(c, f)$. For a solution S and $f \in S$, $g \notin S$, we define $C(S, f, g) = \sum d(c, g)$ over $\{c \in V(f) \mid \phi_S(c) = g\}$, that is, the sum of connections of clients that switched from f to g. Note that as long as $f \in F$ remains open, there is no need to change connections of the clients in $V(f)$. We can express the difference of connection costs after closing S as

$$\Delta(S) = \sum_{f \in S} \sum_{c \in V(f)} d(c, F \setminus S) - \sum_{f \in S} C(f) = \sum_{f \in S} \sum_{g \in F \setminus S} C(S, f, g) - \sum_{f \in S} C(f).$$

We have $cost(S) = \sum_{f \in F} C(f) + \Delta(S)$, therefore the optimal solution closes set S of size ℓ minimizing $\Delta(S)$.

The crucial observation is that any small set of closed facilities S can be associated with a small set of open facilities that are relevant for serving the clients from $\bigcup_{f \in S} V(f)$. Intuitively, if $C(S, f, g) = \mathcal{O}(\frac{\varepsilon}{\ell^2}) \cdot cost(S)$ for all $f \in S$, then we can afford replacing ℓ such facilities g with others that are not too far away.

Definition 1. *The ε-support of a solution $S \subseteq F$, $|S| = \ell$, referred to as ε-SUPP(S), is the set of all open facilities g (i.e., $g \notin S$) satisfying one of the following conditions:*

1. *there is $f \in S$ such that g minimizes distance $d(f, g)$ among all open facilities,*
2. *there is $f \in S$ such that $C(S, f, g) > \frac{\varepsilon}{6\ell^2} \cdot cost(S)$.*

We break ties in condition (1) according to the same rule as in the definition of $V(f)$, so there is a single g satisfying condition (1) for each f.

Lemma 1. *For a solution S of size ℓ, we have $|\varepsilon\text{-SUPP}(S)| \leq 6 \cdot \ell^3/\varepsilon + \ell$.*

Proof. We get at most ℓ facilities from condition (1). Since the sets of clients being served by different $g \in F \setminus S$ are disjoint and $\sum_{g \in F} C(S, f, g) \leq cost(S)$, we obtain at most $6 \cdot \ell^3/\varepsilon$ facilities from condition (2).

Even though we will not compute the set $\varepsilon\text{-SUPP}(Opt)$ directly, we are going to work with partitions $F = A \uplus B$, such that $Opt \subseteq A$ and $\varepsilon\text{-SUPP}(Opt) \subseteq B$. Such a partition already gives us a valuable hint. By looking at each facility $f \in A$ separately, we can deduce that if $f \in Opt$ and some other facility g belongs to A (so it cannot belong to $\varepsilon\text{-SUPP}(Opt)$) then in some cases g must also belong to Opt. More precisely, if $g \in A$ is closer to f than the closest facility in B, then g must be closed, as otherwise it would violate condition (1). Furthermore, suppose that $h \in A$ serves clients from $V(f)$ (assuming f is closed) of total cost at least $\frac{\varepsilon}{6\ell^2} \cdot cost(S)$. If we keep h open and close some other facilities, this relation is preserved and having h in A violates condition (2). We formalize this idea with the notion of *required sets*, given by the following procedure, supplied additionally with a real number D, which can be regarded as the guessed value of $cost(Opt)$.

Algorithm 1. COMPUTE-REQUIRED-SET$(A, B, f, \varepsilon, \ell, D)$ (assume $f \in A$ and $A \cap B = \emptyset$)

1: $s_f \leftarrow \min_{y \in B} d(f, y)$
2: $R_f \leftarrow \{g \in A \mid d(f, g) < s_f\}$ (including f)
3: **while** $\exists g \in A : C(R_f, f, g) > \frac{\varepsilon}{3\ell^2} \cdot D$ **do**
4: $\quad R_f \leftarrow R_f \cup \{g\}$
5: **end while**
6: **return** R_f

Lemma 2. *Let $Opt \subseteq F$ be the optimal solution. Suppose $F = A \uplus B$, $f \in Opt \subseteq A$, $\varepsilon\text{-SUPP}(Opt) \subseteq B$, and $cost(Opt) \leq 2D$. Then the set R_f returned by the routine* COMPUTE-REQUIRED-SET$(A, B, f, \varepsilon, \ell, D)$ *satisfies $R_f \subseteq Opt$.*

Proof. Let y_f be the facility in B that is closest to f. Due to condition (1) in Definition 1, all facilities $g \in A$ satisfying $d(f, g) < d(f, y_f)$ must be closed in the optimal solution, so we initially add them to R_f. We keep invariant $R_f \subseteq Opt$, so for any $g \in F \setminus Opt$ it holds that $C(Opt, f, g) \geq C(R_f, f, g)$. Whenever there is $g \in A$ satisfying $C(R_f, f, g) > \frac{\varepsilon}{3\ell^2} \cdot D$, we get

$$C(Opt, f, g) \geq C(R_f, f, g) > \frac{\varepsilon}{3\ell^2} \cdot D \geq \frac{\varepsilon}{6\ell^2} \cdot cost(Opt).$$

Since g does not belong to $\varepsilon\text{-SUPP}(Opt) \subseteq B$, then by condition (2) it must be closed. Hence, adding g to R_f preserves the invariant.

Before proving the main technical lemma, we need one more simple observation, in which we exploit the fact that the function d is indeed a metric.

Lemma 3. *Suppose* $c \in V(f_0)$ *and* $d(f_0, f_1) \leq d(f_0, f_2)$. *Then* $d(c, f_1) \leq 3 \cdot d(c, f_2)$.

Proof. An illustration is given in Fig. 1. Since c belongs to the Voronoi cell of f_0, we have $d(c, f_0) \leq d(c, f_2)$. By the triangle inequality

$$\begin{aligned} d(c, f_1) &\leq d(c, f_0) + d(f_0, f_1) \leq d(c, f_0) + d(f_0, f_2) \\ &\leq d(c, f_0) + d(c, f_0) + d(c, f_2) \leq 3 \cdot d(c, f_2). \end{aligned}$$

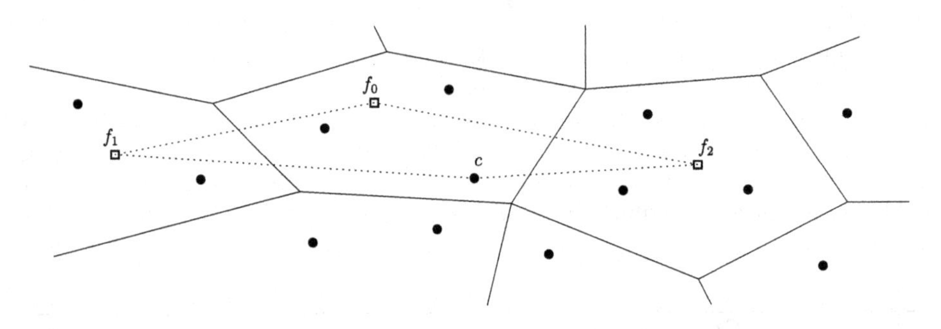

Fig. 1. An example of a Voronoi diagram with squares representing facilities and dots being clients. Lemma 3 states that even if $d(c, f_1) > d(c, f_2)$ for $c \in V(f_0)$ and $d(f_0, f_1) \leq d(f_0, f_2)$, then $d(c, f_1)$ cannot be larger than $3 \cdot d(c, f_2)$.

Lemma 4. *Suppose we are given a partition* $F = A \uplus B$, *such that* $Opt \subseteq A$, ε-SUPP$(Opt) \subseteq B$, *and a number* $D \in \mathbb{R}_{>0}$, *such that* $cost(Opt) \in [D, 2D]$. *Then we can find a solution* $S \subseteq A$, *such that* $cost(S) \leq (1 + \varepsilon) \cdot cost(Opt)$, *in polynomial time.*

Proof. We compute the set R_f = COMPUTE-REQUIRED-SET$(A, B, f, \varepsilon, \ell, D)$ for each facility $f \in A$. The subroutine from Algorithm 1 clearly runs in polynomial time. Furthermore, for each $f \in A$ we compute its *marginal cost* of closing

$$m_f = \sum_{c \in V(f)} d(c, F \setminus R_f) - C(f).$$

If $|R_f| > \ell$ then f cannot belong to any solution consistent with the partition (A, B) and in this case we set $m_f = \infty$. Since the marginal cost depends only on f, we can greedily choose ℓ facilities from A that minimize m_f – we refer to this set as S.

We first argue that $\sum_{f \in F} C(f) + \sum_{f \in S} m_f$ is at most the cost of the optimal solution. By greedy choice we have that $\sum_{f \in S} m_f \leq \sum_{f \in Opt} m_f$. We have assumed $cost(Opt) \leq 2D$ so by Lemma 2 we get that if $f \in Opt$, then $R_f \subseteq Opt$. The set of facilities $F \setminus Opt$ that can serve clients from $V(f)$ is a subset of $F \setminus R_f$ and the distances can only increase, thus for $f \in Opt$ we have $m_f \leq \sum_{c \in V(f)} d(c, F \setminus Opt) - C(f)$. We conclude that $\sum_{f \in F} C(f) + \sum_{f \in S} m_f$ is upper bounded by

$$\sum_{f \in F} C(f) + \sum_{f \in Opt} \sum_{c \in V(f)} d(c, F \setminus Opt) - \sum_{f \in Opt} C(f) = cost(Opt). \tag{1}$$

The second argument is that after switching benchmark from the marginal cost to the true cost of closing S, we will additionally pay at most εD. These quantities differ when for a facility $f \in S$ we have 'connected' some clients from $V(f)$ to $g \in S \setminus R_f$ when computing m_f. More precisely, we want to show that for each $f \in S$ we have

$$\sum_{c \in V(f)} d(c, F \setminus S) \leq \sum_{c \in V(f)} d(c, F \setminus R_f) + \frac{\varepsilon D}{\ell}. \tag{2}$$

By the construction of R_f, whenever $g \in S \setminus R_f$ we are guaranteed that there exists a facility $y \in B$ such that $d(f, g) \geq d(f, y)$ and, moreover, $C(R_f, f, g) \leq \frac{\varepsilon}{3\ell^2} \cdot D$. We can reroute all such clients c to the closest open facility and we know it is not further than $d(c, y)$. By Lemma 3 we know that $d(c, y) \leq 3 \cdot d(c, g)$ so rerouting those clients costs at most $\frac{\varepsilon}{\ell^2} \cdot D$. Since there are at most ℓ such facilities $g \in S \setminus R_f$, we have proved Formula (2). Combining this with bound from (1) implies that $cost(S) \leq cost(Opt) + \varepsilon D$. As we have assumed $D \leq cost(Opt)$, the claim follows.

In order to apply Lemma 4, we need to find a partition $F = A \uplus B$ satisfying $Opt \subseteq A$ and ε-SUPP$(Opt) \subseteq B$. Since ε-SUPP$(Opt) = \mathcal{O}(\ell^3/\varepsilon)$, we can do this via randomization. Consider tossing a biased coin for each facility independently: with probability $\frac{\varepsilon}{\ell^3}$ we place it in A, and with remaining probability in B. The probability of obtaining a partitioning satisfying $Opt \subseteq A$ and ε-SUPP$(Opt) \subseteq S$ equals $(\frac{\varepsilon}{\ell^3})^\ell$ times $(1 - \frac{\varepsilon}{\ell^3})^{O(\frac{\ell^3}{\varepsilon})} = \Omega(1)$. Therefore $2^{O(\ell \log(\ell/\varepsilon))}$ trials give a constant probability of sampling a correct partitioning. In order to derandomize this process, we take advantage of the following construction which is a folklore corollary from the framework of color-coding [2]. As we are not aware of any self-contained proof of this claim in the literature, we provide it for completeness.

Lemma 5. *For a set U of size n, there exists a family $\mathcal{H}$ of partitions $U = A \uplus B$ such that $|\mathcal{H}| = 2^{O(\ell \log(\ell + r))} \log n$ and for every pair of disjoint sets $A_0, B_0 \subseteq U$ with $|A_0| \leq \ell$, $|B_0| \leq r$, there is $(A, B) \in \mathcal{F}$ satisfying $A_0 \subseteq A, B_0 \subseteq B$. The family $\mathcal{H}$ can be constructed in time $2^{O(\ell \log(\ell + r))} n \log n$.*

Proof. Let use denote $[n] = \{1, 2, \ldots, n\}$ and identify $U = [n]$. We rely on the following theorem: for any integers n, k there exists a family $\mathcal{F}$ of functions $f : [n] \to [k^2]$, such that $|\mathcal{F}| = k^{\mathcal{O}(1)} \log n$ and for each $X \subseteq [n]$ of size k there is a function $f \in \mathcal{F}$ which is injective on X; moreover, $\mathcal{F}$ can be constructed in time $k^{\mathcal{O}(1)} n \log n$ [14, Theorem 5.16].

We use this construction for $k = \ell + r$. Next, consider the family $\mathcal{G}$ of all functions $g : [(\ell + r)^2] \to \{0, 1\}$ such that $|g^{-1}(0)| \leq \ell$. Clearly, $|\mathcal{G}| \leq (\ell + r)^{2\ell}$. The family $\mathcal{H}$ is given by taking all compositions $\{h = g \circ f \mid g \in \mathcal{G}, f \in \mathcal{F}\}$ and setting $(A_h, B_h) = (h^{-1}(0), h^{-1}(1))$. We have $|\mathcal{H}| \leq |\mathcal{G}| \cdot |\mathcal{F}| = 2^{O(\ell \log(\ell + r))} \log n$. Let us consider any pair of disjoint subsets $A_0, B_0 \subseteq [n]$ with $|A_0| \leq \ell$, $|B_0| \leq r$. There exists $f \in \mathcal{F}$ injective on $A_0 \cup B_0$ and $g \in \mathcal{G}$ that maps $f(A_0)$ to 0 and $f(B_0)$ to 1, so $A_0 \subseteq A_{g \circ f}, B_0 \subseteq B_{g \circ f}$.

Theorem 4. *The* CO-ℓ-MEDIAN *problem admits a deterministic* $(1+\varepsilon)$*-approximation algorithm running in time* $2^{\mathcal{O}(\ell \log(\ell/\varepsilon))} \cdot n^{\mathcal{O}(1)}$ *for any constant* $\varepsilon > 0$.

Proof. We apply Lemma 5 for $U = F$, ℓ being the parameter, and $r = 6 \cdot \ell^3/\varepsilon + \ell$, which upper bounds the size of ε-SUPP(Opt) (Lemma 1). The family $\mathcal{H}$ contains a partition $F = A \uplus B$ satisfying $Opt \subseteq A$ and ε-SUPP(Opt) $\subseteq B$. Next, we need to find D, such that $cost(Opt) \in [D, 2D]$. We begin with any polynomial-time α-approximation algorithm for k-MEDIAN ($\alpha = \mathcal{O}(1)$) to get an interval $[X, \alpha X]$, which contains $cost(Opt)$. We cover this interval with a constant number of intervals of the form $[X, 2X]$ and one of these provides a valid value of D. We invoke the algorithm from Lemma 4 for each such triple (A, B, D) and return a solution with the smallest cost.

4 Hardness of Capacitated co-ℓ-Median

In this section we show that, unlike CO-ℓ-MEDIAN, its capacitated counterpart does not admit a parameterized approximation scheme.

We shall reduce from the MAX k-COVERAGE problem, which was also the source of lower bounds for k-MEDIAN in the polynomial-time regime [21] and when parameterized by k [12]. However, the latter reduction is not longer valid when we consider a different parameterization for k-MEDIAN, as otherwise we could not obtain Theorem 1. Therefore, we need to design a new reduction, that exploits the capacity constraints and translates the parameter k of an instance of MAX k-COVERAGE into the parameter ℓ of an instance of CAPACITATED CO-ℓ-MEDIAN. To the best of our knowledge, this is the first hardness result in which the capacities play a role and allow us to obtain a better lower bound.

We rely on the following strong hardness result. Note that this result is a strengthening of [12], which only rules out $f(k) \cdot n^{k^{poly(1/\delta)}}$-time algorithm. This suffices to rule out a parameterized approximation scheme for CAPACITATED CO-ℓ-MEDIAN, but not for a strong running time lower bound of the form $f(\ell) \cdot n^{o(\ell)}$.

Theorem 5 ([28])**.** *Assuming Gap-ETH, there is no* $f(k) \cdot n^{o(k)}$*-time algorithm that can approximate* MAX k-COVERAGE *to within a factor of* $(1 - 1/e + \delta)$ *for any function* f *and any constant* $\delta > 0$*. Furthermore, this holds even when every input subset is of the same size and with a promise that there exists* k *subsets that covers each element exactly once.*

We can now prove our hardness result for CAPACITATED CO-ℓ-MEDIAN.

Theorem 6. *Assuming Gap-ETH, there is no $f(\ell) \cdot n^{o(\ell)}$-time algorithm that can approximate* Capacitated co-ℓ-Median *to within a factor of $(1+2/e-\epsilon)$ for any function f and any constant $\epsilon > 0$.*

Proof. Let $U, T_1, \ldots, T_n$ be an instance of Max k-Coverage. We create an instance (F, C) of Capacitated co-ℓ-Median as follows.

- For each subset T_i with $i \in [n]$, create a facility f_i^{set} with capacity $|T_i|$. For each element $u \in U$, create a facility $f_u^{element}$ with capacity $|U|+2$.
- For every $i \in [n]$, create $|T_i|$ clients $c_{i,1}^{set}, \ldots, c_{i,|T_i|}^{set}$. For each $j \in [|T_i|]$, we define the distance from $c_{i,j}^{set}$ to the facilities by

$$\begin{aligned} d(c_{i,j}^{set}, f_i^{set}) &= 0, \\ d(c_{i,j}^{set}, f_u^{element}) &= 1 && \forall u \in T_i, \\ d(c_{i,j}^{set}, f_{i'}^{set}) &= 2 && \forall i' \neq i, \\ d(c_{i,j}^{set}, f_u^{element}) &= 3 && \forall u \notin T_i. \end{aligned}$$

- For every element $u \in U$, create $|U|+1$ clients $c_{u,1}^{element}, \ldots, c_{u,|U|+1}^{element}$ and, for each $j \in [|U|+1]$, define the distance from $c_{u,j}^{element}$ to the facilities by

$$\begin{aligned} d(c_{u,j}^{element}, f_u^{element}) &= 0, \\ d(c_{u,j}^{element}, f_i^{set}) &= 1 && \forall T_i \ni u, \\ d(c_{u,j}^{element}, f_{u'}^{element}) &= 2 && \forall u' \neq u, \\ d(c_{u,j}^{element}, f_i^{set}) &= 3 && \forall T_i \not\ni u. \end{aligned}$$

- Let $\ell = k$.

Suppose that we have an $f(\ell) \cdot n^{o(\ell)}$-time $(1+2/e-\epsilon)$-approximation algorithm for Capacitated co-ℓ-Median. We will use it to approximate Max k-Coverage instance with $|T_1| = \cdots = |T_n| = |U|/k$ with a promise that there exists k subsets that covers each element exactly once, as follows. We run the above reduction to produce an instance (F, C) and run the approximation algorithm for Capacitated co-ℓ-Median; let $S \subseteq F$ be the produced solution. Notice that S may not contain any element-facility, as otherwise there would not even be enough capacity left to serve all clients. Hence, $S = \{f_{i_1}^{set}, \ldots, f_{i_k}^{set}\}$. We claim that $T_{i_1}, \ldots, T_{i_k}$ is an $(1-1/e+\epsilon/2)$-approximate solution for Max k-Coverage.

To see that $T_{i_1}, \ldots, T_{i_k}$ is an $(1-1/e+\epsilon/2)$-approximate solution for Max k-Coverage, notice that the cost of closing $\{f_{i_1}^{set}, \ldots, f_{i_k}^{set}\}$ is exactly $|T_{i_1} \cup \cdots \cup T_{i_k}| + 3 \cdot |U \setminus (T_{i_1} \cup \cdots \cup T_{i_k})|$ because each element-facility $f_u^{element}$ can only serve one more client in addition to $c_{u,1}^{element}, \ldots, c_{u,|U|+1}^{element}$. (Note that we may assume without loss of generality that $f_u^{element}$ serves $c_{u,1}^{element}, \ldots, c_{u,|U|+1}^{element}$.) Moreover, there are exactly $|T_{i_1}| + \cdots + |T_{i_k}| = |U|$ clients left to be served after the closure of $\{f_{i_1}^{set}, \ldots, f_{i_k}^{set}\}$. Hence, each element-facility $f_u^{element}$ with $u \in T_{i_1} \cup \cdots \cup T_{i_k}$

can serve a client of distance one from it. All other element-facilities will have to serve a client of distance three from it. This results in the cost of exactly $|T_{i_1} \cup \cdots \cup T_{i_k}| + 3 \cdot |U \setminus (T_{i_1} \cup \cdots \cup T_{i_k})|$. Now, since we are promised that there exists k subsets that uniquely covers the universe U, the optimum of Capacitated co-ℓ-Median must be $|U|$. Since our (assumed) approximation algorithm for Capacitated co-ℓ-Median has approximation factor $(1 + 2/e - \epsilon)$, we must have $|T_{i_1} \cup \cdots \cup T_{i_k}| + 3 \cdot |U \setminus (T_{i_1} \cup \cdots \cup T_{i_k})| \leq |U| \cdot (1 + 2/e - \epsilon)$, which implies that $|T_{i_1} \cup \cdots \cup T_{i_k}| \geq |U| \cdot (1 - 1/e + \epsilon/2)$. Hence, the proposed algorithm is an $f(k) \cdot n^{o(k)}$-time algorithm that approximates Max k-Coverage to within a factor of $(1 - 1/e + \epsilon/2)$, which by Theorem 5 contradicts Gap-ETH.

5 Conclusions and Open Problems

We have presented a parameterized approximation scheme for co-ℓ-Median and shown that its capacitated version does not admit such a scheme. It remains open whether Capacitated co-ℓ-Median admits any constant-factor FPT approximation. Obtaining such a result might be an important step towards getting a constant-factor polynomial-time approximation, which is a major open problem.

Another interesting question concerns whether one can employ the framework of lossy kernelization [26] to get a polynomial size approximate kernelization scheme (PSAKS) for co-ℓ-Median, which would be a strengthening of our main result. In other words, can we process an instance $\mathcal{I}$ in polynomial time to produce an equivalent instance $\mathcal{I}'$ of size $poly(\ell)$ so that solving $\mathcal{I}'$ would provide a $(1 + \varepsilon)$-approximation for $\mathcal{I}$?

References

1. Adamczyk, M., Byrka, J., Marcinkowski, J., Meesum, S.M., Wlodarczyk, M.: Constant-factor FPT approximation for capacitated k-median. In: 27th Annual European Symposium on Algorithms, ESA 2019. Leibniz International Proceedings in Informatics (LIPIcs), vol. 144, pp. 1:1–1:14. Schloss Dagstuhl-Leibniz-Zentrum fuer Informatik, Dagstuhl, Germany (2019). https://doi.org/10.4230/LIPIcs.ESA.2019.1
2. Alon, N., Yuster, R., Zwick, U.: Color-coding. J. ACM **42**(4), 844–856 (1995). https://doi.org/10.1145/210332.210337
3. Arya, V., Garg, N., Khandekar, R., Meyerson, A., Munagala, K., Pandit, V.: Local search heuristics for k-median and facility location problems. SIAM J. Comput. **33**(3), 544–562 (2004). https://doi.org/10.1137/S0097539702416402
4. Bhattacharyya, A., Ghoshal, S., Karthik C.S., Manurangsi, P.: Parameterized intractability of even set and shortest vector problem from gap-ETH. In: 45th International Colloquium on Automata, Languages, and Programming, ICALP 2018, Prague, Czech Republic, 9–13 July 2018, pp. 17:1–17:15 (2018). https://doi.org/10.4230/LIPIcs.ICALP.2018.17

5. Byrka, J., Pensyl, T., Rybicki, B., Srinivasan, A., Trinh, K.: An improved approximation for k-median, and positive correlation in budgeted optimization. In: Proceedings of the 26th Annual ACM-SIAM Symposium on Discrete Algorithms, pp. 737–756. SIAM (2015). https://doi.org/10.1145/2981561
6. Byrka, J., Rybicki, B., Uniyal, S.: An approximation algorithm for uniform capacitated k-median problem with $1+\epsilon$ capacity violation. In: Proceedings of the 18th International Conference on Integer Programming and Combinatorial Optimization, IPCO 2016, Liège, Belgium, 1–3 June 2016, pp. 262–274 (2016). https://doi.org/10.1007/978-3-319-33461-5_22
7. Chalermsook, P., et al.: From gap-ETH to FPT-inapproximability: clique, dominating set, and more. In: 2017 IEEE 58th Annual Symposium on Foundations of Computer Science (FOCS), pp. 743–754 (October 2017). https://doi.org/10.1109/FOCS.2017.74
8. Charikar, M., Chekuri, C., Goel, A., Guha, S.: Rounding via trees: deterministic approximation algorithms for group Steiner trees and k-median. In: Proceedings of the 30th Annual ACM Symposium on the Theory of Computing, Dallas, Texas, USA, 23–26 May 1998, pp. 114–123 (1998). https://doi.org/10.1145/276698.276719
9. Charikar, M., Guha, S., Tardos, É., Shmoys, D.B.: A constant-factor approximation algorithm for the k-median problem. In: Proceedings of the 31st annual ACM Symposium on Theory of Computing, pp. 1–10. ACM (1999). https://doi.org/10.1145/301250.301257
10. Chitnis, R., Feldmann, A.E., Manurangsi, P.: Parameterized approximation algorithms for bidirected Steiner network problems. In: 26th Annual European Symposium on Algorithms, ESA 2018. Schloss Dagstuhl-Leibniz-Zentrum fuer Informatik (2018). https://doi.org/10.4230/LIPIcs.ESA.2018.20
11. Chitnis, R., Feldmann, A.E., Suchý, O.: A tight lower bound for planar Steiner orientation. Algorithmica (2019). https://doi.org/10.1007/s00453-019-00580-x
12. Cohen-Addad, V., Gupta, A., Kumar, A., Lee, E., Li, J.: Tight FPT approximations for k-median and k-Means. In: 46th International Colloquium on Automata, Languages, and Programming, ICALP 2019. Leibniz International Proceedings in Informatics (LIPIcs), vol. 132, pp. 42:1–42:14. Schloss Dagstuhl-Leibniz-Zentrum fuer Informatik, Dagstuhl, Germany (2019). https://doi.org/10.4230/LIPIcs.ICALP.2019.42
13. Cohen-Addad, V., Li, J.: On the fixed-parameter tractability of capacitated clustering. In: Baier, C., Chatzigiannakis, I., Flocchini, P., Leonardi, S. (eds.) 46th International Colloquium on Automata, Languages, and Programming, ICALP 2019. Leibniz International Proceedings in Informatics (LIPIcs), vol. 132, pp. 41:1–41:14. Schloss Dagstuhl-Leibniz-Zentrum fuer Informatik, Dagstuhl, Germany (2019). https://doi.org/10.4230/LIPIcs.ICALP.2019.41
14. Cygan, M., et al.: Parameterized Algorithms. Springer, Cham (2015). https://doi.org/10.1007/978-3-319-21275-3
15. Cygan, M., Hajiaghayi, M., Khuller, S.: LP rounding for k-centers with non-uniform hard capacities. In: 2012 IEEE 53rd Annual Symposium on Foundations of Computer Science (FOCS), pp. 273–282. IEEE (2012). https://doi.org/10.1109/FOCS.2012.63
16. Dehne, F., Fellows, M., Fernau, H., Prieto, E., Rosamond, F.: NONBLOCKER: parameterized algorithmics for MINIMUM DOMINATING SET. In: Wiedermann, J., Tel, G., Pokorný, J., Bieliková, M., Štuller, J. (eds.) SOFSEM 2006. LNCS, vol. 3831, pp. 237–245. Springer, Heidelberg (2006). https://doi.org/10.1007/11611257_21

17. Demirci, H.G., Li, S.: Constant approximation for capacitated k-median with $(1+\epsilon)$-capacity violation. In: 43rd International Colloquium on Automata, Languages, and Programming, ICALP 2016, Rome, Italy, 11–15 July 2016, pp. 73:1–73:14 (2016). https://doi.org/10.4230/LIPIcs.ICALP.2016.73
18. Dinur, I.: Mildly exponential reduction from gap 3SAT to polynomial-gap label-cover. Electron. Colloquium Comput. Complex. (ECCC) **23**, 128 (2016)
19. Fakcharoenphol, J., Rao, S., Talwar, K.: A tight bound on approximating arbitrary metrics by tree metrics. In: Proceedings of the 35th Annual ACM Symposium on Theory of Computing, San Diego, CA, USA, 9–11 June 2003, pp. 448–455 (2003). https://doi.org/10.1145/780542.780608
20. Feldmann, A.E., Karthik, C., Lee, E., Manurangsi, P.: A survey on approximation in parameterized complexity: hardness and algorithms. Algorithms **13**(6), 146 (2020). https://doi.org/10.3390/a13060146
21. Guha, S., Khuller, S.: Greedy strikes back: improved facility location algorithms. J. Algorithms **31**(1), 228–248 (1999). https://doi.org/10.1006/jagm.1998.0993
22. Gupta, A., Lee, E., Li, J.: An FPT algorithm beating 2-approximation for k-cut. In: Proceedings of the 29th Annual ACM-SIAM Symposium on Discrete Algorithms, pp. 2821–2837. Society for Industrial and Applied Mathematics (2018). https://doi.org/10.1137/1.9781611975031.179
23. Korupolu, M.R., Plaxton, C.G., Rajaraman, R.: Analysis of a local search heuristic for facility location problems. J. Algorithms **37**(1), 146–188 (2000). https://doi.org/10.1006/jagm.2000.1100
24. Li, S.: On uniform capacitated k-median beyond the natural LP relaxation. In: Proceedings of the 26th Annual ACM-SIAM Symposium on Discrete Algorithms (SODA), pp. 696–707. SIAM (2015). https://doi.org/10.1145/2983633
25. Li, S.: Approximating capacitated k-median with $(1+\epsilon)k$ open facilities. In: Proceedings of the 27th Annual ACM-SIAM Symposium on Discrete Algorithms (SODA), pp. 786–796. SIAM (2016). https://doi.org/10.1137/1.9781611974331.ch56
26. Lokshtanov, D., Panolan, F., Ramanujan, M.S., Saurabh, S.: Lossy kernelization. In: Proceedings of the 49th Annual ACM SIGACT Symposium on Theory of Computing, STOC 2017, pp. 224–237. Association for Computing Machinery, New York, NY, USA (2017). https://doi.org/10.1145/3055399.3055456
27. Lokshtanov, D., Ramanujan, M.S., Saurabh, S., Zehavi, M.: Parameterized complexity and approximability of directed odd cycle transversal. In: Proceedings of the 2020 ACM-SIAM Symposium on Discrete Algorithms, SODA 2020, Salt Lake City, UT, USA, 5–8 January 2020, pp. 2181–2200 (2020). https://doi.org/10.1137/1.9781611975994.134
28. Manurangsi, P.: Tight running time lower bounds for strong inapproximability of maximum k-coverage, unique set cover and related problems (via t-wise agreement testing theorem). In: Proceedings of the 14th Annual ACM-SIAM Symposium on Discrete Algorithms, pp. 62–81. SIAM (2020). https://doi.org/10.1137/1.9781611975994.5
29. Manurangsi, P., Raghavendra, P.: A birthday repetition theorem and complexity of approximating dense CSPs. In: ICALP, pp. 78:1–78:15 (2017). https://doi.org/10.4230/LIPIcs.ICALP.2017.78
30. van Rooij, J.M.: Exact Exponential-Time Algorithms for Domination Problems in Graphs. BOXpress (2011)
31. Xu, Y., Möhring, R.H., Xu, D., Zhang, Y., Zou, Y.: A constant FPT approximation algorithm for hard-capacitated k-means. Optim. Eng. **21**(3), 709–722 (2020). https://doi.org/10.1007/s11081-020-09503-0

Explorable Uncertainty in Scheduling with Non-uniform Testing Times

Susanne Albers[1] and Alexander Eckl[1,2](✉)

[1] Department of Informatics, Technical University of Munich, Boltzmannstr. 3, 85748 Garching, Germany
albers@in.tum.de, alexander.eckl@tum.de

[2] Advanced Optimization in a Networked Economy, Technical University of Munich, Arcisstraße 21, 80333 Munich, Germany

Abstract. The problem of scheduling with testing in the framework of explorable uncertainty models environments where some preliminary action can influence the duration of a task. In the model, each job has an unknown processing time that can be revealed by running a test. Alternatively, jobs may be run untested for the duration of a given upper limit. Recently, Dürr et al. [4] have studied the setting where all testing times are of unit size and have given lower and upper bounds for the objectives of minimizing the sum of completion times and the makespan on a single machine. In this paper, we extend the problem to non-uniform testing times and present the first competitive algorithms. The general setting is motivated for example by online user surveys for market prediction or querying centralized databases in distributed computing. Introducing general testing times gives the problem a new flavor and requires updated methods with new techniques in the analysis. We present constant competitive ratios for the objective of minimizing the sum of completion times in the deterministic case, both in the non-preemptive and preemptive setting. For the preemptive setting, we additionally give a first lower bound. We also present a randomized algorithm with improved competitive ratio. Furthermore, we give tight competitive ratios for the objective of minimizing the makespan, both in the deterministic and the randomized setting.

Keywords: Online scheduling · Explorable uncertainty · Competitive analysis · Single machine · Sum of completion times · Makespan

1 Introduction

In scheduling environments, uncertainty is a common consideration for optimization problems. Commonly, results are either based on worst case considerations or a random distribution over the input. These approaches are known as robust

Work supported by Deutsche Forschungsgemeinschaft (DFG), GRK 2201 and by the European Research Council, Grant Agreement No. 691672, project APEG.

C. Kaklamanis and A. Levin (Eds.): WAOA 2020, LNCS 12806, pp. 127–142, 2021.
https://doi.org/10.1007/978-3-030-80879-2_9

optimization and stochastic optimization, respectively. However, it is often the case that unknown information can be attained through investing some additional resources, e.g. time, computing power or money. In his seminal paper, Kahan [11] has first introduced the notion of explorable or queryable uncertainty to model obtaining additional information for a problem at a given cost during the runtime of an algorithm. Since then, these kind of problems have been explored in different optimization contexts, for example in the framework of combinatorial, geometric or function value optimization tasks.

Recently, Dürr et al. [4] have introduced a model for scheduling with testing on a single machine within the framework of explorable uncertainty. In their approach, a number of jobs with unknown processing times are given. Testing takes one unit of time and reveals the processing time. If a job is executed untested, the time it takes to run the job is given by an upper bound. The novelty of their approach lies in having tests executed directly on the machine running the jobs as opposed to considering tests separately.

In view of this model, a natural extension is to consider non-uniform testing times to allow for a wider range of problems. Dürr et al. state that for certain applications it is appropriate to consider a broader variation on testing times and leave this question up for future research.

Situations where a preliminary action, operation or test can be executed before a job are manifold and include a wide range of real-life applications. In the following, we discuss a small selection of such problems and emphasize cases with heterogeneous testing requirements. Consider first a situation where an online user survey can help predict market demand and production times. The time needed to produce the necessary amount of goods for the given demand is only known after conducting the survey. Depending on its scope and size, the invested costs for the survey may vary significantly.

As a second example, we look at distributed computing in a setting with many distributed local databases and one centralized master server. At the local stations, only estimates of some data values are stored; in order to obtain the true value one must query the master server. It depends on the distance and connection quality from any localized database to the master how much time and resources this requires. Olston and Widom [14] have considered this setting in detail.

Another possible example is the acquisition of a house through an agent giving us more information about its value, location, condition, etc., but demanding a price for her services. This payment could vary based on the price of the house, the amount of work of the agent or the number of competitors.

In their paper, Dürr et al. [4] mention fault diagnosis in maintenance and medical treatment, file compression for transmissions, and running jobs in an alternative fast mode whose availability can be determined through a test. Generally, any situation involving diverse cost and duration estimates, like e.g. in construction work, manufacturing or insurance, falls into our category of possible applications.

In view of all these examples, we investigate non-uniform testing in the scope of explorable uncertainty on a single machine as introduced by [4]. We study whether algorithms can be extended to this non-uniform case and if not, how we can find new methods for it.

1.1 Problem Statement

We consider n jobs to be scheduled on a single machine. Every job j has an unknown processing time p_j and a known upper bound u_j. It holds $0 \leq p_j \leq u_j$ for all j. Each job also has a testing time $t_j \geq 0$. A job can either be executed untested, which takes time u_j, or be tested and then executed, which takes a total time of $t_j + p_j$. Note that a tested job does not necessarily have to be executed right after its test, it may be delayed arbitrarily while the algorithm tests or executes other jobs.

Since only the upper bounds are initially known to the algorithm, the task can be viewed as an online problem with an adaptive adversary. The actual processing times p_j are only realized after job j has been tested by the algorithm. In the randomized case, the adversary knows the distribution of the random input parameters of an algorithm, but not their outcome.

We denote the completion time of a job j as C_j and primarily consider the objective of minimizing the total sum of completion times $\sum_j C_j$. As a secondary objective, we also investigate the simpler goal of minimizing the makespan $\max_j C_j$. We use competitive analysis to compare the value produced by an algorithm with an optimal offline solution.

Clearly, in the offline setting where all processing times are known, an optimal schedule can be determined directly: If $t_j + p_j \leq u_j$ then job j is tested, otherwise it is run untested. For the sum of completion times, the jobs are therefore scheduled in order of non-decreasing $\min(t_j + p_j, u_j)$. Any algorithm for the online problem not only has to decide whether to test a given job or not, but also in which order to run all tests and executions of both untested and tested jobs. For a solution to the makespan objective, the ordering of the jobs does not matter and an optimal offline algorithm decides the testing by the same principle as above.

1.2 Related Work

Our setting is directly based on the problem of scheduling uncertain jobs on a single machine with explorable processing times, introduced by Dürr et al. [4] in 2018. They only consider the special case where $t_j \equiv 1$ for all jobs. For deterministic algorithms, they give a lower bound of 1.8546 and an upper bound of 2. In the randomized case, they give a lower bound of 1.6257 and a 1.7453-competitive algorithm. For several deterministic special case instances, they provide upper bounds closer to the best possible ratio of 1.8546. Additionally, tight algorithms for the objective of minimizing the makespan are given for both the deterministic and randomized cases.

Testing and executing jobs on a single machine can be viewed as part of the research area of *queryable uncertainty* or *explorable uncertainty*. The first seminal paper on dealing with uncertainty by querying parts of the input was published in 1991 by Kahan [11]. In his paper, Kahan considers a set of elements with uncertain values that lie in a closed interval. He explores approximation guarantees for the number of queries necessary to obtain the maximum and median value of the uncertain elements.

Since then, there has been a large amount of research concerned with the objective of minimizing the number of queries to obtain a solution. A variety of numerical, geometric and combinatorial problems have been studied in this framework, the following is a selection of some of these publications: Next to Kahan, Feder et al. [8], Khanna and Tan [12], and Gupta et al. [10] have also considered the objective of determining different function values, in particular the k-smallest value and the median. Bruce et al. [2] have analysed geometric tasks, specifically the Maximal Points and Convex Hull problems. They have also introduced the notion of *witness sets* as a general concept for queryable uncertainty, which was then generalized by Erlebach et al. [6]. Olston and Widom [14] researched caching problems while allowing for some inaccuracy in the objective function. Other studied combinatorial problems include minimum spanning tree [6,13], shortest path [7], knapsack [9] and boolean trees [3]. See also the survey by Erlebach and Hoffmann [5] for an overview over research in this area.

A related type of problems within optimization under uncertainty are settings where the cost of the queries is a direct part of the objective function. Most notably, the paper by Dürr et al. [4] falls into this category. There, the tests necessary to obtain additional information about the runtime of the jobs are executed on the same machine as the jobs themselves. Other examples include Weitzman's original Pandora's Box problem [17], where n independent random variables are probed to maximize the highest revealed value. Every probing incurs a price directly subtracted from the objective function. Recently, Singla [16] introduced the 'price of information' model to describe receiving information in exchange for a probing price. He gives approximation ratios for various well-known combinatorial problems with stochastic uncertainty.

1.3 Contribution

In this paper, we provide the first algorithms for the more general scheduling with testing problem where testing times can be non-uniform. Consult Table 1 for an overview of results for both the non-uniform and uniform versions of the problem. All ratios provided without citation are introduced in this paper. The remaining results are presented in [4].

For the problem of scheduling uncertain jobs with non-uniform testing times on a single machine, our results are the following: A deterministic 4-competitive algorithm for the objective of minimizing the sum of completion times and a randomized 3.3794-competitive algorithm for the same objective. If we allow preemption - that is, to cancel the execution of a job at any time and start

Table 1. Overview of results

Objective type	General tests	Uniform tests	Lower bound
$\sum C_j$ - deterministic	4	2 [4]	1.8546 [4]
$\sum C_j$ - randomized	3.3794	1.7453 [4]	1.6257 [4]
$\sum C_j$ - determ. preemptive	$2\varphi \approx 3.2361$	-	1.8546
$\max C_j$ - deterministic	$\varphi \approx 1.6180$	φ [4]	φ [4]
$\max C_j$ - randomized	$\frac{4}{3}$	$\frac{4}{3}$ [4]	$\frac{4}{3}$ [4]

working on a different job - then we can improve the deterministic case to be 2φ-competitive. Here, $\varphi \approx 1.6180$ is the golden ratio.

For the objective of minimizing the makespan, we adopt and extend the ideas of Dürr et al. [4] to provide a tight φ-competitive algorithm in the deterministic case and a tight $\frac{4}{3}$-competitive algorithm in the randomized case.

Our approaches handle non-uniform testing times in a novel fashion distinct from the methods of [4]. As we show in the full version of this paper [1], the idea of scheduling untested jobs with small upper bounds in the beginning of the schedule, which works well in the uniform case, fails to generalize to non-uniform tests. Additionally, describing parameterized worst-case instances becomes intangible in the presence of an arbitrary number of different testing times.

In place of these methods, we compute job completion times by cross-examining contributions of other jobs in the schedule. We determine tests based on the ratio between the upper bound and the given test time and pay specific attention to sorting the involved executions and tests in an suitable way.

The paper is structured as follows: Sects. 2 and 3 examine the deterministic and randomized cases respectively. Various algorithms are presented and their competitive ratios proven. We extend the optimal results for the objective of minimizing the makespan from the uniform case to general testing times in Sect. 4. Finally, we conclude with some open problems.

2 Deterministic Setting

In this section, we introduce our basic algorithm and prove deterministic upper bounds for the non-preemptive as well as the preemptive case. The basic structure introduced in Sect. 2.1 works as a framework for other algorithms presented later. We give a detailed analysis of the deterministic algorithm and prove that it is 4-competitive if parameters are chosen accordingly. In Sect. 2.2 we prove that an algorithm for the preemptive case is 3.2361-competitive and that no preemptive algorithm can have a ratio better than 1.8546.

2.1 Basic Algorithm and Proof of 4-Competitiveness

We now present the elemental framework of our algorithm, which we call (α, β)-*SORT*. As input, the algorithm has two real parameters, $\alpha \geq 1$ and $\beta \geq 1$.

Algorithm 1: (α, β)-SORT

```
1  T ← ∅, N ← ∅, σ_j ≡ 0;
2  foreach j ∈ [m] do
3  |   if u_j ≥ α t_j then
4  |   |   add j to T;
5  |   |   set σ_j ← β t_j;
6  |   else
7  |   |   add j to N;
8  |   |   set σ_j ← u_j;
9  |   end
10 end
11 while N ∪ T ≠ ∅ do
12 |   choose j_min ∈ argmin_{j∈N∪T} σ_j;
13 |   if j_min ∈ N then
14 |   |   execute j_min untested;
15 |   |   remove j_min from N;
16 |   else if j_min ∈ T then
17 |   |   if j_min not tested then
18 |   |   |   test j_min;
19 |   |   |   set σ_{j_min} ← p_{j_min};
20 |   |   else
21 |   |   |   execute j_min;
22 |   |   |   remove j_min from T;
23 |   |   end
24 end
```

The algorithm is divided into two phases. First, we decide for each job whether we test this job or not based on the ratio $\frac{u_j}{t_j}$. This gives us a partition of $[m]$ into the disjoint sets $T = \{j \in [m] : \text{ALG tests } j\}$ and $N = \{j \in [m] : \text{ALG runs } j \text{ untested}\}$. In the second phase, we always attend to the job $j_{\min}$ with the current smallest *scaling time* σ_j. The scaling time is the time needed for the next step of executing j:

- If j is in N, then $\sigma_j = u_j$.
- If j is in T and has not been tested, then $\sigma_j = \beta t_j$.
- If j is in T and has already been tested, then $\sigma_j = p_j$.

Note that in the second case above, we 'stretch' the scaling time by multiplying with $\beta \geq 1$. The intention behind this stretching is that testing a job, unlike executing it, does not immediately lead to a job being completed. Therefore the parameter β artificially lowers the relevance of testing in the ordering of our algorithm. Note that the actual time needed for testing remains t_j.

In the following, we show that the above algorithm achieves a provably good competitive ratio. The parameters are kept general in the proof and are then optimized in a final step. We present the computations with general parameters for a clearer picture of the proof structure, which we will reuse in later sections.

In the final optimization step it will turn out that setting $\alpha = \beta = 1$ yields a best-possible competitive ratio of 4.

Theorem 1. *The* $(1,1)$*-SORT algorithm is* 4*-competitive for the objective of minimizing the sum of completion times.*

Proof. For the purpose of estimating the algorithmic result against the optimum, let $\rho_j := \min(u_j, t_j + p_j)$ be the optimal running time of job j. Without loss of generality, we order the jobs s.t. $\rho_1 \geq \ldots \geq \rho_n$. Hence the objective value of the optimum is

$$\mathrm{OPT} = \sum_{j=1}^{n} j \cdot \rho_j \tag{1}$$

Additionally, let

$$p_j^A := \begin{cases} t_j + p_j & \text{if } j \in T, \\ u_j & \text{if } j \in N, \end{cases} \tag{2}$$

be the *algorithmic running time* of j, i.e. the time the algorithm spends on running job j.

We start our analysis by comparing p_j^A to the optimal runtime ρ_j for a single job, summarized in the following Proposition:

Proposition 1. *(a)* $\forall j \in T\colon t_j \leq \rho_j,\ p_j \leq \rho_j$
(b) $\forall j \in T\colon p_j^A \leq \left(1 + \frac{1}{\alpha}\right) \rho_j$
(c) $\forall j \in N\colon p_j^A \leq \alpha \rho_j$

Part (a) directly estimates testing and running times of tested jobs against the values of the optimum. We will use this extensively when computing the completion time of the jobs. The proof of parts (b) and (c) is very similar to the proof of Theorem 14 in [4] for uniform testing times. We refer to the full version [1] for a complete write-down of the proof. Note that instead of considering a single bound, we split the upper bound of the algorithmic running time p_j^A into different results for tested (b) and untested jobs (c). This allows us to differentiate between different cases in the proof of Lemma 1 in more detail. We will often make use of this Proposition to upper bound the algorithmic running time in later sections.

To obtain an estimate of the completion time C_j, we consider the *contribution* $c(k, j)$ of all jobs $k \in [n]$ to C_j. We define $c(k, j)$ to be the amount of time the algorithm spends scheduling job k before the completion of j. Obviously it holds that $c(k, j) \leq p_k^A$. The following central lemma computes an improved upper bound on the contribution $c(k, j)$, using a rigorous case distinction over all possible configurations of k and j:

Lemma 1 (Contribution Lemma). *Let* $j \in [n]$ *be a given job. The completion time of* j *can be written as*

$$C_j = \sum_{k \in [n]} c(k, j).$$

Additionally, for the contribution of k to j it holds that

$$c(k,j) \leq \max\left(\left(1+\frac{1}{\beta}\right)\alpha, 1+\frac{1}{\alpha}, 1+\beta\right)\rho_j.$$

Refer to the full version [1] for the proof. Depending on whether j and k are tested or not, the lemma computes various upper bounds on the contribution using estimates from Proposition 1. Finally, the given bound on $c(k,j)$ is achieved by taking the maximum over the different cases.

Recall that the jobs are ordered by non-increasing optimal execution times ρ_j, which by Proposition 1 are directly tied to the algorithmic running times. Hence, the jobs k with small indices are the 'bad' jobs with possibly large running times. For jobs with $k \leq j$ we therefore use the independent upper bound from the Contribution Lemma. Jobs with large indices $k > j$ are handled separately and we directly estimate them using their running time p_k^A.

By Lemma 1 and Proposition 1(b),(c) we have

$$\begin{aligned}
C_j &= \sum_{k>j} c(k,j) + \sum_{k\leq j} c(k,j) \\
&\leq \sum_{k>j} p_k^A + \sum_{k\leq j} \max\left(\left(1+\frac{1}{\beta}\right)\alpha, 1+\frac{1}{\alpha}, 1+\beta\right)\rho_j \\
&= \sum_{k>j} \max\left(\alpha, 1+\frac{1}{\alpha}\right)\rho_k + \max\left(\left(1+\frac{1}{\beta}\right)\alpha, 1+\frac{1}{\alpha}, 1+\beta\right) j\cdot\rho_j.
\end{aligned}$$

Finally, we sum over all jobs j:

$$\begin{aligned}
\sum_{j=1}^{n} C_j &= \sum_{j=1}^{n}\sum_{k=j+1}^{n} \max\left(\alpha, 1+\frac{1}{\alpha}\right)\rho_k \\
&\quad + \sum_{j=1}^{n} \max\left(\left(1+\frac{1}{\beta}\right)\alpha, 1+\frac{1}{\alpha}, 1+\beta\right) j\cdot\rho_j \\
&= \max\left(\alpha, 1+\frac{1}{\alpha}\right)\sum_{j=1}^{n}(j-1)\rho_j \\
&\quad + \max\left(\left(1+\frac{1}{\beta}\right)\alpha, 1+\frac{1}{\alpha}, 1+\beta\right)\sum_{j=1}^{n} j\cdot\rho_j \\
&\leq \underbrace{\left(\max\left(\alpha, 1+\frac{1}{\alpha}\right) + \max\left(\left(1+\frac{1}{\beta}\right)\alpha, 1+\frac{1}{\alpha}, 1+\beta\right)\right)}_{=:f(\alpha,\beta)}\sum_{j=1}^{n} j\cdot\rho_j \\
&= f(\alpha,\beta)\cdot \mathrm{OPT}
\end{aligned}$$

Minimizing $f(\alpha,\beta)$ on the domain $\alpha,\beta \geq 1$ yields optimal parameters $\alpha = \beta = 1$ and a value of $f(1,1) = 4$. We conclude that $(1,1)$-SORT is 4-competitive.

The parameter selection $\alpha = 1$, $\beta = 1$ is optimal for the closed upper bound formula we obtained in our proof. It is possible and somewhat likely that a different parameter choice leads to better overall results for the algorithm. In the optimal makespan algorithm (see Sect. 4) the value of α is higher, suggesting that $\alpha = 1$, which leads to testing all non-trivial jobs, might not be the best choice. The problem structure and the approach by Dürr et al. [4] also motivate setting β to some higher value than 1. For our proof, setting parameters like we did is optimal.

In the full version of the paper [1], we take advantage of this somewhat unexpected parameter outcome to prove that $(1, 1)$-SORT cannot be better than 3-competitive. Additionally, we show that for *any* choice of parameters, (α, β)-SORT is not better than 2-competitive.

2.2 A Deterministic Algorithm with Preemption

The goal of this section is to show that if we allow jobs to be preempted there exists a 3.2361-competitive algorithm. In his book on Scheduling, Pinedo [15] defines preemption as follows: "The scheduler is allowed to interrupt the processing of a job (preempt) at any point in time and put a different job on the machine instead."

The idea for our algorithm in the preemptive setting is based on the so-called *Round Robin* rule, which is used frequently in preemptive machine scheduling [15, Chapters 3.7, 5.6, 12.4]. The scheduling time frame is divided into very small equal-sized units. The Round Robin algorithm then cycles through all jobs, tending to each job for exactly one unit of time before switching to the next. It ensures that at any time the amount every job has been processed only differs by at most one time unit [15].

The Round Robin algorithm is typically applied when job processing times are completely unknown. In our setting, we are actually given some upper bounds for our processing times and may invest testing time to find out the actual values. Despite having more information, it turns out that treating all job processing times as unknown in a Round Robin setting gives a provably good result. The only way we employ upper bounds and testing times is again to decide which jobs will be tested and which will not. We again do this at the beginning of our schedule for all given jobs. The rule to decide testing is exactly the same as in the first phase of Algorithm 1: If $u_j/t_j \geq \alpha$, then test j, otherwise run j untested. Again, α is a parameter that is to be determined. It will turn out that setting $\alpha = \varphi$ gives the best result.

The pseudo-code for the *Golden Round Robin* algorithm is given in Algorithm 2.

Essentially, the algorithm first decides for all jobs whether to test them and then runs a regular Round Robin scheme on the algorithmic testing time p_j^A, which is defined as in (2).

Theorem 2. *The Golden Round Robin algorithm is* 3.2361*-competitive in the preemptive setting for the objective of minimizing the sum of completion times. This analysis is tight.*

Algorithm 2: Golden Round Robin

```
1  T ← ∅, N ← ∅, σ_j ≡ 0;
2  foreach j ∈ [m] do
3      if u_j ≥ φt_j then
4          add j to T;
5          set σ_j ← t_j;
6      else
7          add j to N;
8          set σ_j ← u_j;
9      end
10 end
11 while ∃j ∈ [m] not completely scheduled do
12     run Round Robin on all jobs using σ_j as their processing time;
13     let j_min be the first job to finish during the current execution;
14     if j_min ∈ T and j_min tested but not executed then
15         set σ_{j_min} ← p_{j_min} and keep j_min in the Round Robin rotation;
16     end
17 end
```

We only provide a sketch of the proof here, the complete proof can be found in the full version of the paper [1].

Proof (Proof sketch). We set $\alpha = \varphi$ and use Proposition 1(b),(c) to bound the algorithmic running time p_j^A of a job j by its optimal running time ρ_j.

$$p_j^A \leq \varphi \rho_j.$$

We then compute the contribution of a job k to a fixed job j by grouping jobs based on their finishing order in the schedule. This allows us to estimate the completion time of job j:

$$C_j \leq \sum_{k>j} p_k^A + j \cdot p_j^A$$

Finally, we sum over all jobs to receive $\text{ALG} \leq 2\varphi \cdot \text{OPT}$.

To show that the analysis is tight, we provide an example where the algorithmic solution has a value of $2\varphi \cdot \text{OPT}$ if we let the number of jobs approach infinity.

The following theorem additionally provides a lower bound for the deterministic preemptive setting, giving us a first simple lower bound for this case. The proof is based on the lower bound provided in [4] for the deterministic non-preemptive case. We again defer the proof to the full version [1].

Theorem 3. *No algorithm in the preemptive deterministic setting can be better than* 1.8546*-competitive.*

3 Randomized Setting

In this section we introduce randomness to further improve the competitive ratio of Algorithm 1. There are two natural places to randomize: when deciding which jobs to test and the decision about the ordering of the jobs. These decisions directly correspond to the parameters α and β.

Making α randomized, for instance, could be achieved by defining α as a random variable with density function $f_\alpha : [1, \infty] \to \mathbb{R}_0^+$ and testing j if and only if $r_j := u_j/t_j \geq \alpha$. Then the probability for testing j would be given by $p = \int_1^{r_j} f_\alpha(x)dx$. Using a random variable α like this would make the analysis unnecessarily complicated, therefore we directly consider the probability p without defining a density, and let p depend on r_j. This additionally allows us to compute the probability of testing *independently* for each job.

Introducing randomness for β is even harder. The choice of β influences multiple jobs at the same time, therefore independence is hard to establish. Additionally, β appears in the denominator of our analysis frequently, hindering computations using expected values. We therefore forgo using randomness for the β-parameter and focus on α in this paper. We encourage future research to try their hand at making β random.

We give a short pseudo-code of our randomized algorithm in Algorithm 3. It is given a parameter-function $p(r_j)$ and a parameter β, both of which are to be determined later.

Algorithm 3: Randomized-SORT

```
1  T ← ∅, N ← ∅, σ_j ≡ 0;
2  foreach j ∈ [m] do
3  |   add j to T with probability p(r_j) and set σ_j ← βt_j;
4  |   otherwise add it to N and set σ_j ← u_j;
5  end
6  while N ∪ T ≠ ∅ do
7  |   choose j_min ∈ argmin_{j∈N∪T} σ_j;
8  |   if j_min ∈ N then
9  |   |   execute j_min untested;
10 |   |   remove j_min from N;
11 |   else if j_min ∈ T then
12 |   |   if j_min not tested then
13 |   |   |   test j_min;
14 |   |   |   set σ_{j_min} ← p_{j_min};
15 |   |   else
16 |   |   |   execute j_min;
17 |   |   |   remove j_min from T;
18 |   |   end
19 end
```

Theorem 4. *Randomized-SORT is* 3.3794*-competitive for the objective of minimizing the sum of completion times.*

Proof. Again, we let $\rho_1 \geq \ldots \geq \rho_n$ denote the ordered optimal running time of jobs $1, \ldots, n$. The optimal objective value is given by (1). Fix jobs j and k. For easier readability, we write p instead of $p(r_j)$. Since the testing decision is now done randomly, the algorithmic running time p_j^A as well as the contribution $c(k, j)$ are now random variables. It holds

$$p_j^A = \begin{cases} t_j + p_j & \text{with probability } p \\ u_j & \text{with probability } 1 - p \end{cases}$$

For the values of $c(k, j)$ we consult the case distinctions from the proof of the Contribution Lemma 1. If $j \in N$, one can easily determine that $c(k, j) \leq (1 + 1/\beta)u_j$ for all cases. Note that for this we did not need to use the final estimates with parameter α from the case distinction. Therefore this upper bound holds deterministically as long as we assume $j \in N$. By extension it also trivially holds for the expectation of $c(k, j)$:

$$E[c(k, j) \mid j \text{ untested}] \leq (1 + 1/\beta)u_j.$$

Doing the same for the case distinction of $j \in T$, we get

$$E[c(k, j) \mid j \text{ tested}] \leq \max\left((1 + \beta)t_j, \left(1 + \frac{1}{\beta}\right)p_j, t_j + p_j\right).$$

For the expected value of the contribution we have by the law of total expectation:

$$\begin{aligned} E[c(k, j)] &= E[c(k, j) \mid j \text{ untested}] \cdot Pr[j \text{ untested}] \\ &\quad + E[c(k, j) \mid j \text{ tested}] \cdot Pr[j \text{ tested}] \\ &\leq \left(1 + \frac{1}{\beta}\right)u_j \cdot (1 - p) + \max\left((1 + \beta)t_j, \left(1 + \frac{1}{\beta}\right)p_j, t_j + p_j\right) \cdot p \end{aligned}$$

Note that this estimation of the expected value is independent of any parameters of k. That means, for fixed j we estimate the contribution to be the same for all jobs with small parameter $k \leq j$. Of course, as before, for the jobs with large parameter $k > j$ we may also alternatively directly use the algorithmic runtime of k:

$$E[c(k, j)] \leq E[p_k^A].$$

Putting the above arguments together, we use the Contribution Lemma and linearity of expectation to estimate the completion time of j:

$$\begin{aligned} E[C_j] &= \sum_{j=1}^{n} E[c(k, j)] \\ &\leq \sum_{k>j} E[p_k^A] + \sum_{k \leq j} E[c(k, j)]. \end{aligned}$$

For the total objective value of the algorithm we receive again using linearity of expectation:

$$
\begin{aligned}
E\left[\sum_{j=1}^{n} C_j\right] &\leq \sum_{j=1}^{n}(j-1)E[p_j^A] + \sum_{j=1}^{n} j \cdot E[c(k,j)] \\
&\leq \sum_{j=1}^{n}(j-1)(u_j \cdot (1-p) + (t_j + p_j) \cdot p) \\
&\quad + \sum_{j=1}^{n} j\left(\left(1+\frac{1}{\beta}\right) u_j \cdot (1-p)\right. \\
&\quad \left. + \max\left((1+\beta)t_j, \left(1+\frac{1}{\beta}\right) p_j, t_j + p_j\right) \cdot p\right) \\
&\leq \sum_{j=1}^{n} j \cdot \lambda_j(\beta, p),
\end{aligned}
$$

where we define

$$
\begin{aligned}
\lambda_j(\beta, p) := &\left(u_j + \left(1+\frac{1}{\beta}\right) u_j\right) \cdot (1-p) \\
&+ \left(t_j + p_j + \max\left((1+\beta)t_j, \left(1+\frac{1}{\beta}\right) p_j, t_j + p_j\right)\right) \cdot p.
\end{aligned}
$$

Having computed this first estimation for the objective of the algorithm, we now consider the ratio $\lambda_j(\beta, p)/\rho_j$ as a standalone. If we can prove an upper bound for this ratio, the same holds as competitive ratio for our algorithm.

Hence the goal is to choose parameters β and p, where p can depend on j, s.t. $\lambda_j(\beta, p)/\rho_j$ is as small as possible. In the best case, we want to compute

$$
\min_{\beta \geq 1, p \in [0,1]} \max_j \frac{\lambda_j(\beta, p)}{\rho_j}.
$$

Lemma 2. *There exist parameters $\hat{\beta} \geq 1$ and $\hat{p} \in [0, 1]$ s.t.*

$$
\max_j \frac{\lambda_j(\hat{\beta}, \hat{p})}{\rho_j} \leq 3.3794.
$$

The choice of parameters is given in the proof of the lemma, which can be found in the full version of our paper [1]. During the proof we use computer-aided computations with Mathematica. The Mathematica code can be found in the full version.

To conclude the proof of the theorem, we write

$$
E\left[\sum_{j=1}^{n} C_j\right] \leq \sum_{j=1}^{n} j \cdot \lambda_j(\hat{\beta}, \hat{p}) \leq 3.3794 \sum_{j=1}^{n} j \cdot \rho_j = 3.3794 \cdot \mathrm{OPT}.
$$

4 Optimal Results for Minimizing the Makespan

In this section, we consider the objective of minimizing the makespan of our schedule. It turns out that we are able to prove the same tight algorithmic bounds for this objective function as Dürr et al. in the unit-time testing case, both for deterministic and randomized algorithms. The decisions of the algorithms only depend on the ratio $r_j = u_j/t_j$. Refer to the full version [1] for the proofs.

Theorem 5. *The algorithm that tests job j iff $r_j \geq \varphi$ is φ-competitive for the objective of minimizing the makespan. No deterministic algorithm can achieve a smaller competitive ratio.*

Theorem 6. *The randomized algorithm that tests job j with probability $p = 1-1/(r_j^2-r_j+1)$ is $4/3$-competitive for the objective of minimizing the makespan. No randomized algorithm can achieve a smaller competitive ratio.*

5 Conclusion

In this paper, we introduced the first algorithms for the problem of scheduling with testing on a single machine with general testing times that arises in the context of settings where a preliminary action can influence cost, duration or difficulty of a task. For the objective of minimizing the sum of completion times, we presented a 4-approximation for the deterministic case, and a 3.3794-approximation for the randomized case. If preemption is allowed, we can improve the deterministic result to 3.2361. We also considered the objective of minimizing the makespan, for which we showed tight ratios of 1.618 and 4/3 for the deterministic and randomized cases, respectively.

Our results open promising avenues for future research, in particular tightening the gaps between our ratios and the lower bounds given by the unit case. Based on various experiments using different adversarial behaviour and multiple testing times it seems hard to force the algorithm to make mistakes that lead to worse ratios than those proven in [4] for the unit case. We conjecture that in order to achieve better lower bounds, the adversary must make live decisions based on previous choices of the algorithm, in particular depending on how much the algorithm has already tested, run or deferred jobs up to a certain point.

Further interesting directions for future work are the extension of the problem to multiple machines to consider scheduling problems like open shop, flow shop, or other parallel machine settings.

References

1. Albers, S., Eckl, A.: Explorable uncertainty in scheduling with non-uniform testing times (2020). https://arxiv.org/abs/2009.13316
2. Bruce, R., Hoffmann, M., Krizanc, D., Raman, R.: Efficient update strategies for geometric computing with uncertainty. Theor. Comput. Syst. **38**(4), 411–423 (2005). https://doi.org/10.1007/s00224-004-1180-4

3. Charikar, M., Fagin, R., Guruswami, V., Kleinberg, J., Raghavan, P., Sahai, A.: Query strategies for priced information. J. Comput. Syst. Sci. **64**(4), 785–819 (2002). https://doi.org/10.1006/jcss.2002.1828
4. Dürr, C., Erlebach, T., Megow, N., Meißner, J.: Scheduling with explorable uncertainty. In: Karlin, A.R. (ed.) 9th Innovations in Theoretical Computer Science Conference (ITCS 2018). Leibniz International Proceedings in Informatics (LIPIcs), vol. 94, pp. 30:1–30:14. Schloss Dagstuhl-Leibniz-Zentrum fuer Informatik, Dagstuhl, Germany (2018). https://doi.org/10.4230/LIPIcs.ITCS.2018.30
5. Erlebach, T., Hoffmann, M.: Query-competitive algorithms for computing with uncertainty. Bull. EATCS (116), 22–39 (2015)
6. Erlebach, T., Hoffmann, M., Krizanc, D., Mihal'ák, M., Raman, R.: Computing minimum spanning trees with uncertainty. In: Albers, S., Weil, P. (eds.) 25th International Symposium on Theoretical Aspects of Computer Science. Leibniz International Proceedings in Informatics (LIPIcs), vol. 1, pp. 277–288. Schloss Dagstuhl-Leibniz-Zentrum fuer Informatik, Dagstuhl, Germany (2008). https://doi.org/10.4230/LIPIcs.STACS.2008.1358
7. Feder, T., Motwani, R., O'Callaghan, L., Olston, C., Panigrahy, R.: Computing shortest paths with uncertainty. J. Algorithms **62**(1), 1–18 (2007). https://doi.org/10.1016/j.jalgor.2004.07.005
8. Feder, T., Motwani, R., Panigrahy, R., Olston, C., Widom, J.: Computing the median with uncertainty. SIAM J. Comput. **32**(2), 538–547 (2003). https://doi.org/10.1137/S0097539701395668
9. Goerigk, M., Gupta, M., Ide, J., Schöbel, A., Sen, S.: The robust knapsack problem with queries. Comput. Oper. Res. **55**, 12–22 (2015). https://doi.org/10.1016/j.cor.2014.09.010
10. Gupta, M., Sabharwal, Y., Sen, S.: The update complexity of selection and related problems. In: Chakraborty, S., Kumar, A. (eds.) IARCS Annual Conference on Foundations of Software Technology and Theoretical Computer Science (FSTTCS 2011). Leibniz International Proceedings in Informatics (LIPIcs), vol. 13, pp. 325–338. Schloss Dagstuhl-Leibniz-Zentrum fuer Informatik, Dagstuhl, Germany (2011). https://doi.org/10.4230/LIPIcs.FSTTCS.2011.325
11. Kahan, S.: A model for data in motion. In: Proceedings of the Twenty-Third Annual ACM Symposium on Theory of Computing, pp. 265–277. STOC 1991. Association for Computing Machinery, New York, NY, USA (1991). https://doi.org/10.1145/103418.103449
12. Khanna, S., Tan, W.C.: On computing functions with uncertainty. In: Proceedings of the Twentieth ACM SIGMOD-SIGACT-SIGART Symposium on Principles of Database Systems, pp. 171–182. PODS 2001. Association for Computing Machinery, New York, NY, USA (2001). https://doi.org/10.1145/375551.375577
13. Megow, N., Meißner, J., Skutella, M.: Randomization helps computing a minimum spanning tree under uncertainty. SIAM J. Comput. **46**(4), 1217–1240 (2017). https://doi.org/10.1137/16M1088375
14. Olston, C., Widom, J.: Offering a precision-performance tradeoff for aggregation queries over replicated data. In: 26th International Conference on Very Large Data Bases (VLDB 2000), pp. 144–155. VLDB 2000. Morgan Kaufmann Publishers Inc., San Francisco, CA, USA (2000)
15. Pinedo, M.L.: Scheduling: Theory, Algorithms, and Systems. Springer International Publishing, 5 edn. (2016). https://doi.org/10.1007/978-1-4614-2361-4

16. Singla, S.: The price of information in combinatorial optimization. In: Proceedings of the Twenty-Ninth Annual ACM-SIAM Symposium on Discrete Algorithms, pp. 2523–2532. SODA 2018, Society for Industrial and Applied Mathematics, USA (2018)
17. Weitzman, M.L.: Optimal search for the best alternative. Econometrica **47**(3), 641–654 (1979). https://doi.org/10.2307/1910412

Memoryless Algorithms for the Generalized k-server Problem on Uniform Metrics

Dimitris Christou[1(✉)], Dimitris Fotakis[1], and Grigorios Koumoutsos[2]

[1] National Technical University of Athens (NTUA), Athens, Greece
fotakis@cs.ntua.gr
[2] Université libre de Bruxelles (ULB), Bruxelles, Belgium

Abstract. We consider the generalized k-server problem on uniform metrics. We study the power of memoryless algorithms and show tight bounds of $\Theta(k!)$ on their competitive ratio. In particular we show that the *Harmonic Algorithm* achieves this competitive ratio and provide matching lower bounds. Combined with the $\approx 2^{2^k}$ doubly-exponential bound of Chiplunkar and Vishwanathan for the more general setting of memoryless algorithms on uniform metrics with different weights, this shows that the problem becomes exponentially easier when all weights are equal.

1 Introduction

The k-server problem is one of the most fundamental and extensively studied problems in the theory of online algorithms. In this problem, we are given a metric space of n points and k mobile servers located at points of the metric space. At each step, a request arrives at a point of a metric space and must be served by moving a server there. The goal is to minimize the total distance travelled by the servers.

The k-server problem generalizes various online problems, most notably the paging (caching) problem, which corresponds to the k-server problem on uniform metric spaces. Paging, first studied in the seminal work of Sleator and Tarjan [26], is well-understood: the competitive ratio is k for deterministic algorithms and $H_k = \Theta(\log k)$ for randomized; those algorithms and matching lower bounds are folklore results for online algorithms [1,22,26].

The k-server problem in general metric spaces is much deeper and intriguing. In a landmark result, Koutsoupias and Papadimitriou [18] showed that the *Work Function Algorithm* (WFA) [18] is $(2k-1)$-competitive, which is almost optimal for deterministic algorithms since the competitive ratio is at least k [21]. For

The full version of this paper can be found at: https://arxiv.org/abs/2007.08669.
Dimitris Fotakis is supported by the Hellenic Foundation for Research and Innovation (H.F.R.I.) under the "First Call for H.F.R.I. Research Projects to support Faculty members and Researchers' and the procurement of high-cost research equipment grant", project BALSAM, HFRI-FM17-1424. Grigorios Koumoutsos was supported by by Fonds de la Recherche Scientifique-FNRS Grant no MISU F 6001.

C. Kaklamanis and A. Levin (Eds.): WAOA 2020, LNCS 12806, pp. 143–158, 2021.
https://doi.org/10.1007/978-3-030-80879-2_10

randomized algorithms, it is believed that an $O(\log k)$-competitive algorithm is possible; despite several breakthrough results over the last decade [2,9,10,20], this conjecture still remains open.

Memoryless Algorithms. One drawback of the online algorithms achieving the best-known competitive ratios for the k-server problem is that they are computationally inefficient. For example, the space used by the WFA is proportional to the number of different configurations of the servers, i.e., $\binom{n}{k}$, which makes the whole approach quite impractical.

This motivates the study of trade-offs between the competitive ratio and computational efficiency. A starting point in this line of research is to determine the competitive ratio of *memoryless algorithms:* a memoryless algorithm decides the next move based solely on the current configuration of the servers and the given request.

Memoryless algorithms for the k-server problem have been extensively studied (see e.g., [8,17] for detailed surveys). The most natural memoryless algorithm is the *Harmonic Algorithm*, which moves each server with probability inversely proportional to its distance from the requested point. It is known that its competitive ratio is $O(2^k \cdot \log k)$ and $\Omega(k^2)$ [5]. It is conjectured that in fact the Harmonic Algorithm is $\frac{k(k+1)}{2} = O(k^2)$-competitive; this remains a long-standing open problem. For special cases such as uniform metrics and resistive metric spaces, an improved competitive ratio of k can be achieved and this is the best possible for memoryless algorithms [13].

We note that the study of memoryless algorithms for the k-server problem is of interest only for randomized algorithms; it is easy to see that any deterministic memoryless algorithm is not competitive. Throughout this paper, we adopt the standard benchmark to evaluate randomized memoryless algorithms, which is comparing them against an *adaptive online adversary*, unless stated otherwise. For a detailed discussion on the different adversary models and relations between them, see [6,8].

The Generalized k-server Problem. In this work, we focus on the generalized k-server problem, a far-reaching extension of the k-server problem, introduced by Koutsoupias and Taylor [19]. Here, each server s_i lies in a different metric space M_i and a request is a tuple $(r_1, \dots, r_k)$, where $r_i \in M_i$; to serve it, some server s_i should move to point r_i. The standard k-server problem is the very special case where all metric spaces are identical, i.e., $M_i = M$ and all requests are of the form $(r, r, \dots, r)$. Other well-studied special cases of the generalized k-server problem are (i) the weighted k-server problem [3,14], where all metrics are scaled copies of a fixed metric M, i.e., $M_i = w_i M$ and all requests of the form $r = (r, r, \dots, r)$ and (ii) the CNN problem [16,19], where all metrics M_i are real lines.

Previous Work. The generalized k-server problem has a much richer structure than the classical k-server problem and is much less understood. For general metric spaces, no $f(k)$- competitive algorithms are known, except for the special case of $k = 2$ [23–25]. For $k \geq 3$, competitive algorithms are known only for the following special cases:

1. Uniform Metrics: All metric spaces $M_1, \ldots, M_k$ are uniform (possibly with different number of points), with the same pairwise distance, say 1.
2. Weighted Uniform Metrics: All metrics are uniform, but they have different weights; the cost of moving in metric M_i is w_i.

Perhaps surprisingly, those two cases are very different. For deterministic algorithms Bansal et al. [4] obtained algorithms with (almost) optimal competitive ratio. For uniform metrics their algorithm is $(k \cdot 2^k)$-competitive, while the best possible ratio is at least $2^k - 1$ [19]. For weighted uniform metrics, they obtained a $2^{2^{k+3}}$-competitive algorithm (by extending an algorithm of Fiat and Ricklin [14] for weighted k-server on uniform metrics), while the lower bound for the problem is $2^{2^{k-4}}$ [3].

We note that for uniform metrics, if memory is allowed and we compare against oblivious adversaries, competitive randomized algorithms are known: Bansal et al. [4] designed a $O(k^3 \log k)$-competitive randomized algorithm with memory; this was recently improved to $O(k^2 \log k)$ by Bienkowski et al. [7].

Memoryless Algorithms for Generalized k-server. Recently Chiplunkar and Vishnawathan [11] studied randomized memoryless algorithms in weighted uniform metrics. They showed tight doubly exponential ($\approx 1.6^{2^k}$) bounds on the competitive ratio. Interestingly, the memoryless algorithm achieving the optimal bound in this case is different from the Harmonic Algorithm.

Since the weighted uniform case seems to be much harder than the uniform case, it is natural to expect that a better bound can be achieved by memoryless algorithms in uniform metrics. Moreover, in weighted uniform metrics the competitive ratios of deterministic algorithms (with memory) and randomized memoryless algorithms are essentially the same. Recall that a similar phenomenon occurs for the paging problem (standard k-server on uniform metrics) where both deterministic and randomized memoryless algorithms have a competitive ratio of k. Thus, it is natural to guess that for uniform metrics, a competitive ratio of order 2^k (i.e., same as the deterministic competitive ratio) can be achieved by memoryless algorithms.

1.1 Our Results

In this work we study the power of memoryless algorithms for the generalized k-server problem in uniform metrics and we determine the exact competitive ratio by obtaining tight bounds.

First, we determine the competitive ratio of the Harmonic Algorithm on uniform metrics.

Theorem 1. *The Harmonic Algorithm for the generalized k-server problem on uniform metrics is $(k \cdot \alpha_k)$-competitive, where α_k is the solution of the recurrence relation $\alpha_k = 1 + (k-1)\alpha_{k-1}$, with $\alpha_1 = 1$.*

It is not hard to see that $\alpha_k = \Theta((k-1)!)$, therefore the competitive ratio of the Harmonic Algorithm is $O(k!)$. This shows that indeed, uniform metrics allow for substantial improvement on the performance compared to weighted uniform metrics where there is a doubly-exponential lower bound.

To obtain this result, we analyse the Harmonic Algorithm using Markov Chains and random walks, based on the *Hamming distance* between the configuration of the algorithm and the adversary, i.e., the number of metric spaces where they have their servers in different points. Based on this, we then provide a proof using a potential function, which essentially captures the expected cost of the algorithm until it reaches the same configuration as the adversary. The proof is in Sect. 2.

Next we show that the upper bound of Theorem 1 is tight, by providing a matching lower bound.

Theorem 2. *The competitive ratio of any randomized memoryless algorithm for the generalized k-server problem on uniform metrics is at least* $k \cdot \alpha_k$.

Here the analysis differs, since the Hamming distance is not the right metric to capture the "distance" between the algorithm and the adversary: assume that all their servers are at the same points, except one, say server s_i. Then, in the next request, the algorithm will reach the configuration of the adversary with probability p_i; clearly, if p_i is large, the algorithm is in a favourable position, compared to the case where p_i is small.

This suggests that the structure of the algorithm is not solely characterized by the number of different servers (i.e., Hamming distance) between the algorithm and the adversary, but also the labels of the servers matter. For that reason, we need to focus on the subset of different servers, which gives a Markov Chain on 2^k states. Unfortunately, analyzing such chains in a direct way can be done only for easy cases like $k = 2$ or $k = 3$. For general values of k, we find an indirect way to characterize the solution of this Markov Chain. A similar approach was taken by Chiplunkar and Vishwanathan [11] for weighted uniform metrics; we use some of the properties they showed, but our analysis differs since we need to make use of the special structure of our problem to obtain our bounds.

In fact, we are able to show that any memoryless algorithm other than the Harmonic has competitive ratio strictly larger than $k \cdot \alpha_k$. We describe the details in Sect. 3.

On the positive side, our results show that improved guarantees can be achieved compared to the weighted uniform case. On the other hand, the competitive ratio of memoryless algorithms ($\Theta(k!)$) is asymptotically worse than the deterministic competitive ratio of $2^{O(k)}$. This is somewhat surprising, since (as discussed above) in most uniform metric settings of k-server and generalizations, the competitive ratio of deterministic algorithms (with memory) and randomized memoryless is (almost) the same.

1.2 Notation and Preliminaries

Memoryless Algorithms. A memoryless algorithm for the generalized k-server problem receives a request $r = (r_1, \ldots, r_k)$ and decides which server to move based only on its current configuration $q = (q_1, \ldots, q_k)$ and r. For the case of uniform metrics, a memoryless algorithm is fully characterized by a probability distribution $p = (p_1, \ldots, p_k)$; whenever it needs to move a server, it uses server s_i of metric M_i with probability p_i. Throughout the paper we assume for convenience (possibly by relabeling the metrics) that given a memoryless algorithm we have that $p_1 \geq p_2 \geq \ldots \geq p_k$. We also assume that $p_i > 0$ for all i; otherwise it is trivial to show that the algorithm is not competitive.

The Harmonic Algorithm. In the context of generalized k-server on uniform metrics, the Harmonic Algorithm is a memoryless algorithm which moves at all metric spaces with equal probability, i.e., $p_i = 1/k$, for all $i \in [k]$.

The Harmonic Recursion. We now define the recursion that will be used to get the competitive ratio of the Harmonic Algorithm, which we refer to as the *harmonic recursion* in the context of this work and do some basic observations that will be useful throughout the paper.

Definition 1 (Harmonic recursion). *The harmonic recursion $\alpha_\ell \in \mathbb{N}$ satisfies the recurrence $\alpha_\ell = 1 + (\ell - 1)\alpha_{\ell-1}$ for $\ell > 1$, and $\alpha_1 = 1$.*

Based on the definition, we make the following observation:

Observation 1. *The harmonic recursion is strictly increasing, i.e., $\alpha_{\ell+1} > \alpha_\ell$ for any $\ell \in \mathbb{N}$.*

Also it is easy to show that α_ℓ has a closed form, given by

$$\alpha_\ell = (\ell - 1)! \sum_{i=0}^{\ell-1} \frac{1}{i!}. \tag{1}$$

Based on this closed form, we get the following:

Observation 2. *For any $\ell \in \mathbb{N}$, it holds that $(\ell - 1)! \leq \alpha_\ell \leq e(\ell - 1)!$.*

This observation also shows that for any $\ell \in \mathbb{N}$, we have $\alpha_\ell = \Theta((\ell - 1)!)$.

2 Upper Bound

In this section we prove Theorem 1. More precisely, we use a potential function argument to show that for any request sequence, the expected cost of the Harmonic Algorithm is at most $k \cdot \alpha_k$ times the cost of the adversary.

Organization. In Sect. 2.1, we define a potential between the Harmonic Algorithm's and the adversary's configurations that is inspired by random walks on a special type of Markov Chains [15] we refer to as the "*Harmonic Chain*". The required background of Markov Chains is presented in the Appendix of the full version. Then, in Sect. 2.2 we will use this potential to prove the upper bound of Theorem 1 with a standard potential-based analysis.

2.1 Definition of the Potential Function

We begin by presenting the intuition behind the definition of our potential function. Our first observation is that since (i) the metrics are uniform with equal weights and (ii) the Harmonic Algorithm does not distinguish between metrics since it has equal probabilities $\frac{1}{k}$, it makes sense for the potential between two configurations $p, q \in [n]^k$ to depend only on their Hamming distance and not on the labels of their points. In order to come up with an appropriate potential, we need to understand how the Hamming distance between the Harmonic Algorithm's and the adversary's configurations evolves over time.

Imagine that the adversary moves to an "optimal" configuration of his choice and then it serves requests until the Harmonic Algorithm reaches this configuration as well. Since the adversary must serve all the requests using a server from its configuration, we know that for each request $r = (r_1, \dots, r_k)$, at least one of the requested points r_i should coincide with the i-th server of the adversary. In that case, with probability $\frac{1}{k}$ the Harmonic Algorithm moves in metric M_i, thus it decreases his Hamming distance from the adversary by 1. On the other hand, assume that ℓ servers of the algorithm coincide with the ones of the adversary. Then, with probability $\frac{\ell}{k}$ it would increase its Hamming distance from the optimal configuration by 1. This shows that the evolution of the Hamming distance between the Harmonic Algorithm's and the adversary's configurations is captured by a random walk on the following Markov Chain that we refer to as the *Harmonic Chain*.

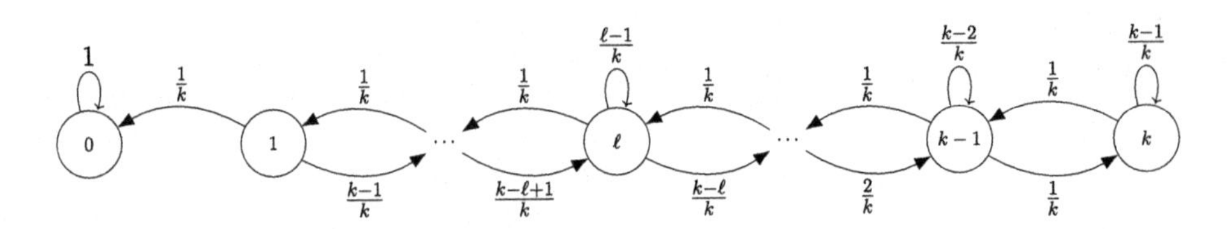

Fig. 1. The Harmonic Chain - Here, the states of the chain denote the Hamming distance between the configurations of the Harmonic Algorithm and the adversary.

In the scenario we described above, the expected movement cost of the Harmonic Algorithm until it reaches the adversary's configuration with an initial Hamming distance of ℓ would be the *Expected Extinction Time* (ETT) [15] $h(\ell)$ of this chain, that is $h(\ell) = \mathbb{E}[N|X_0 = \ell]$ where N denotes a random variable defined as $N = min_{\tau \geq 0}\{X_\tau = 0\}$ and X_t denotes the state of the Harmonic Chain at time t. Intuitively, $h(k)$ should immediately give an upper bound on the competitive ratio of the Harmonic Algorithm. In the full version, we study the Harmonic Chain and prove the following Theorem:

Theorem 3. *For any initial state* $\ell \in \{0, 1, \ldots, k\}$*, the EET of the Harmonic Chain is given by*

$$h(\ell) = k \sum_{i=k-\ell+1}^{k} \alpha_i$$

Proof. By using conditional probabilities on the EET of the Harmonic Chain, we get that for any $\ell \in \{1, 2, \ldots, k\}$,

$$h(\ell) = 1 + \frac{1}{k}h(\ell - 1) + \frac{k - \ell}{k}h(\ell + 1) + (1 - \frac{1}{k} - \frac{k - \ell}{k})h(\ell).$$

This yields a second-order recurrence relation we need to solve for $h(\ell)$. A formal proof is given in the full version, where we derive the Theorem from the EET of the more general class of Markov Chains called Birth-Death Chains. □

From Theorem 3 and Observations 1, 2 we immediately get that $h(\ell)$ is (strictly) increasing and $h(\ell) = \Theta(k!)$ $\forall \ell \in \{1, \ldots, k\}$. Furthermore, we make the following observation:

Observation 3. *For any* $\ell \in \{1, 2, \ldots, k\}$*,* $\frac{h(\ell)}{\ell} \leq h(1)$ *in the Harmonic Chain, with the equality holding only for* $\ell = 1$*.*

Proof. Fix any $\ell \in \{2, 3, \ldots, k\}$. Then:

$$\frac{h(\ell)}{\ell} = \frac{k\alpha_k + k\sum_{i=k-\ell}^{k-2} \alpha_{i+1}}{\ell} < \frac{k\alpha_k + k(\ell - 1)\alpha_{k-1}}{\ell} < \frac{k\alpha_k + (\ell - 1)k\alpha_k}{\ell} = k\alpha_k.$$

where both inequalities hold from Observation 1 which states that α_i is *strictly* increasing and the Observation follows from $h(1) = k\alpha_k$ by definition. □

Suppose that the adversary moves ℓ servers whenever the algorithm reaches its configuration and then it doesn't move until the algorithm reaches its new configuration. Intuitively, the competitive ratio would be $\frac{h(\ell)}{\ell}$ which is maximized for $\ell = 1$ by Observation 3. This means that $h(1) = k \cdot \alpha_k$ is an upper bound for the competitive ratio of the Harmonic Algorithm. Using this important intuition, we define the potential between two configurations of Hamming distance ℓ as $h(\ell)$. Formally,

Definition 2 (Potential Function). *The potential between two configurations* $p, q \in [n]^k$ *is defined as*

$$\phi(p, q) = h(d_H(p, q)).$$

2.2 Bounding the Competitive Ratio

In this section, we will prove the upper bound of Theorem 1 by using the potential we defined in Sect. 2.1. We consider a sequence of $T \in \mathbb{N}$ requests such that $r^t \in [n]^k$ $\forall t \in [T]$. Let $q^t, \mathcal{A}^t \in [n]^k$ be used to denote the Harmonic Algorithm's and the adversary's configurations after serving request r^t respectively. Also, let

$q^0 = \mathcal{A}^0$ be the initial configuration of the instance. For any time $t \in [T]$, we first let the adversary specify the request r^t and serve it, possibly depending on the algorithm's previous configuration q^{t-1}. We will prove that when the adversary moves x servers the increase in potential is at most $k \cdot \alpha_k \cdot x$ and when the Harmonic Algorithm moves one server, the expected decrease in potential is at least 1. Then, using these properties, we will prove Theorem 1.

To simplify the analysis, we make the following observation for the potential function.

Observation 4. *For any $\ell, \ell' \in \{0, 1, \ldots, k\}$ such that $\ell < \ell'$ it holds that*

$$h(\ell') - h(\ell) = k \sum_{i=\ell}^{\ell'-1} \alpha_{k-i}$$

Proof. By telescoping we have

$$h(\ell') - h(\ell) = \sum_{i=\ell}^{\ell'-1} (h(i+1) - h(i)) = \sum_{i=\ell}^{\ell'-1} (k \sum_{j=k-i}^{k} \alpha_j - k \sum_{j=k-i+1}^{k} \alpha_j) = k \sum_{i=\ell}^{\ell'-1} \alpha_{k-i}$$

where the second equality holds by the definition of the potential. □

Using this observation, we are now ready to prove the following lemmata:

Lemma 1 (Adversary Moves). *For any $t \in \{1, \ldots, T\}$ it holds that*

$$\phi(q^{t-1}, \mathcal{A}^t) - \phi(q^{t-1}, \mathcal{A}^{t-1}) \leq k \cdot \alpha_k \cdot d_H(\mathcal{A}^t, \mathcal{A}^{t-1}).$$

Proof. Let $\ell^{t-1} = d_H(q^{t-1}, \mathcal{A}^{t-1})$ and $\ell^t = d_H(q^{t-1}, \mathcal{A}^t)$. Clearly, $\ell^{t-1}, \ell^t \in \{0, 1, \ldots, k\}$. Since the potential $h(\ell)$ is strictly increasing on ℓ, if $\ell^t \leq \ell^{t-1}$ then this means that the adversary's move didn't increase the potential and then the Lemma follows trivially. Thus, we only need to prove the Lemma for $0 \leq \ell^{t-1} < \ell^t \leq k$. We have:

$$h(\ell^t) - h(\ell^{t-1}) = k \sum_{i=\ell^{t-1}}^{\ell^t - 1} \alpha_i \leq (\ell^t - \ell^{t-1}) k \alpha_k \tag{2}$$

where the equality is given from Observation 4 and the inequality from the fact that the recursion α_ℓ is increasing. Thus, we have proven that $\phi(q^{t-1}, \mathcal{A}^t) - \phi(q^{t-1}, \mathcal{A}^{t-1}) \leq (\ell^t - \ell^{t-1}) \cdot k \cdot \alpha_k$. To conclude the proof of the Lemma, by the triangle inequality of the Hamming distance we have

$$d_H(q^{t-1}, \mathcal{A}^{t-1}) + d_H(\mathcal{A}^{t-1}, \mathcal{A}^t) \geq d_H(q^{t-1}, \mathcal{A}^t)$$

which gives $\ell^t - \ell^{t-1} \leq d_H(\mathcal{A}^{t-1}, \mathcal{A}^t)$. Combined with (2), we get the Lemma. □

Lemma 2 (Harmonic Moves). *For any $t \in \{1, \ldots, T\}$ it holds that*

$$\mathbb{E}[\phi(q^{t-1}, \mathcal{A}^t) - \phi(q^t, \mathcal{A}^t)] \geq d_H(q^{t-1}, q^t).$$

Proof. If the Harmonic Algorithm serves the request, then $q^t = q^{t-1}$ and the Lemma follows trivially. Otherwise, by definition, it moves to a configuration q^t such that $d_H(q^{t-1}, q^t) = 1$. Let $\ell^{t-1} = d_H(q^{t-1}, \mathcal{A}^t)$ and $\ell^t = d_H(q^t, \mathcal{A}^t)$. Also, let $C = |\{i : \mathcal{A}_i^t = r_i^t\}|$, i.e., the number of the adversary's servers that could serve the current request. By definition, $\mathcal{A}^t$ must serve r^t which gives $C \geq 1$. Furthermore, q^{t-1} doesn't serve the request but $\mathcal{A}^t$ does, and thus $\ell^{t-1} \geq 1$.

Recall that the Harmonic Algorithm randomly moves at a metric with equal probabilities in order to serve a request. If it moves in any of the C metrics where the adversary serves the request, we get $\ell^t = \ell^{t-1} - 1$ and the potential decreases with probability $\frac{C}{k}$. If it moves on any of the $k - \ell^{t-1}$ metrics where $\mathcal{A}_i^t = q_i^{t-1}$, we get $\ell^t = \ell^{t-1} + 1$ and the potential increases with probability $\frac{k-\ell^{t-1}}{k}$. In any other case, we have $\ell^t = \ell^{t-1}$ and the potential doesn't change. To simplify the notation, we define $j \in \{1, \dots, k\}$ as $j = k - \ell^{t-1} + 1$. We have:

$$\begin{aligned}
\mathbb{E}[\phi(q^{t-1}, \mathcal{A}^t) - \phi(q^t, \mathcal{A}^t)] &= \mathbb{E}[h(\ell^{t-1}) - h(\ell^t)] \\
&= \frac{C}{k}(h(\ell^{t-1}) - h(\ell^{t-1} - 1)) \\
&\quad + \frac{k - \ell^{t-1}}{k}(h(\ell^{t-1}) - h(\ell^{t-1} + 1)) \\
&= C\alpha_j - (j-1)\alpha_{j-1} \\
&= C\alpha_j - (\alpha_j - 1) = (C-1)\alpha_j + 1 \\
&\geq 1 = d_h(q^{t-1}, q^t)
\end{aligned}$$

where the first equality follows from the definition of the potential, the second equality from the possible changes in the Hamming distance between the algorithm and the adversary, the third equality follows from Observation 4 and the definition of j, the fourth equality follows from the definition of the recursion α_ℓ and the inequality follows from $C \geq 1$. □

Proof of Theorem 1. We are now ready to prove Theorem 1. By combining lemmata 1 and 2, we get that for any $t \in \{1, \dots, T\}$, the expected difference in potential is

$$\mathbb{E}[\Delta\phi^t] = \mathbb{E}[\phi(q^t, \mathcal{A}^t) - \phi(q^{t-1}, \mathcal{A}^{t-1})] \leq k\alpha_k d_H(\mathcal{A}^t, \mathcal{A}^{t-1}) - d_H(q^{t-1}, q^t)$$

Now, let $\text{ADV} = \sum_{t=1}^T d_H(\mathcal{A}^t, \mathcal{A}^{t-1})$ be used to denote the total cost of the adversary and $\text{ALG} = \sum_{t=1}^T d_H(q^t, q^{t-1})$ be used to denote the expected cost of the Harmonic Algorithm. Summing over all $t \in \{1, 2, \dots, T\}$ we finally get

$$\mathbb{E}[\sum_{t=1}^T \Delta\phi^t] = \mathbb{E}[\phi(q^T, \mathcal{A}^T) - \phi(q^0, \mathcal{A}^0)] \leq k\alpha_k \cdot \text{ADV} - \text{ALG}$$

and since $\mathcal{A}^0 = q^0$ (i.e., $\phi(q^0, \mathcal{A}^0) = 0$) and $\phi(q^T, \mathcal{A}^T) \geq 0$, we get that $\text{ALG} \leq k \cdot \alpha_k \cdot \text{ADV}$, which concludes the proof of Theorem 1.

3 Lower Bound

In this section we prove Theorem 2. More precisely, we construct an adaptive online adversarial request sequence against any memoryless algorithm and prove that its competitive ratio is lower bounded by the solution of a linear system of 2^k equations. Since solving this system directly is possible only for easy cases like $k = 2$ or $k = 3$, we show how to get a lower bound for the solution (similarly to the approach taken by Chiplunkar and Vishwanathan [11] for weighted uniform metric spaces) and thus for the competitive ratio of any memoryless algorithm.

Organization. In Sect. 3.1 we formally define the adversarial request sequence and the intuition behind it. In Sect. 3.2 we state the linear system of equations that results from our request sequence and prove a lower bound on its solution. This leads to the proof of Theorem 2.

3.1 Constructing the Adversarial Instance

Before we state the adversarial instance, it is useful to give the intuition behind it. It is natural to construct an adversary that moves only when it has the same configuration with the algorithm.

In fact, we construct an adversary that moves in only one metric space: the one that the algorithm uses with the smallest probability (ties are broken arbitrarily). Recall that in the analysis of the Harmonic Algorithm from Sect. 2, the competitive ratio is also maximized when in each "phase" the adversary starts with only one different server than the algorithm and does not move until the configurations (of algorithm and adversary) match (Observation 3).

Let ALG be any memoryless online algorithm and ADV be the adversary. Consider a "phase" to be a part of the request sequence where in the beginning the configurations of ALG and ADV coincide and it ends when ALG matches the configuration of ADV. Since ADV must serve all requests, in each request r one point r_i is such that the adversary's server in metric M_i occupies it; we say that the i-th position of ADV is revealed in such a request. Thus every request will reveal to the algorithm exactly one of the positions of the adversary's servers in some metric space M_i. The main idea behind our lower bound instance is that, in each request, out of the metric spaces that servers of ALG and ADV differ, we reveal to the algorithm the position of ADV in the metric that ALG serves with the highest probability; this implies that whenever ALG and ADV differ by only one server, this will be in metric M_k. Intuitively, this way we exploit best the "asymetries" in the distribution of ALG (this is formalized in Lemma 3).

The Instance. Recall that any memoryless algorithm for the generalized k-server problem on uniform metric spaces is fully characterized by a probability distribution $p = [p_1, p_2, \ldots, p_k]$ over the k-metric spaces $M_1, M_2, \ldots, M_k$. W.l.o.g., we can assume that $p_1 \geq p_2 \geq \cdots \geq p_k$. Let $q^t, \mathcal{A}^t$ be used to denote the configurations of the algorithm and the adversary after serving request r^t respectively. Also, let $q^0 = \mathcal{A}^0$ be used to denote the initial configuration of both the algorithm and the adversary. We will now construct the request sequence. For $t = 1, 2, \ldots, T$:

1. Observe q^{t-1}, i.e., the algorithm's current configuration.
2. If $q^{t-1} = \mathcal{A}^{t-1}$, then:
   - $\mathcal{A}^t = [q_1^0, q_2^0, \ldots, q_{k-1}^0, Z]$ for some $Z \in [n]$ such that $Z \neq \mathcal{A}_k^{t-1} = q_k^{t-1}$.

   otherwise:
   - $\mathcal{A}^t = \mathcal{A}^{t-1}$.
3. Determine $m = \min(\{j : q_j^{t-1} \neq \mathcal{A}_j^t\})$.
4. Pick any $r^t \in [n]^k$ such that $r_m^t = \mathcal{A}_m^t$ and $r_j^t \notin \{\mathcal{A}_j^t, q_j^{t-1}\}$ $\forall j \in [k] \setminus \{m\}$.

Note that for step 4, we need to have at least $n \geq 3$ points in order to pick a point that isn't occupied by neither the algorithm's nor the adversary's servers. As we explain in Sect. 4, this is a necessary requirement; if all metrics have $n = 2$ points, then the competitive ratio of the Harmonic Algorithm is $O(2^k)$ and therefore a lower bound of order $k!$ is not possible. Furthermore, this adversary is clearly online adaptive since it constructs the request r^t and the configuration $\mathcal{A}^t$ after observing the algorithm's configuration q^{t-1}.

As an example of our instance, for $k = 4$, let $\mathcal{A}^{t-1} = [0, 0, 0, 0]$ and $q^{t-1} = [1, 0, 0, 1]$ for some t. Clearly, the algorithm and the adversary have different servers in metric M_1 and M_4. From step 3, $m = \min(1, 4) = 1$, i.e., M_1 is the metric space that the algorithm serves with highest probability out of the metric spaces that it and the adversary have their servers in different points. Then, from step 4, $r^t = [0, 2, 2, 2]$ (actually, the selection of the last three coordinates is arbitrary as long as neither the algorithm nor the adversary have their server on this point).

Notice that ADV moves one server in metric space M_k whenever it has the same configuration with ALG. On the other hand, ALG never serves request r^t with configuration q^{t-1} and thus moves at every time step. This means that the competitive ratio of ALG is lower bounded by the (expected) number of requests it takes for it to reach configuration of ADV.

3.2 Proving the Lower Bound

Our Approach. We define the *state* of the algorithm at time t as $S^t = \{i : q_i^t \neq \mathcal{A}_i^t\}$, i.e., the subset of metric spaces with different servers between the algorithm and the adversary. In this context, $h(S)$ is used to denote the expected number of requests it takes for the algorithm to reach the adversary's configuration, i.e. state $\emptyset$, starting from some state $S \subseteq [k]$. From the request sequence we defined, $h(\{k\})$ is a lower bound for the competitive ratio of any memoryless algorithm.

By observing how the state S of the algorithm (and by extension $h(S)$) evolves under the request sequence, we can write down a linear system of 2^k equations on the 2^k variables $h(S)$ $\forall S \subseteq [k]$. In fact, these equations give the EET of a random walk in a Markov Chain of 2^k states. We then prove a lower bound on $h(\{k\})$ and thus the competitive ratio of any memoryless algorithm.

Notice that for the given instance, if we were analyzing the Harmonic Algorithm, then the Hamming distance between it and the adversary would be captured by the Harmonic Chain and we would immediately get that $h(\{k\}) = k \cdot \alpha_k$.

Analysis. Fix any two different configurations $q, \mathcal{A}$ for the algorithm and the adversary that are represented by state $S = \{i : q_i \neq \mathcal{A}_i\} \neq \emptyset$ with $\min(S) = m$. Then, we know that for the next request r we have constructed it holds that $r_m = \mathcal{A}_m \neq q_m$ and $r_j \notin \{q_j, \mathcal{A}_j\}$ for any $j \in [k] \setminus \{m\}$. Recall that the memoryless algorithm will randomly move to some state M_j and move to a different configuration $q' = [q_1, \ldots, q_{j-1}, r_j, q_{j+1}, \ldots, q_k]$ that is captured by state S'. We distinguish between the following three cases:

1. If $j \notin S$, then this means that $q_j = \mathcal{A}_j$ and $q'_j = r_j \neq \mathcal{A}_j$ and thus $S' = S \cup \{j\}$.
2. If $j = m$, then $q_j \neq \mathcal{A}_j$ and $q'_j = \mathcal{A}_j = r_m$ and thus $S' = S \setminus \{m\}$.
3. If $j \in S \setminus \{m\}$ then $q_j \neq \mathcal{A}_j$ and $q'_j \neq \mathcal{A}_j$ and thus $S' = S$.

Since $h(S)$ denotes the expected number of steps until the state of the algorithm becomes $\emptyset$ starting from S, from the above cases we have that for any state $S \neq \emptyset$:

$$h(S) = 1 + p_m \cdot h(S \setminus \{m\}) + \sum_{j \notin S} p_j \cdot h(S \cup \{j\}) + \sum_{j \in S \setminus \{m\}} p_j \cdot h(S), \quad m = \min(S).$$

Combined we the fact that obviously $h(\emptyset) = 0$ and $\sum_{j=1}^{k} p_j = 1$, we get the following set of 2^k linear equations with 2^k variables:

$$\begin{gathered} h(\emptyset) = 0 \\ p_m(h(S) - h(S \setminus \{m\})) = 1 + \sum_{j \notin S} p_j(h(S \cup \{j\}) - h(S)), \forall S \neq \emptyset, m = \min(S) \end{gathered} \tag{3}$$

Normally, we would like to solve this linear system to compute $h(\{k\})$ and this would be the proven lower bound for the memoryless algorithm. However, even for $k = 4$ it is hopeless to find closed form expressions for the solutions of this system. Interestingly, similar equations were studied by Chiplunkar and Vishnawathan [11] for the weighted uniform metric case. In their study, they showed a monotonicity property on the solutions of their linear system that directly transfers to our setting and is stated in Lemma 3 below. Using this, combined with the special structure of our problem, we show how to derive a lower bound of $k \cdot \alpha_k$ for $h(\{k\})$ instead of solving (3) to directly compute it.

Lemma 3. *For any $S \subseteq [k]$ with $i, j \in S$ such that $i < j$ (and thus $p_i \geq p_j$), the solutions of linear system* (3) *satisfy*

$$h(S) - h(S \setminus \{j\}) \geq \frac{p_i}{p_j}(h(S) - h(S \setminus \{i\}))$$

The proof is deferred to the full version. Let us first see the intuition behind the inequality of Lemma 3. Let S be the subset of metric spaces where the servers of ALG and ADV occupy different points: then, in the next move, the expected time to match ADV decreases the most, if ALG matches first the jth server of the adversary (i.e., the "state" changes from S to $S\setminus\{j\}$) where j is the metric with the smallest the probability p_j. This explains why in our adversarial instance we choose to reveal to ALG the location of ADV in the metric it serves with the highest probability: this makes sure that the decrease in the expected time to reach ADV is minimized.

Using Lemma 3, we can now prove the following:

Lemma 4. *For any $S \subseteq [k]$ with $S \neq \emptyset$ and $i \in S$, the solutions of linear system* (3) *satisfy*

$$p_i(h(S) - h(S \setminus \{i\})) \geq 1 + \sum_{j \notin S} p_j(h(S \cup \{j\}) - h(S))$$

Proof. Fix any non-empty set $S \subseteq [k]$ and any $i \in S$. Let $m = \min(S) \leq i$. Then, by Lemma 3 we have

$$p_i(h(S) - h(S \setminus \{i\})) \geq p_m(h(S) - h(S \setminus \{m\}))$$

Since $m = \min(S)$, and we study the solution of linear system (3), we have

$$p_m(h(S) - h(S \setminus \{m\})) = 1 + \sum_{j \notin S} p_j(h(S \cup \{j\}) - h(S))$$

and the lemma follows. □

We are now ready to prove the main theorem of this section.

Theorem 4. *The solution of linear system* (3) *satisfies*

$$h(\{k\}) \geq \frac{\alpha_k}{p_k}$$

Proof. In order to prove the theorem, it suffices to show that for any $S \subseteq [k]$ such that $S \neq \emptyset$ and $i \in S$, it holds that

$$p_i(h(S) - h(S \setminus \{i\})) \geq \alpha_{k-|S|+1}$$

Then, by setting $S = \{k\}$ ($|S| = 1$) and $i = k \in S$, we get $p_k(h(\{k\}) - h(\emptyset)) \geq \alpha_k$, and since $h(\emptyset) = 0$ by definition, the Theorem follows. It remains to prove the desired property. This can be shown by induction on the size of S.

Base case: If $|S| = k$ (this means that $S = [k]$) then for any $i \in S$, by Lemma 4 we have

$$p_i(h(S) - h(S \setminus \{i\})) = 1 = \alpha_1 = \alpha_{k-|S|+1}.$$

Inductive hypothesis: Suppose that for any $S \subseteq [k]$ with $|S| = \ell > 1$ and any $i \in S$, we have

$$p_i(h(S) - h(S \setminus \{i\})) \geq \alpha_{k-\ell+1}.$$

Inductive step: Let $S \subseteq [k]$ be any set with $|S| = \ell - 1 > 0$ and $i \in S$ be any element of this set. By Lemma 4, we have that

$$p_i(h(S) - h(S \setminus \{i\})) \geq 1 + \sum_{j \notin S} p_j(h(S \cup \{j\}) - h(S))$$

Now, for any $j \notin S$ we can use the hypothesis on the set $S \cup \{j\}$ with size ℓ. Thus, we have

$$p_j(h(S \cup \{j\}) - h(S)) \geq \alpha_{k-\ell+1} = \alpha_{k-|S|}$$

for any $j \notin S$. Combining, we get

$$p_i(h(S) - h(S \setminus \{i\})) \geq 1 + (k - |S|)\alpha_{k-|S|} = \alpha_{k-|S|+1}.$$

□

Proof of Theorem 2. Since $p_1 \geq p_2 \geq \cdots \geq p_k$, we have that $p_k \leq \frac{1}{k}$. Thus, by Theorem 4 we have that $h(\{k\}) \geq k \cdot \alpha_k$ for any distribution. Since $h(\{k\})$ is a lower bound for any memoryless algorithm, the Theorem follows.

Corollary 1. *The Harmonic Algorithm is the only memoryless algorithm with a competitive ratio of* $k \cdot \alpha_k$.

Proof. By Theorem 4, the competitive ratio of the Harmonic Algorithm is at least $k \cdot \alpha_k$ and combined with the upper bound of Theorem 1 we get that the Harmonic Algorithm is $(k \cdot \alpha_k)$-competitive. Assuming $p_1 \geq \cdots \geq p_k$, any other memoryless algorithm will have $p_k < \frac{1}{k}$. Thus, by Theorem 4 its competitive ratio will be lower bounded by $h(\{k\}) > k \cdot \alpha_k$ which is strictly worse that the competitive ratio of the Harmonic Algorithm. □

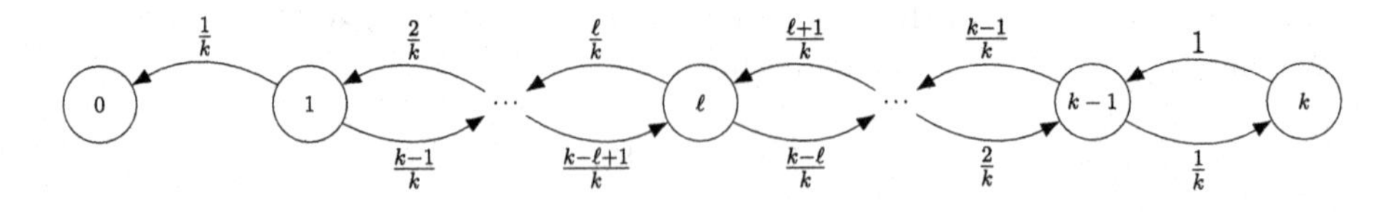

Fig. 2. The evolution of the Harmonic Algorithm's Hamming distance from the adversary when $n = 2$.

4 Concluding Remarks

We provided tight bounds on the competitive ratio of randomized memoryless algorithms for generalized k-server in uniform metrics. Combining our results with the work of Chiplunkar and Vishwanathan [11], the power of memoryless algorithms in uniform and weighted uniform metrics is completely characterized. It might be interesting to determine the power of memoryless algorithms for other metric spaces such as e.g., weighted stars. However we note that memoryless algorithms are not competitive on arbitrary metric spaces, even for $k = 2$; this was shown by Chrobak and Sgall [12] and Koutsoupias and Taylor [19] independently. We conclude with the following side remark.

Metrics with $n = 2$ Points. In our lower bound instance from Sect. 3 we require that all metric spaces have at least $n \geq 3$ points. We observe that this is necessary, and that if all metric spaces have $n = 2$ points, the Harmonic Algorithm is $O(2^k)$-competitive, thus a lower bound of $k \cdot \alpha_k$ can not be achieved. The underlying reason is the following: in the Harmonic Chain described in Sect. 2, while being at state ℓ (i.e., having ℓ servers different than the adversary), the algorithm moves to state $\ell - 1$ with probability $1/k$ and remains in the same state with probability $\frac{\ell-1}{k}$. This happens because if $n \geq 3$, then given the algorithm's configuration q and the adversary's configuration $a \neq q$, we can construct a request r such that $r_i \neq q_i$ and $r_i \neq a_i$ in $\ell - 1$ metric spaces. However if $n = 2$, the algorithm moves only for $r = \bar{q}$ (i.e., r is the algorithm's anti-configuration) and thus $a_i \neq q_i$ implies that $r_i = a_i$ and if the algorithm moves in M_i, then it reduces the number of different servers to $\ell - 1$. Thus the Markov Chain used to analyse this instance becomes the following:

Then, as we show in the full version, for this random walk, $h(\ell) = O(2^k)$ for any $1 \leq \ell \leq k$ and using a similar technique as in Sect. 2 we can prove that the Harmonic Algorithm is $O(2^k)$-competitive.

References

1. Achlioptas, D., Chrobak, M., Noga, J.: Competitive analysis of randomized paging algorithms. Theor. Comput. Sci. **234**(1–2), 203–218 (2000)
2. Bansal, N., Buchbinder, N., Madry, A., Naor, J.: A polylogarithmic-competitive algorithm for the k-server problem. J. ACM **62**(5), 40 (2015)
3. Bansal, N., Eliáš, M., Koumoutsos, G.: Weighted k-server bounds via combinatorial dichotomies. In: 58th IEEE Annual Symposium on Foundations of Computer Science (FOCS), pp. 493–504 (2017)
4. Bansal, N., Eliáš, M., Koumoutsos, G., Nederlof, J.: Competitive algorithms for generalized k-server in uniform metrics. In: Proceedings of the Twenty-Ninth Annual ACM-SIAM Symposium on Discrete Algorithms (SODA), pp. 992–1001 (2018)
5. Bartal, Y., Grove, E.: The harmonic k-server algorithm is competitive. J. ACM **47**(1), 1–15 (2000)
6. Ben-David, S., Borodin, A., Karp, R.M., Tardos, G., Wigderson, A.: On the power of randomization in on-line algorithms. Algorithmica **11**(1), 2–14 (1994). https://doi.org/10.1007/BF01294260

7. Bienkowski, M., Jez, l., Schmidt, P.: Slaying hydrae: improved bounds for generalized k-server in uniform metrics. In 30th International Symposium on Algorithms and Computation, ISAAC 2019, pp. 14:1–14:14 (2019)
8. Borodin, A., El-Yaniv, R.: Online Computation and Competitive Analysis. Cambridge University Press, Cambridge (1998)
9. Bubeck, S., Cohen, M.B., Lee, Y.T., Lee, J.R., Madry, A.: k-server via multiscale entropic regularization. In: Proceedings of the 50th Annual ACM SIGACT Symposium on Theory of Computing (STOC), pp. 3–16 (2018)
10. Buchbinder, N., Gupta, A., Molinaro, M., Naor, J.: k-Servers with a smile: online algorithms via projections. In: Proceedings of the Thirtieth Annual ACM-SIAM Symposium on Discrete Algorithms, SODA 2019, pp. 98–116 (2019)
11. Chiplunkar, A., Vishwanathan, S.: Randomized memoryless algorithms for the weighted and the generalized k -server problems. ACM Trans. Algorithms **16**, 1–28 (2019)
12. Chrobak, M., Sgall, J.: The weighted 2-server problem. Theor. Comput. Sci. **324**(2–3), 289–312 (2004)
13. Coppersmith, D., Doyle, P., Raghavan, P., Snir, M.: Random walks on weighted graphs and applications to on-line algorithms. J. ACM **40**(3), 421–453 (1993)
14. Fiat, A., Ricklin, M.: Competitive algorithms for the weighted server problem. Theor. Comput. Sci. **130**(1), 85–99 (1994)
15. Grinstead, C.M., Snell, J.L.: Introduction to Probability, AMS (2003)
16. Iwama, K., Yonezawa, K.: Axis-bound CNN problem. IEICE TRANS **87**, 1–8 (2001)
17. Koutsoupias, E.: The k-server problem. Comput. Sci. Rev. **3**(2), 105–118 (2009)
18. Koutsoupias, E., Papadimitriou, C.H.: On the k-server conjecture. J. ACM **42**(5), 971–983 (1995)
19. Koutsoupias, E., Taylor, D.S.: The CNN problem and other k-server variants. Theor. Comput. Sci. **324**(2–3), 347–359 (2004)
20. Lee, J.R.: Fusible HSTs and the randomized k-server conjecture. In: 59th IEEE Annual Symposium on Foundations of Computer Science, FOCS 2018, pp. 438–449 (2018)
21. Manasse, M.S., McGeoch, L.A., Sleator, D.D.: Competitive algorithms for server problems. J. ACM **11**(2), 208–230 (1990)
22. McGeoch, L.A., Sleator, D.D.: A strongly competitive randomized paging algorithm. Algorithmica **6**(6), 816–825 (1991)
23. Sitters, R.: The generalized work function algorithm is competitive for the generalized 2-server problem. SIAM J. Comput. **43**(1), 96–125 (2014)
24. Sitters, R.A., Stougie, L., de Paepe, W.E.: A competitive algorithm for the general 2-server problem. In: Baeten, J.C.M., Lenstra, J.K., Parrow, J., Woeginger, G.J. (eds.) ICALP 2003. LNCS, vol. 2719, pp. 624–636. Springer, Heidelberg (2003). https://doi.org/10.1007/3-540-45061-0_50
25. Sitters, R.A., Stougie, L.: The generalized two-server problem. J. ACM **53**(3), 437–458 (2006)
26. Sleator, D.D., Tarjan, R.E.: Amortized efficiency of list update and paging rules. Commun. ACM **28**(2), 202–208 (1985)

Tight Bounds on Subexponential Time Approximation of Set Cover and Related Problems

Magnús M. Halldórsson[1(✉)], Guy Kortsarz[2], and Marek Cygan[3]

[1] ICE-TCS, Reykjavik University, Reykjavik, Iceland
mmh@ru.is
[2] Rutgers University, Camden, NJ, USA
guyk@rutgers.edu
[3] University of Warsaw, Warsaw, Poland
mcygan@mimuw.edu.pl

Abstract. We show that SET-COVER on instances with N elements cannot be approximated within $(1-\gamma)\ln N$-factor in time $\exp(N^{\gamma-\delta})$, for any $0 < \gamma < 1$ and any $\delta > 0$, assuming the Exponential Time Hypothesis. This essentially matches the best upper bound known by Cygan, Kowalik, and Wykurz [6] of $(1-\gamma)\ln N$-factor in time $exp(O(N^{\gamma}))$.

The lower bound is obtained by extracting a standalone reduction from LABEL-COVER to SET-COVER from the work of Moshkovitz [18], and applying it to a different PCP theorem than done there. We also obtain a tighter lower bound when conditioning on the Projection Games Conjecture.

We also treat three problems (Directed Steiner Tree, Submodular Cover, and Connected Polymatroid) that strictly generalize SET-COVER. We give a $(1-\gamma)\ln N$-approximation algorithm for these problems that runs in $exp(\tilde{O}(N^{\gamma}))$ time, for any $1/2 \le \gamma < 1$.

Keywords: Subexponential time algorithms · Lower bounds · Set cover

1 Introduction

We show that SET-COVER on instances with N elements cannot be approximated within $(1-\gamma)\ln N$-factor in time $\exp(N^{\gamma-\delta})$, assuming the Exponential Time Hypothesis (ETH), for any $\gamma, \delta > 0$. This essentially matches the best upper bound known by Cygan, Kowalik, and Wykurz [6]. This is obtained by extracting a standalone reduction from LABEL-COVER to SET-COVER from the work of Moshkovitz [18], and applying it to a different PCP theorem than done

The work of Marek Cygan is part of a project TOTAL that has received funding from the European Research Council (ERC) under the European Union Horizon 2020 research and innovation programme (grant agreement No 677651). Magnús Halldórsson is partially supported by Icelandic Research Fund grant 174484-051. Guy Kortsarz is partially supported by NSF grants 1540547 and 1910565.

C. Kaklamanis and A. Levin (Eds.): WAOA 2020, LNCS 12806, pp. 159–173, 2021.
https://doi.org/10.1007/978-3-030-80879-2_11

there. We also obtain a tighter lower bound when conditioning on the Projection Games Conjecture.

We also treat the Directed Steiner Tree (DST) problem that strictly generalizes SET-COVER. The input to DST consists of a directed graph G with costs on edges, a set of terminals, and a designated root r. The goal is to find a subgraph of G that forms an arborescence rooted at r containing all the N terminals and minimizing the cost. We give a $(1-\gamma)\ln N$-approximation algorithm for DST that runs in $exp(\tilde{O}(N^{\gamma}))$ time, for any $\gamma \geq 1/2$. Recall that the $\tilde{O}$-notation hides logarithmic factors.

This algorithm also applies to two other generalizations of SET-COVER. In the SUBMODULAR COVER problem, the input is a set system $(U, \mathcal{C})$ with a cost $c(s)$ on each element s of the universe U. We are given a non-decreasing *submodular* function $f : 2^U \rightarrow \mathbb{R}$ satisfying $f(S+x) - f(S) \geq f(T+x) - f(T)$, for every $S \subseteq T \subseteq V$ and every $x \in U \setminus T$. The objective is to minimize the cost $c(S) = \sum_{s \in S} c(s)$ subject to $f(S) = f(U)$. In the CONNECTED POLYMATROID problem, which generalizes SUBMODULAR COVER, the elements of U are leaves of a tree and both elements and sets have cost. The goal is to select a set $S \subseteq U$ so that $f(S) = f(U)$ and $c(S) + c(T(S))$ is minimized, where $T(S)$ is the unique tree rooted at r spanning S.

1.1 Related Work

Johnson [14] and Lovász [16] showed that a greedy algorithm yields a $1 + \lg N$-approximation of SET-COVER, where N is the number of elements. Chvátal [5] extended it to the weighted version. Slavík [22] refined the bound to $\ln n - \ln\ln n + O(1)$.

Lund and Yannakakis [17] showed that logarithmic factor was essentially best possible for polynomial-time algorithms, unless $NP \subseteq DTIME(n^{polylog(n)})$. Feige [9] gave the precise lower bound that SET-COVER admits no $(1-\epsilon)$-approximation, for any $\epsilon > 0$, with a similar complexity assumption. Assuming the stronger ETH, he shows that $\ln N - c\log\log N$-approximation is not possible for polynomial algorithms, for some $c > 0$. The work of Moshkovitz [18] and Dinur and Steurer [7] combined shows that $(1-\epsilon)\ln N$-approximation is hard modulo $P = NP$. All the inapproximability results relate to two prover interactive proofs via the LABEL-COVER problem.

Recent years have seen increased interest in subexponential time algorithms, including approximation algorithms. A case in point is the *maximum clique* problem that has a trivial $2^{n/\alpha}\mathrm{poly}(n)$-time algorithm that gives a α-approximation, for any $1 \leq \alpha \leq \sqrt{n}$, and Bansal et al. [1] improved the time to $exp(n/\Omega(\alpha \log^2 \alpha))$. Chalermsook et al. [3] showed that this is nearly tight, as α-approximation requires $exp(n^{1-\epsilon}/\alpha^{1+\epsilon})$ time, for any $\epsilon > 0$, assuming ETH.

For SET-COVER, Cygan, Kowalik and Wykurz [6] gave a $(1-\alpha)\ln N$-approximation algorithm that runs in time $2^{O(N^{\alpha})}$, for any $0 < \alpha < 1$. The results of [7,18] imply a $exp(N^{\alpha/c})$-time lower bound for $(1-\alpha)\ln N$-approximation, for some constant $c \geq 3$.

DST can be approximated within N^ϵ-factor, for any $\epsilon > 0$, in polynomial time [4,15]. In quasi-polynomial time, it can be approximated within $O(\log^2 N/\log\log N)$-factor [11], which is also best possible in that regime [11,12]. This was recently extended to CONNECTED POLYMATROID [10]. For SUBMODULAR COVER, the greedy algorithm also achieves a $1 + \ln N$-approximation [23].

1.2 Organization

The paper is organized so as to be accessible at different levels of detail. In Sect. 2, we derive two different hardness results for subexponential time algorithms under different complexity assumptions: ETH and PGC. For this purpose, we only state (but not prove) the hardness reduction, and introduce the LABEL-COVER problem with its key parameters: size, alphabet size, and degrees.

We prove the properties of the hardness reduction in Sect. 2.2, by combining two lemmas extracted from [18]. The proofs of these lemmas are given in Sect. 3.1 and 3.2. To make it easy for the reader to spot the differences with the arguments of [18], we underline the changed conditions or parameters in our presentation.

Finally, the approximation algorithm for Directed Steiner Tree is given in Sect. 4.

2 Hardness of Set Cover

We give our technical results for SET-COVER in this section. Starting with definition of LABEL-COVER in Sect. 2.1, we give a reduction from LABEL-COVER to SET-COVER in Sect. 2.2, and derive specific approximation hardness results in Sect. 2.3. A full proof of the correctness of the reduction is given in the following section.

2.1 Label Cover

The intermediate problem in all known approximation hardness reductions for SET-COVER is the LABEL-COVER problem.

Definition 1. *In the* LABEL-COVER *problem with the projection property (a.k.a., the* Projection Game*), we are given a bipartite graph $G(A, B, E)$, finite alphabets (also called* labels*) Σ_A and Σ_B, and a function $\pi_e : \Sigma_A \to \Sigma_B$ for each edge $e \in E$. A* labeling *is a pair $\varphi_A : A \to \Sigma_A$ and $\varphi_B : B \to \Sigma_B$ of assignments of labels to the vertices of A and B, respectively. An edge $e = (a, b)$ is* covered *(or* satisfied*) by (φ_A, φ_B) if $\pi_e(\varphi_A(a)) = \varphi_B(b)$. The goal in* LABEL-COVER *is to find a labeling (φ_A, φ_B) that covers as many edges as possible.*

The *size* of a label cover instance $\mathcal{G} = (G = (A, B, E), \Sigma_A, \Sigma_B, \Pi = \{\pi_e\}_e)$ is denoted by $n_{\mathcal{G}} = |A| + |B| + |E|$. The *alphabet size* is $\max(|\Sigma_A|, |\Sigma_B|)$. The LABEL-COVER instances we deal with will be *bi-regular*, meaning that all nodes of the same bipartition have the same degree. We refer to the degree of nodes in A (B) as the *A-degree* (*B-degree*), respectively.

LABEL-COVER is a central problem in computational complexity, corresponding to *projection PCPs*, or probabilistically checkable proofs that make 2 queries. A key parameter is the *soundness error*:

Definition 2. *A* LABEL-COVER *construction* $\mathcal{G} = \mathcal{G}_\phi$*, formed from a 3-SAT formula* ϕ*, has* soundness error ϵ *if: a) whenever* ϕ *is satisfiable, there is a labeling of* $\mathcal{G}$ *that covers all edges, and b) when* ϕ *is unsatisfiable, every labeling of* $\mathcal{G}$ *covers at most an* ϵ*-fraction of the edges.*

Note that the construction is a schema that holds for a range of values for ϵ. A LABEL-COVER construction is *almost-linear size* if it is of size $n^{1+o(1)}$, possibly with extra $poly(1/\epsilon)$ factors.

We use the following PCP theorem of Moshkovitz and Raz [19].

Theorem 1 ([19]). *For every* $\epsilon \geq 1/polylog(n)$*,* SAT *on input of size* n *can be reduced to* LABEL-COVER *on a bi-regular graph of degrees* $\text{poly}(1/\epsilon)$*, with soundness error* ϵ*, almost-linear size* $n^{1+o(1)}\text{poly}(1/\epsilon) = n^{1+o(1)}$*, and alphabet size that is exponential in* $poly(1/\epsilon)$*. The reduction can be computed in time linear in the size and alphabet size of the Label-Cover.*

Dinur and Steurer [7] later gave a PCP construction whose alphabet size depends only polynomially on $1/\epsilon$. This is crucial for NP-hardness results, and combined with the reduction of [18], implies the essentially tight bound of $(1-\epsilon)\ln N$-approximation of SET-COVER by poly-time algorithms.

2.2 Set Cover Reduction

We present here a reduction from a generic LABEL-COVER (a two-prover PCP theorem) to the SET-COVER problem. This is extracted from the work of Moshkovitz [18]. The presentation in [18] was tightly linked with the PCP construction of Dinur and Steurer [7] that was used in order to stay within polynomial time. When allowing superpolynomial time, it turns out to be more frugal to apply the older PCP construction of Moshkovitz and Raz [19]. This construction has exponential dependence on the alphabet size, which precludes its use in NP-hardness results. On the other hand, it has nearly-linear dependence on the size of the Label Cover, unlike Dinur-Steurer, and this becomes a dominating factor in subexponential reductions.

Our main technical contribution is then to provide a standalone reduction from LABEL-COVER to SET-COVER that allows specific PCP theorems to be plugged in.

We say that a reduction that originates in SAT achieves *approximation gap* ρ if there is a value a such that SET-COVER instances originating in satisfiable formulas have a set cover of size at most a, while instances originating in unsatisfiable formulas have all set covers of size greater than $\rho \cdot a$.

Theorem 2. *Let* $\gamma > 0$ *and* $0 < \delta < \gamma$*. There is a reduction from* LABEL-COVER *to* SET-COVER *with the following properties. Let* $\mathcal{G}$ *be a bi-regular*

LABEL-COVER of almost-linear size n_0, soundness error parameter ϵ, B-degree poly$(1/\epsilon)$, and alphabet size $\sigma_A(\epsilon)$. Then for each $\gamma > 0$, $\mathcal{G}$ is reduced to a SET-COVER instance $\mathcal{SC} = \mathcal{SC}_{\mathcal{G},\gamma,\delta}$ with approximation gap $(1-\gamma)\ln N$, $N = \tilde{O}(n_0^{1/(\gamma-\delta)})$ elements, and $M = \tilde{O}(n_0) \cdot \sigma_A(polylog(n))$ sets. The time of the reduction is linear in the size of $\mathcal{SC}$.

2.3 Approximation Hardness Results

When it comes to hardness results for subexponential time algorithms the standard assumption is the *Exponential Time Hypothesis (ETH)*. ETH asserts that the 3-SAT problem on n variables and m clauses cannot be solved in $2^{o(n)}$-time. Impagliazzo, Paturi and Zane [13] showed that any 3-SAT instance can be sparsified in $2^{o(n)}$-time to an instance with $m = O(n)$ clauses. When we refer to SAT input of size n, we mean 3-CNF formula on n variables and $O(n)$ clauses. Thus, ETH together with the sparsification lemma [2] implies the following:

Conjecture 1. (ETH) There is no $2^{o(n)}$-time algorithm that decides SAT on inputs of size n.

We need only a weaker version: There is some $\zeta > 0$ such that there is no $exp(n^{1-\zeta})$-time algorithm to decide SAT.

We are now ready for our main result.

Theorem 3. *Let $0 < \gamma < 1$ and $0 < \delta < \gamma$. Assuming ETH, there is no $(1-\gamma)\ln N$-approximation algorithm of SET-COVER with N elements and M sets that runs in time $exp(N^{\gamma-\delta}) \cdot poly(M)$.*

Proof. We show how such an algorithm can be used to decide SAT in subexponential time, contradicting ETH. Given a SAT instance of size n, apply Theorem 1 with $\epsilon = \Theta(1/\log^2 n)$, to obtain LABEL-COVER instance $\mathcal{G}$ of size $n_0 = n^{1+o(1)}$ and alphabet size $\sigma_A(\epsilon) = exp(poly(1/\epsilon)) = exp(polylog(n))$. Next apply Theorem 2 to obtain a SET-COVER instance $\mathcal{SC} = \mathcal{SC}_{\mathcal{G}}$ with $M = \tilde{O}(n_0)\sigma_A(polylog(n)) = exp(polylog(n))$ sets and $N = \tilde{O}(n_0^{1/\gamma}) = n^{1/\gamma+o(1)}$ elements, and approximation gap $(1-\gamma)\ln N$.

Suppose there is a $(1-\gamma)\ln N$-approximation algorithm of SET-COVER running in time $exp(N^{\gamma-\delta}) \cdot poly(M)$. Since it achieves this approximation, it can decide the satisfiability of ϕ. Since $N^{\gamma-\delta} = n^{(1/\gamma+o(1))\cdot(\gamma-\delta)} \leq n^{1-\delta'}$, for some $\delta' > 0$, the running time contradicts ETH.

Note: We could also allow the algorithm greater than polynomial complexity in terms of M without changing the implication. In fact, the complexity can be as high as $exp(M^{\delta_0})$, for some small constant δ_0.

Still Tighter Bound Under Stronger Assumptions. Moshkovitz [18] proposed a conjecture on the parameters of possible Label Cover constructions. We require a particular version with almost linear size and low degree.

Definition 3 (The Projection Games Conjecture (PGC)). *3-SAT of inputs of size n can be reduced to* LABEL-COVER *of almost-linear size, alphabet size poly$(1/\epsilon)$, and bi-regular degrees poly$(1/\epsilon)$, where ϵ is the soundness error parameter.*

The key difference of PGC from known PCP theorems is the alphabet size. PGC is considered quite plausible and has been used to prove conditional hardness results for a number of problems [18,20].

By assuming PGC, we improve the dependence on M, the number of sets.

Theorem 4. *Let $0 < \gamma < 1$ and $0 < \delta < \gamma$. Assuming PGC and ETH, there is no $(1-\gamma)\ln N$-approximation, nor a $O(\log M)$-approximation, of* SET-COVER *with N elements and M sets that runs in time $exp(N^{\gamma-\delta}M^{1-\delta})$.*

Namely, both the approximation factor and the time complexity can depend more strongly on the number of sets in the SET-COVER instance. The only result known in terms of M is a folklore $\sqrt{M}$-approximation in polynomial time.

Proof. We can proceed in the same way as in the proof of Theorem 3, but starting from the conjectured LABEL-COVER given by PGC, in which the alphabet size is polynomial in ϵ. We then obtain a set cover instance $\mathcal{SC}_{\mathcal{G}}$ that differs only in that we now have $M = |A||\Sigma_A| = n^{1+o(1)} = N^{\gamma+o(1)}$. So, a $c\log M$-approximation, with constant $c > 0$, implies a $c\gamma\ln N$-approximation, which is smaller than $(1-\gamma)\ln N$-approximation when $c\gamma < (1-\gamma)$ or $\gamma < 1/(2c)$. Also, $exp(M^{1-\delta}) = exp(n^{1-\delta'})$, for some $\delta' > 0$, and hence such a running time again breaks ETH.

3 Proof of the Set Cover Reduction

We extract here a sequence of two reductions from the work of Moshkovitz. By untangling them from the LABEL-COVER construction, we can use them for our standalone SET-COVER reduction.

Moshkovitz [18] chooses some of the parameters of the lemmas so as to fit the purpose of proving NP-hardness of approximation. As a result, the size of the intermediate LABEL-COVER instance generated grows to be a polynomial of degree larger than 1. This leads to weaker hardness results for sub-exponential time algorithms than what we desire. We indicate therefore how we can separate a key parameter to maintain nearly-linear size label covers.

A key tool in her argument is the concept of agreement soundness error.

Definition 4 (List-agreement soundness error). *Let $\mathcal{G}$ be a* LABEL-COVER *for deciding the satisfiability of a Boolean formula ϕ. Let φ_A assign each A-vertex ℓ alphabet symbols. We say that the A-vertices* totally disagree *on a vertex $b \in B$ if, there are no two neighbors $a_1, a_2 \in A$ of b for which there exist $\sigma_1 \in \varphi_A(a_1), \sigma_2 \in \varphi_A(a_2)$ such that $\pi_{(a_1,b)}(\sigma_1) = \pi_{(a_2,b)}(\sigma_2)$.*

We say that $\mathcal{G}$ has list-agreement soundness error *(ℓ, ϵ) if, for unsatisfiable ϕ, for any assignment $\varphi_A : A \to \binom{\sigma_A}{\ell}$, the A-vertices are in total disagreement on at least $1-\epsilon$ fraction of the B-vertices.*

The reduction of Theorem 2 is obtained by stringing together two reductions: from LABEL-COVER to a modified LABEL-COVER with a low agreement soundness error, and from that to SET-COVER.

The first one is laid out in the following lemma that combines Lemmas 4.4 and 4.7 of [18]. The proof is given in the upcoming subsection.

Lemma 1. *Let $D \geq 2$ be a prime power, q be a power of D, $\ell > 1$, and $\epsilon_0 > 0$. There is a polynomial reduction from a* LABEL-COVER *with soundness error $\epsilon_0^2 D^2$ and B-degree q to a* LABEL-COVER *with list-agreement soundness error $(\ell, 2\epsilon_0 D^2 \cdot \ell^2)$ and B-degree D. The reduction preserves alphabets, and the size is increased by poly(q)-factor.*

Moshkovitz also gave a reduction from LABEL-COVER with small agreement soundness error to SET-COVER approximation. We extract a more general parameterization than is stated explicitly around Claim 4.10 in [18]. The proof, with minor changes from [18], is given for completeness in Sect. 3.2.

Lemma 2 ([18], rephrased). *Let $\mathcal{G}' = (G' = (A', B', E'), \Sigma_A, \Sigma_B, \Pi')$ be a bi-regular* LABEL-COVER *instance with soundness parameter ϵ for deciding the satisfiability of a boolean formula ϕ. Let D be the B'-degree. For every α with $2/D < \alpha < 1$,*[1] *and any $u \geq (D^{O(\log D)} \log |\Sigma_{B'}|)^{1/\alpha}$, there is a reduction from $\mathcal{G}$ to a* SET-COVER *instance $\mathcal{SC} = \mathcal{SC}_{\mathcal{G}'}$ with a certain choice of ϵ that attains the following properties:*

1. Completeness: *If all edges of G' can be covered, then $\mathcal{SC}$ has a set cover of size $|A'|$.*
2. Soundness: *If $\mathcal{G}'$ has list-agreement soundness error (ℓ, α), where $\ell = D(1 - \alpha) \ln u$, then every set cover of $\mathcal{SC}$ is of size more than $|A'|(1 - 2\alpha) \ln u$.*
3. *The number N of elements of $\mathcal{SC}$ is $|B'| \cdot u$ and the number M of sets is $|A'| \cdot |\Sigma_A|$.*
4. *The time for the reduction is polynomial in $|A'|, |B'|, |\Sigma_A|, |\Sigma_B|$ and u.*

Given these lemmas, we can now prove Theorem 2, which we restate for convenience.

Theorem 2 (restated). *Let $\gamma > 0$ and $0 < \delta < \gamma$. There is a reduction from* LABEL-COVER *to* SET-COVER *with the following properties. Let $\mathcal{G}$ be a bi-regular* LABEL-COVER *of almost-linear size n_0, soundness error parameter ϵ, B-degree poly($1/\epsilon$), and alphabet size $\sigma_A(\epsilon)$. Then for each $\gamma > 0$, $\mathcal{G}$ is reduced to a* SET-COVER *instance $\mathcal{SC} = \mathcal{SC}_{\mathcal{G},\gamma,\delta}$ with approximation gap $(1 - \gamma) \ln N$, $N = \tilde{O}(n_0^{1/(\gamma-\delta)})$ elements, and $M = \tilde{O}(n_0) \cdot \sigma_A(polylog(n))$ sets. The time of the reduction is linear in the size of $\mathcal{SC}$.*

Proof (Proof of Theorem 2). Let $\alpha = 2\delta$, D be a prime power at least $2/\alpha$, and $\gamma' = (\gamma - \delta)/(1 - \delta)$. Let $n_1 = \tilde{O}(n_0)$ be the size of the instance that is formed by

[1] Moshkovitz doesn't explicitly relate α and D, but indicates that α be small and D "sufficiently large".

Lemma 1 on $\mathcal{G}$ with $\epsilon = \Theta(1/\log^4 n)$. Let $u = n_1^{(1-\gamma')/\gamma'}$ and note that $\log u = \Theta(\log n)$. Let $\ell = D(1-\alpha)\ln u = \Theta(\log n)$, $\epsilon = \alpha^2 D^{-2}\ell^{-4}/4 = \Theta(1/\log^4 n)$, and $\epsilon_0 = \sqrt{\epsilon}/D = \alpha/(2D^2\ell^2) = \Theta(1/\log^2 n)$. Let q be the B-degree of $\mathcal{G}$.

We apply Lemma 1 to $\mathcal{G}$ (with D, q, ℓ and ϵ_0) and obtain a LABEL-COVER instance $\mathcal{G}' = (G' = (A', B', E'), \Sigma_A, \Sigma_B, \Pi)$ with B-degree D and size n_1, with alphabets unchanged. The list-agreement soundness error of $\mathcal{G}'$ is $(\ell, 2\epsilon_0\ell^2 D^2)$, or (ℓ, α).

We can verify that $\mathcal{G}'$, α, and D and u satisfy the prerequisites of Lemma 2, which yields a SET-COVER instance $\mathcal{SC}$=$\mathcal{SC}_{\mathcal{G}}$ with $N = |B'| \cdot u = n_1 \cdot n_1^{(1-\gamma')/\gamma'} = n_1^{1/\gamma'} = \tilde{O}(n_0^{(1-\delta)/(\gamma-\delta)}) = \tilde{O}(n_0^{1/(\gamma-\delta)})$ elements and $M = |A'| \cdot |\Sigma_A|$ sets.

If ϕ is satisfiable, then $\mathcal{SC}$ has set cover of size $|A'|$, while if it is unsatisfiable, then every set cover of $\mathcal{SC}$ has size more than $|A'|(1-2\alpha)\ln u = |A'|(1-\delta)\ln u$. Note that $\ln u = \ln(N/n_1) = (1-\gamma')\ln N$. Hence, the approximation gap is $(1-\delta)\ln u = (1-\delta)(1-\gamma')\ln N = (1-\gamma)\ln N$.

3.1 Proof of Lemma 1

We give here a full proof of Lemma 1, based on [18], with minor modification.

When the labeling assigns a single label to each node, i.e., when $\ell = 1$, the list-agreement soundness error reduces to *agreement soundness error*, which is otherwise defined equivalently. Moshkovitz first showed how to reduce a LABEL-COVER with small soundness error to one with a small agreement soundness error. The lemma stated here is unchanged from [18] except that Moshkovitz used the parameter name n instead of our parameter q.[2]

Lemma 3 (Lemma 4.4 of [18]). *Let $D \geq 2$ be a prime power and let q be a power of D. Let $\epsilon_0 > 0$. There is a polynomial reduction from a LABEL-COVER with soundness error $\epsilon_0^2 D^2$ and B-degree q to a LABEL-COVER with agreement soundness error $2\epsilon_0 D^2$ and B-degree D. The reduction preserves alphabets, and the size is increased by poly(q)-factor.*

We have underlined the parts that changed because of using q as parameter instead of n. The proof is based on the following combinatorial lemma, whose proof we omit. We note that the set U here is different from the one used in Lemma 2 (but we retained the notation to remain faithful to [18]).

Lemma 4 (Lemma 4.3 of [18]). *For $0 < \epsilon < 1$, for a prime power D, and q that is a power of D, there is an explicit construction of a regular bipartite graph $H = (U, V, E)$ with $|U| = q$, V-degree D, and $|V| \leq q^{O(1)}$ that satisfies the following. For every partition $U_1, \ldots, U_\ell$ of U into sets such that $|U_i| \leq \epsilon|U|$ for $i = 1, 2, \ldots, \ell$, the fraction of vertices $v \in V$ with more than one neighbor in any single set U_i, is at most ϵD^2.*

[2] This invited confusion, since n was also used to denote the size of the LABEL-COVER (like we do here).

Again, we used the parameter name q, rather than n as in [18]. We show how to take a LABEL-COVER with standard soundness and convert it to a LABEL-COVER instance with total disagreement soundness, by combining it with the graph from Lemma 4.

Proof (Proof of Lemma 3). Let $\mathcal{G} = (G = (A, B, E), \Sigma_A, \Sigma_B, \Pi)$ be the original LABEL-COVER. Let $H = (U, V, E_H)$ be the graph from Lemma 4, where q, D and ϵ are as given in the current lemma. Let us use U to enumerate the neighbors of a B-vertex, i.e., there is a function $E^{\leftarrow} : B \times U \to A$ that given a vertex $b \in B$ and $u \in U$ gives us the A-vertex which is the u neighbor (in G) of b.

We create a new LABEL-COVER $\mathcal{G}' = (G = (A, B \times V, E'), \Sigma_A, \Sigma_B, \Pi')$. The intended assignment to every vertex $a \in A$ is the same as its assignment in the original instance. The intended assignment to a vertex $\langle b, v\rangle \in B \times V$ is the same as the assignment to b in the original game. We put an edge $e' = (a, \langle b, v\rangle)$ if $E^{\leftarrow}(b, u) = a$ and $(u, v) \in E_H$. We define $\pi_{e'} \equiv \pi_{(a,b)}$.

If there is an assignment to the original instance that satisfies c fraction of its edges, then the corresponding assignment to the new instance satisfies c fraction of its edges.

Suppose there is an assignment for the new instance $\varphi_A : A \to \Sigma_A$ in which more than $2\epsilon D^2$ fraction of the vertices in $B \times V$ do not have total disagreement.

Let us say that $b \in B$ is "good" if for more than an ϵD^2 fraction of the vertices in $\{b\} \times V$, the A-vertices do not totally disagree. Note that the fraction of good $b \in B$ is at least ϵD^2.

Focus on a good $b \in B$. Consider the partition of U into $|\Sigma_B|$ sets, where the set corresponding to $\sigma \in \Sigma_B$ is:

$$U_\sigma = \{u \in U | a = E^{\leftarrow}(b, u) \wedge e = (a, b) \in E_G \wedge \pi_e(\varphi_A(a)) = \sigma\} .$$

By the goodness of b and the property of H, there must be $\sigma \in \Sigma_B$ such that $|U_\sigma| > \epsilon |U|$. We call σ the "champion" for b.

We define an assignment $\varphi_B : B \to \Sigma_B$ that assigns good vertices b their champions, and other vertices b arbitrary values. The fraction of edges that φ_A, φ_B satisfy in the original instance is at least $\epsilon^2 D^2$.

Moshkovitz then shows that small agreement soundness error translates to the list version. The proof is unchanged from [18].

Lemma 5 (Lemma 4.7 of [18]). *Let $\ell \geq 1$, $0 < \epsilon' < 1$. A* LABEL-COVER *with agreement soundness error ϵ' has list-agreement soundness error $(\ell, \epsilon'\ell^2)$.*

Proof. Assume by the way of contradiction that the LABEL-COVER instance has an assignment $\hat{\varphi}_A : A \to \binom{\sigma_A}{\ell}$ such that on more than $\epsilon'\ell^2$-fraction of the B-vertices, the A-vertices do not totally disagree. Define an assignment $\varphi_A : A \to \Sigma_A$ by assigning every vertex $a \in A$ a symbol picked uniformly at random from the ℓ symbols in $\hat{\varphi}_A(a)$. If a vertex $b \in B$ has two neighbors $a_1, a_2 \in A$ that agree on b under the list assignment $\hat{\varphi}_A$, then the probability that they agree on b under the assignment φ_A is at least $1/\ell^2$. Thus, under φ_A, the expected fraction of the B-vertices that have at least two neighbors that agree on them, is more than ϵ'. In particular, there exists an assignment to the A-vertices, such that more than ϵ' fraction of the B-vertices have two neighbors that agree on them. This contradicts the agreement soundness.

Lemma 1 follows directly from combining Lemmas 3 and 5.

3.2 Proof of Lemma 2

We give here a proof of Lemma 2, following closely the exposition of [18], with minor modifications.

Feige [9] introduced the concept of *partition systems* (also known as *anti-universal sets* [21]) which is key to tight inapproximibility results for SET-COVER. It consists of a universe along with a collection of partitions. Each partition covers the universe, but any cover that uses at most one set out of each partition, is necessarily large. The idea is to form the reduction so that if the SAT instance is satisfiable, then one can use a single partition to cover the universe, while if it is unsatisfiable, then one must use sets from different partitions, necessarily resulting in a large set cover.

Naor, Schulman, and Srinivasan [21] gave the following combinatorial construction (which as appears as Lemma 4.9 of [18]) that derandomizes one introduced by Feige [9].

Lemma 6 ([21]). *For natural numbers m, D and $0 < \alpha < 1$, with $\alpha \geq 2/D$, and for all $u \geq (D^{O(\log D)} \log m)^{1/\alpha}$, there is an explicit construction of a universe U of size u and partitions $\mathcal{P}_1, \ldots, \mathcal{P}_m$ of U into D sets that satisfy the following: there is no cover of U with $\ell = D \ln |U|(1-\alpha)$ sets $S_{i_1}, \ldots, S_{i_\ell}$, $1 \leq i_1 < \cdots < i_\ell \leq m$, such that set S_{i_j} belongs to partition $\mathcal{P}_{i_j}$.*

Naor et al. [21] state the result in terms of the relation $u \geq (D/(D-1))^\ell \ell^{O(\log \ell} \log m$. Note that for $\ell = D \ln u(1-\alpha)$, we have $(D/(D-1))^\ell \approx u^{(1-\alpha)D/(D-1)} \approx u^{1-\alpha+1/D}$, and hence we need D to be sufficiently large.

The following reduction follows Moshkovitz [18], which in turns is along the lines of Feige [9].

Lemma 2 (restated). *Let $\mathcal{G}' = (G' = (A', B', E'), \Sigma_A, \Sigma_B, \Pi')$ be a bi-regular* LABEL-COVER *instance with soundness parameter ϵ for deciding the satisfiability of a boolean formula ϕ. Let D be the B'-degree. For every α with $2/D < \alpha < 1$,*[3] *and any $u \geq (D^{O(\log D)} \log |\Sigma_{B'}|)^{1/\alpha}$, there is a reduction from $\mathcal{G}$ to a* SET-COVER *instance $\mathcal{SC}=\mathcal{SC}_{\mathcal{G}'}$ with a certain choice of ϵ that attains the following properties:*

1. *Completeness: If all edges of G can be covered, then $\mathcal{SC}$ has a set cover of size $|A'|$.*
2. *Soundness: If $\mathcal{G}'$ has list-agreement soundness error (ℓ, α), where $\ell = D(1-\alpha) \ln u$, then every set cover of $\mathcal{SC}$ is of size more than $|A'|(1-2\alpha) \ln u$.*

[3] Moshkovitz doesn't explicitly relate α and D, but indicates that α be small and D "sufficiently large".

3. *The number N of elements of $\mathcal{SC}$ is $|B'| \cdot u$ and the number M of sets is $|A'| \cdot |\Sigma_A|$.*
4. *The time for the reduction is polynomial in $|A'|, |B'|, |\Sigma_A|, |\Sigma_B|$ and u.*

Proof. Let α and u be values satisfying the statement of the theorem. Let $m = |\Sigma_B|$ and let D be the B-degree of $\mathcal{G}$. Apply Lemma 6 with m, D and u, obtaining a universe U of size u and partitions $\mathcal{P}_{\sigma_1}, \dots, \mathcal{P}_{\sigma_m}$ of U. We index the partitions by the symbols $\sigma_1, \dots, \sigma_m$ of Σ_B. The elements of the SET-COVER instances are $B \times U$. Equivalently, each vertex $b \in B$ has a copy of the universe U. Covering this universe corresponds to satisfying the edges that touch b. There are m ways to satisfy the edges that touch b—one for every possible assignment $\sigma \in \Sigma_B$ to b. The different partitions covering u correspond to those different assignments.

For every vertex $a \in A$ and an assignment $\sigma \in \Sigma_A$ to a we have a set $S_{a,\sigma}$ in the SET-COVER instance. Taking $S_{a,\sigma}$ to the cover corresponds to assigning σ to a. Notice that a cover might consist of several sets of the form $S_{a,\cdot}$ for the same $a \in A$, which is the reason we consider list agreement. The set $S_{a,\sigma}$ is a union of subsets, one for every edge $e = (a, b)$ touching a. If e is the i-th edge coming into b $(1 \leq i \leq D)$, then the subset associated with e is $\{b\} \times S$, where S is the i-th subset of the partition $P_{\phi_e(\sigma)}$.

Completeness follows from taking the set cover corresponding to each of the A-vertices and its satisfying assignments.

To prove soundness, assume by contradiction that there is a set cover C of $\mathcal{SC}_{\mathcal{G}}$ of size at most $|A| \ln |U|(1 - 2\alpha)$. For every $a \in A$, let s_a be the number of sets in C of the form $S_{a,\cdot}$. Hence, $\sigma_{a \in A} s_a = |C|$. For every $b \in B$, let s_b be the number of sets in C that participate in covering $\{b\} \times U$. Then, denoting the A-degree of G by D_A,

$$\sum_{b \in B} s_b = \sum_{a \in A} s_a D_A \leq D_A |A| \ln |U|(1 - 2\alpha) = D|B| \ln |U|(1 - 2\alpha) .$$

In other words, on average over the $b \in B$, the universe $\{b\} \times U$ is covered by at most $D \ln |U|(1 - 2\alpha)$ sets. Therefore, by Markov's inequality, the fraction of $b \in B$ whose universe $\{b\} \times U$ is covered by at most $D \ln |U|(1 - \alpha) = \ell$ sets is at least α. By the contrapositive of Lemma 6 and our construction, for such $b \in B$, there are two edges $e_1 = (a_1, b), e_2 = (a_2, b) \in E$ with $S_{a_1,\sigma_1}, S_{a_2,\sigma_2} \in C$ where $\pi_{e_1}(\sigma_1) = \pi_{e_2}(\sigma_2)$.

We define assignment $\hat{\varphi}_A : A \to \binom{\sigma_A}{\ell}$ to the A-vertices as follows. For every $a \in A$, pick ℓ different symbols $\sigma \in \Sigma_A$ from those with $S_{a,\sigma} \in C$ (add arbitrary symbols if there are not enough). As we showed, for at least α-fraction of the $b \in B$, the A-vertices will not totally disagree. Hence, the soundness property follows.

4 Approximation Algorithm for Directed Steiner Tree

Recall that in DST, the input consists of a directed graph G with costs $c(e)$ on edges, a collection X of terminals, and a designated root $r \in V$. The goal is to find a subgraph of G that forms an arborescence $T_{opt} = T(r, X)$ rooted at r containing all the terminals and minimizing the cost $c(T(r, X)) = \sum_{e \in T(r,X)} c(e)$. Let $N = |X|$ denote the number of terminals and n the number of vertices.

Observe that one can model SET-COVER as a special case of DST on a 3-level acyclic digraph, with a universal root on top, nodes representing sets as internal layer, and the elements as leaves. The cost of an edge coming into a node corresponds to the cost of the corresponding element or set.

Our algorithm consists of "guessing" a set C of intermediate nodes of the optimal tree. After computing the optimal tree on top of this set, we use this set as the source of roots for a collection of trees to cover the terminals. This becomes a set cover problem, where we map each set selected to a tree of restricted size with a root in C. Our algorithm then reduces to applying the classic greedy set cover algorithm on this instance induced by the "right" set C. Because of the size restriction, the resulting approximation has a smaller constant factor.

We may assume each terminal is a leaf, by adding a leaf as a child of a non-leaf terminal and transfer the terminal function to that leaf. If a tree contains a terminal, we say that the tree *covers* the terminal.

Let $\ell(T)$ denote the number of terminals in a tree T. Let T_v denote the subtree of tree T rooted at node v. For node v and child w of v, let T_{vw} be the subtree of T formed by $T_w \cup \{vw\}$, i.e., consisting of the subtree of T rooted at w along with the edge to w's parent (v).

Definition 5. *A set $C \subset V$ is a ϕ-*core *of a tree T if there is a collection of edge-disjoint subtrees $T_1, T_2, \dots$ of T such that: a) the root of each tree T_i is in C, b) every terminal in T is contained in exactly one tree T_i, and c) each T_i contains at most ϕ terminals, $\ell(T_i) \leq \phi$.*

Lemma 7. *Every tree T contains a ϕ-core of size at most $\lceil \ell(T)/\phi \rceil$, for any ϕ.*

Proof. The proof is by induction on the number of terminals in the tree. The root is a core when $\ell(T) \leq \phi$. Let v be a vertex with $\ell(T_v) > \phi$ but whose children fail that inequality. Let C' be a ϕ-core of $T' = T \setminus T_v$ promised by the induction hypothesis, and let $C = C' \cup \{v\}$.

For each child w of v, the subtree T_{vw} contains at most ϕ terminals. Together they cover uniquely the terminals in T_v, and satisfy the other requirements of the definition of a ϕ-core for T_v. Thus C is a ϕ-core for T. Since $\ell(T') \leq \ell(T) - \phi$, the size of C satisfies $|C| = |C'| + 1 \leq \lceil \ell(T')/\phi \rceil + 1 \leq \lceil (\ell(T) - \phi)/\phi \rceil + 1 = \lceil \ell(T)/\phi \rceil$.

A core implicitly suggests a set cover instance, with sets of size at most ϕ, formed by the terminals contained in each of the edge-disjoint subtrees. Our algorithm is essentially based on running a greedy set cover algorithm on that instance.

Let $R(v)$ denote the set of nodes reachable from v in G.

Definition 6. *Let $S \subseteq V$, $U \subset X$, and let ϕ be a parameter. Then S induces a ϕ-bounded* SET-COVER *instance $(U, \mathcal{C}^{\phi}_{S,U})$ with $\mathcal{C}^{\phi}_{S,U} = \{Y \subseteq U : |Y| \leq$

ϕ and $\exists v \in S, Y \subseteq R(v)\}$. Namely, a subset Y of at most ϕ terminals in U is in $\mathcal{C}^{\phi}_{S,U}$ iff there is a v-rooted subtree containing Y.

We relate set cover solutions of $\mathcal{C}_{C,X}$ to DST solutions of G with the following lemmas.

For a node r_0 and set F, let $T(r_0, F)$ be the tree of minimum cost that is rooted by r_0 and contains all the nodes of F. For sets F and S, let $T(S, F)$ be the tree of minimum cost that contains all the nodes of F and is rooted by some node in S.

Lemma 8. *Let $\mathcal{S}$ be a set cover of $\mathcal{C}^{\phi}_{S,X}$ of cost $c(\mathcal{S})$. Then, we can form a valid* DST *solution $T_{\mathcal{S}}$ by combining $T(r, S)$ with the trees $T(S, F)$, for each $F \in \mathcal{S}$. The cost of $T_{\mathcal{S}}$ is at most $c(T_{\mathcal{S}}) \leq c(T(r, S)) + c(\mathcal{S})$.*

Proof. $T_{\mathcal{S}}$ contains all the terminals since the sets in $\mathcal{S}$ cover X. It contains an r-rooted arborescence since $T(r, S)$ contains a path from r to all nodes in S, and the other subtrees contain a path to each terminal from some node in S. The cost bound follows from the definition of the weights of sets in $\mathcal{C}^{\phi}_{S,X}$. The actual cost could be less, if the trees share edges or have multiple paths, in which case some superfluous edges can be shed.

Let $OPT_{SC}(\mathcal{C})$ be the weight of an optimal set cover of a set system $\mathcal{C}$.

Lemma 9. *Let C be a ϕ-core of T_{opt}. The cost of an optimal* DST *of G equals $OPT_{SC}(\mathcal{C}^{\phi}_{C,X})$ plus the cost of an optimal r-rooted tree with C as terminals: $c(T_{Opt}) = OPT_{SC}(\mathcal{C}^{\phi}_{C,X}) + c(T(r, C))$*

Proof. The subtree of T_{Opt} induced by C and the root r has cost $c(T(r, C))$. The rest of the tree consists of the subtrees T_{vw}, for each $v \in C$ and child w of v. T_{vw} contains at most ϕ terminals, so the corresponding set is contained in $\mathcal{C}^{\phi}_{C,X}$. Together, these subtrees contain all the terminals, so the corresponding set collection covers $\mathcal{C}^{\phi}_{C,X}$. Thus, $c(T_{Opt}) \geq c(T(r, C)) + OPT_{SC}(\mathcal{C}_{C,X})^{\phi}$. By Lemma 8, the inequality is tight.

The *density* of a set F in $\mathcal{C}S, X^{\phi}$ is $\min_{s \in S} c(T(s, F))/|F|$: the cost of the optimal tree containing F averaged over the nodes in F.

Given a root and a fixed set S of nodes as leaves, an optimal cost tree $T(r, S)$ can be computed in time $poly(n)2^{|S|}$ by a (non-trivial) algorithm of Dreyfus and Wagner [8].

Lemma 10. *A minimum density set in $\mathcal{C}^{\phi}_{S,X}$ can be found in time $n^{O(\max(\phi, N/\phi))}$.*

Proof. There are at most $2n^{\phi}$ subsets of at most ϕ terminals and at most N/ϕ choices for a root from the set C. Given a potential root r_0 and candidate core S, the algorithm of [8] computes $T(r_0, S)$ in time $poly(n)2^{2N/\phi}$.

Our algorithm for DST is based on guessing the right ϕ-core C, and then computing a greedy set cover of $\mathcal{C}^{\phi}_{S,X}$ by repeatedly applying Lemma 10. More precisely, we try all possible subsets $S \subset V$ of size at most $2N/\phi$ as a ϕ-core (of T_{opt}) and for each such set do the following. Set U initially as X, representing the uncovered terminals. Find a min-density set Z of $\mathcal{C}^{\phi}_{S,U}$ and a corresponding optimal cost tree (with some root in S), remove Z from U and repeat until U is empty. We then compute $T(r,S)$ and combine it with all the computed subtrees into a single tree T_S. The solution output, T_{Alg}, is the T_S of smallest total cost, over all the candidate cores S.

Theorem 5. *Let $\gamma \geq 1/2$ be a parameter, $\phi = N^{1-\gamma}$, and let C be a ϕ-core of T_{opt}. Then the greedy set cover algorithm applied to $\mathcal{C}^{\phi}_{C,X}$ yields a $1+\ln\phi$-approximation of* DST. *Namely, our algorithm is a $(1-\gamma)\ln n$-approximation of* DST. *The running time is $n^{O(max(\phi,N/\phi))} = exp(\tilde{O}(N^{\gamma}))$.*

Proof. Let Gr be the size of the greedy set cover of $\mathcal{C}^{\phi}_{C,X}$ and $O = OPT_{SC}(\mathcal{C}^{\phi}_{C,X})$. Since the cardinality of the largest set in $\mathcal{C}^{\phi}_{C,X}$ is at most ϕ, it follows by the analysis of Chvátal [5] that $Gr \leq (1+\ln\phi)Opt_{SC}(\mathcal{C}^{\phi}_{C,X})$. Thus, letting $t_c = c(T(r,C))$,

$$c(T_{Alg}) \leq t_C + Gr \leq t_C + (1+\ln\phi)O \leq (1+\ln\phi)(t_C + O) = (1+\ln\phi)c(T_{Opt}) \ .$$

applying Lemma 8 in the first inequality and Lemma 9 in the (final) equality. Observe that $\ln\phi = (1-\gamma)\ln n$. For each candidate core S we find a min-density set at most n times. There are $\binom{n}{N/\phi} \leq n^{N/\phi}$ candidate cores and the cost for each is $n \cdot n^{O(\min(\phi,N/\phi))}$, by Lemma 10. Hence, the total cost is $n^{N/\phi} \cdot n^{O(\max(\phi,N/\phi))} = n^{O(N/\phi)} = exp(\tilde{O}(N^{\gamma}))$ using that $\phi = N^{1-\gamma} \leq N/\phi$.

Now we observe that the same theorem applies to the CONNECTED POLYMATROID problem. Since the function is both submodular and *increasing*, for every collection of pairwise disjoint sets $\{S_i\}_{i=1^k}$, it holds that $\sum_{i=1}^{k} f(S_i) \geq f(\bigcup_{i=1}^{k} S_i)$. Thus, for a given $\gamma \geq 1/2$, at iteration i there exists a collection S_i of terminals so that $f(S_i)/c(S_i) \geq f(U)/c(U)$. We can guess S_i in time $\exp(N^{\gamma} \cdot \log n)$ and its set of Steiner vertices X_i in time $O(3^{N^{\gamma}})$. Using the algorithm of [8], we can find a tree of density at most opt/N^{γ}. The rest of the proof is identical.

Acknowledgment. We thank Bundit Laekhanukit for helpful comments and discussions.

References

1. Bansal, N., Chalermsook, P., Laekhanukit, B., Nanongkai, D., Nederlof, J.: New tools and connections for exponential-time approximation. CoRR, abs/1708.03515 (2017)

2. Calabro, C., Impagliazzo, R., Paturi, R.: A duality between clause width and clause density for SAT. In: CCC, pp. 252–260, (2006)
3. Chalermsook, P., Laekhanukit, B., Nanongkai, D.: Independent set, induced matching, and pricing: connections and tight (subexponential time) approximation hardnesses. In: FOCS (2013)
4. Charikar, M., et al.: Approximation algorithms for directed Steiner problems. J. Algorithms **33**(1), 73–91 (1999)
5. Chvátal, V.: A greedy heuristic for the set-covering problem. Math. Oper. Res. **4**(3), 233–235 (1979)
6. Cygan, M., Kowalik, L., Wykurz, M.: Exponential-time approximation of weighted set cover. Inf. Process. Lett. **109**(16), 957–961 (2009)
7. Dinur, I., Steurer, D.: Analytical approach to parallel repetition. CoRR, abs/1305.1979 (2013)
8. Dreyfus, S.E., Wagner, R.A.: The Steiner problem in graphs. Networks **1**(3), 195–207 (1971)
9. Feige, U.: A threshold of $\ln n$ for approximating set cover. J. ACM **45**(4), 634–652 (1998)
10. Ghuge, R., Nagarajan, V.: Quasi-polynomial algorithms for submodular tree orienteering and other directed network design problems. In: SODA, pp. 1039–1048. SIAM (2020)
11. Grandoni, F., Laekhanukit, B., Li, S.: $O(\log^2 k/\log\log k)$-approximation algorithm for directed Steiner tree: a tight quasi-polynomial-time algorithm. In: STOC, pp. 253–264 (2019)
12. Halperin, E., Krauthgamer, R.: Polylogarithmic inapproximability. In: STOC, pp. 585–594 (2003)
13. Impagliazzo, R., Paturi, R., Zane, F.: Which problems have strongly exponential complexity? In: FOCS, pp. 653–663 (1998)
14. Johnson, D.S.: Approximation algorithms for combinatorial problems. J. Comput. Syst. Sci. **9**(3), 256–278 (1974)
15. Kortsarz, G., Peleg, D.: Approximating the weight of shallow Steiner trees. Discrete Appl. Math **93**, 265–285 (1999)
16. Lovász, L.: On the ratio of optimal integral and fractional covers. Discrete Math. **13**, 383–390 (1975)
17. Lund, C., Yannakakis, M.: On the hardness of approximating minimization problems. J. ACM **41**(5), 960–981 (1994)
18. Moshkovitz, D.: The projection games conjecture and the NP-hardness of $\ln n$-approximating set-cover. Theory Comput. **11**(7), 221–235 (2015)
19. Moshkovitz, D., Raz, R.: Sub-constant error probabilistically checkable proof of almost-linear size. Comput. Complex. **19**(3), 367–422 (2010). https://doi.org/10.1007/s00037-009-0278-0
20. Mukhopadhyay, P.: The projection games conjecture and the hardness of approximation of SSAT and related problems. CoRR, abs/1907.05548 (2019)
21. Naor, M., Schulman, L.J., Srinivasan, A.: Splitters and near-optimal derandomization. In: FOCS, pp. 182–191 (1995)
22. Slavík, P.: A tight analysis of the greedy algorithm for set cover. In: STOC, pp. 435–441 (1996)
23. Wolsey, L.A.: An analysis of the greedy algorithm for the submodular set covering problem. Combinatorica **2**, 385–393 (1982). https://doi.org/10.1007/BF02579435

Concave Connection Cost Facility Location and the Star Inventory Routing Problem

Jarosław Byrka and Mateusz Lewandowski[(✉)]

Institute of Computer Science, University of Wrocław, Wrocław, Poland
mlewandowski@cs.uni.wroc.pl

Abstract. We study a variant of the *uncapacitated facility location* (UFL) problem, where connection costs of clients are defined by (client specific) concave nondecreasing functions of the connection distance in the underlying metric. A special case capturing the complexity of this variant is the setting called *facility location with penalties* where clients may either connect to a facility or pay a (client specific) penalty.

We show that the best known approximation algorithms for UFL may be adapted to the concave connection cost setting. The key technical contribution is an argument that the JMS algorithm for UFL may be adapted to provide the same approximation guarantee for the more general concave connection cost variant.

We also study the *star inventory routing with facility location* (SIRPFL) problem that was recently introduced by Jiao and Ravi, which asks to jointly optimize the task of clustering of demand points with the later serving of requests within created clusters. We show that the problem may be reduced to the concave connection cost facility location and substantially improve the approximation ratio for all three variants of SIRPFL.

Keywords: Facility location · Inventory routing · Approximation

1 Introduction

The uncapacitated facility location (UFL) problem has been recognized by both theorists and practitioners as one of the most fundamental problems in combinatorial optimization. In this classical NP-hard problem, we are given a set of facilities F and a set of clients C. We aim to open a subset of facilities and connect each client to the closest opened facility. The cost of opening a facility i is f_i and the cost of connecting the client j to facility i is the distance $d_{j,i}$. The distances d are assumed to define a symmetric metric. We want to minimize the total opening costs and connection costs.

The natural generalization is a variant with penalties. For each client j, we are given its penalty p_j. Now, we are allowed to reject some clients, i.e., leave

Authors were supported by the NCN grant number 2015/18/E/ST6/00456.

C. Kaklamanis and A. Levin (Eds.): WAOA 2020, LNCS 12806, pp. 174–188, 2021.
https://doi.org/10.1007/978-3-030-80879-2_12

them unconnected and pay some fixed positive penalty instead. The objective is to minimize the sum of opening costs, connection costs and penalties. We call this problem facility location with penalties and denote as FLP.

We also study inventory routing problems that roughly speaking deal with scheduling the delivery of requested inventory to minimize the joint cost of transportation and storage subject to the constraint that goods are delivered in time. The approximability of such problems have been studied, see e.g., [13].

Recently, Jiao and Ravi [9] proposed to study a combination of inventory routing and facility location. The obtained general problem appears to be very difficult, therefore they focused on a special case, where the delivery routes are stars. They called the resulting problem the Star Inventory Routing Problem with Facility Location (SIRPFL). Formally, the problem can be described as follows. We are given a set of clients D and facility locations F with opening costs f_i and metric distances d as in the UFL problem. Moreover we are given a time horizon $1, \ldots, T$ and a set of demand points (j, t) with u_t^j units of demand for client $j \in D$ due by day t. Furthermore, we are given holding costs $h_{s,t}^j$ per unit of demand delivered on day s serving (j, t). The goal is to open a set of facilities, assign demand points to facilities, and plan the deliveries to each demand point from its assigned facility. For a single delivery on day t from facility i to client j we pay the distance $d_{j,i}$. The cost of the solution that we want to minimize is the total opening cost of facilities, delivery costs and the holding costs for early deliveries.

The above SIRPFL problem has three natural variants. In the *uncapacitated* version a single delivery can contain unlimited number of goods as opposed to *capacitated* version, where a single order can contain at most U units of demand. Furthermore, the capacitated variant can be *splittable*, where the daily demand can be delivered across multiple visits and the *unsplittable*, where all the demand (j, t) must arrive in a single delivery (for feasibility, the assumption is made that a single demand does not exceed the capacity U).

1.1 Previous Work

The metric UFL problem has a long history of results [4,6,8,12,16]. The current best approximation factor for UFL is 1.488 due to Li [10]. This is done by combining the bifactor[1] (1.11, 1.78)-approximation algorithm by Jain et al. [7] (JMS algorithm) with the LP-rounding algorithm by Byrka and Aardal [1]. The analysis of the JMS algorithm crucially utilizes a *factor revealing LP* by which the upper bound on the approximation ratio of the algorithm is expressed as a linear program. For the lower bounds, Sviridenko [15] showed that there is no better than 1.463 approximation for metric UFL unless P $\neq$ NP.

The above hardness result transfers to the penalty variant as FLP is a generalization of UFL. For approximation, the long line of research [3,5,11,18,19] stopped with the current best approximation ratio of 1.5148 for FLP. It remains

[1] Intuitively a bifactor (λ_f, λ_c) means that the algorithm pays at most λ_f times more for opening costs and λ_c times more for connection cost than the optimum solution.

open[2], whether there is an algorithm for FLP matching the factor for classical UFL without penalties.

For the SIRPFL problem, Jiao and Ravi [9] gave the 12, 24 and 48 approximation algorithms for uncapacitated, capacitated splittable and capacitated unsplittable variants respectively using LP-rounding technique.

1.2 Nondecreasing Concave Connection Costs

We propose to study a natural generalization of the FLP problem called per-client nondecreasing concave connection costs facility location (NCC-FL). The set up is identical as for the standard metric UFL problem, except that the connection cost is now defined as a function of distances. More precisely, for each client j, we have a nondecreasing concave function g_j which maps distances to connection costs. We note the importance of concavity assumption of function g_j. Dropping this assumption would allow to encode the set cover problem similarly to the non-metric facility location, rendering the problem hard to approximate better than within $\log n$.

As we will show, the NCC-FL is tightly related to FLP. We will also argue that an algorithm for NCC-FL can be used as a subroutine when solving SIRPFL problems. Therefore it serves us a handy abstraction that allows to reduce the SIRPFL to the FLP.

Table 1. Summary of improved approximation ratios

	Previous work	Our results
FLP	1.5148 [11]	1.488
NCC-FL	-	1.488
Uncapacitated SIRPFL	12 [9]	1.488
Capacitated splittable SIRPFL	24 [9]	3.236
Capacitated unsplittable SIRPFL	48 [9]	6.029

1.3 Our Results

We give improved approximation algorithms for FLP and all three variants of SIRPFL (see Table 1). Our work closes the current gap between classical facility location and FLP. More precisely, our contributions are as follows:

1. We adapt the JMS algorithm to work for the penalized variant of facility location. The technical argument relies on picking a careful order of the clients in the factor revealing program and an adequate reduction to the factor revealing program without penalties.

[2] Qiu and Kern [14] claimed to close this problem, however they withdrawn their work from arxiv due to a crucial error.

2. Then, we combine the adapted JMS algorithm with LP rounding to give the 1.488-approximation algorithm for FLP. Therefore we match the best known approximation algorithm for UFL.
3. We show a reduction from the NCC-FL to FLP which results in a 1.488-approximation algorithm for NCC-FL.
4. We cast the SIRPFL as the NCC-FL problem, therefore improving approximation factor from 12 to 1.488.
5. For the capacitated versions of SIRPFL we are also able to reduce the approximation factors from 24 (for splittable variant) and 48 (for unsplittable variant) down to 3.236 and 6.029 respectively.

The results from points 2 and 3 are more technical and follow from already known techniques, we therefore only sketch their proofs in Sects. 3 and 4, respectively. The other arguments are discussed in detail.

2 JMS with Penalties

Consider Algorithm 1, a natural analog of the JMS algorithm for penalized version. The only difference to the original JMS algorithm is that we simply freeze the budget α_j of client j whenever it reaches p_j. For brevity, we use notation $[x]^+ = \max\{x, 0\}$.

Algorithm 1. Penalized analog of JMS

1: Set budget $\alpha_j := 0$ for each client j. Declare all clients active and all facilities unopened. At every moment each client j offers some part of its budget to facility i. The amount offered is computed as follows:
   (i) if client j is not connected: $[\alpha_j - d_{i,j}]^+$
   (ii) if client j is connected to some other facility i': $[d_{i',j} - d_{i,j}]^+$
2: While there is any active client:
   – simultaneously and uniformly increase the time t and budgets α_j for all active clients until one of the three following events happen:
      (i) *facility opens*: for some unopened facility i, the total amount of offers from all the clients (active and inactive) is equal to the cost of this facility. In this case open facility i, (re-)connect to it all the clients with positive offer towards i and declare them inactive.
      (ii) *client connects*: for some active client j and opened facility i, the budget $\alpha_j = d_{i,j}$. In this case, connect a client j to facility i and deactivate client j.
      (iii) *potential runs out*: for some active client j, its budget $\alpha_j = p_j$. In this case declare j inactive.
3: Return the set of opened facilities. We pay penalties for clients that did not get connected.

Observe that in the produced solution, clients are connected to the closest opened facility and we pay the penalty if the closest facility is more distant then the penalty.

Note that Algorithm 1 is exactly the same as the one proposed by Qiu and Kern [14]. In the next section we give the correct analysis of this algorithm.

2.1 Analysis: Factor-Revealing Program

We begin by introducing additional variables t_j to Algorithm 1, which does not influence the run of the algorithm, but their values will be crucial for our analysis. Initially set all variables $t_j := 0$. As Algorithm 1 progresses, increase variables t_j simultaneously and uniformly with global time t in the same way as budgets α_j. However, whenever *potential runs out* for client j, we do not freeze variable t_j (as opposed to α_j), but keep increasing it as time t moves on. For such an inactive client j, we will freeze t_j at the earliest time t for which there is an opened facility at distance at most t from j. For other active clients (i.e. the clients that did not run out of potential) we have $t_j = \alpha_j$.

Observe now, that the final budget of a client at the end of the algorithm is equal to $\alpha_j = \min\{t_j, p_j\}$. We will now derive a factor revealing program and show that it upper-bounds the approximation factor.

Theorem 1. *Let $\lambda_f \geq 1$. Let also $\lambda_c = \sup_k o_k$, where o_k is the value of the following optimization program $P(k)$.*

$$\max \frac{\sum_i \min\{t_i, p_i\} - \lambda_f f}{\sum_i d_i} \qquad (P(k))$$

s.t.

$$\sum_{i=1}^{l-1} [\min\{r_{i,l}, p_i\} - d_i]^+ + \sum_{i=l}^{k} [\min\{t_l, p_i\} - d_i]^+ \leq f \qquad l \in [k] \quad (1)$$

$$t_i \leq t_{i+1} \qquad i \in [k-1] \quad (2)$$

$$r_{j,i} \geq r_{j,i+1} \qquad 1 \leq j < i < k \quad (3)$$

$$t_i \leq r_{j,i} + d_i + d_j \qquad 1 \leq j < i \leq k \quad (4)$$

$$r_{i,l} \leq t_i \qquad 1 \leq i < l \leq k \quad (5)$$

$$p_i \geq d_i \qquad 1 \leq i \leq k \quad (6)$$

$$t_i \geq 0, d_i \geq 0, r_{j,i} \geq 0, f \geq 0 \qquad 1 \leq j \leq i \leq k \quad (7)$$

$$p_i \geq 0 \qquad (8)$$

Then, for any solution S with facility cost F_S, connection cost D_S and penalty cost P_S Algorithm 1 returns the solution of cost at most $\lambda_f F_S + \lambda_c(D_S + P_S)$.

Proof. It is easy to see that Algorithm 1 returns a solution of cost equal to the total budget, i.e., $\sum_i \alpha_i = \sum_i \min\{t_i, p_i\}$. To show a bifactor (λ_f, λ_c) we fix λ_f and consider the $\lambda_f \cdot F_S$ of the budget as being spent on opening facilities and ask how large the resulting λ_c can become. Therefore we want to bound $\frac{\sum_i \alpha_i - \lambda_f F_S}{D_S + P_S}$.

Observe that solution S can be decomposed into set X of clients that are left unconnected and a collection $\mathcal{C}$ of stars. Each star $C \in \mathcal{C}$ consist of a single

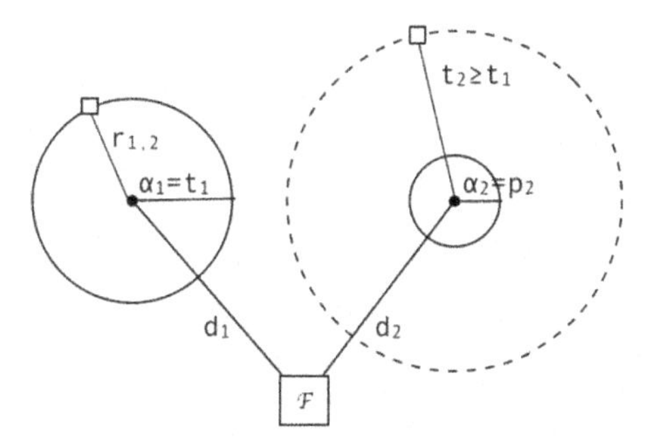

Fig. 1. The ordering of the clients with respect to t_i instead of α_i.

facility and clients connected to this facility (clients for which this facility was closest among opened facilities). Let $d(C)$ and $f(C)$ denote the connection and facility opening cost of the star C respectively. We have to bound the following:

$$\begin{aligned}\frac{\sum_i \alpha_i - \lambda_f F_S}{D_S + P_S} &= \frac{\sum_{C\in\mathcal{C}}\left(\sum_{j\in C}\alpha_j - \lambda_f f(C)\right) + \sum_{j\in X}\alpha_j}{\sum_{C\in\mathcal{C}} d(C) + P_S} \\ &\le \frac{\sum_{C\in\mathcal{C}}\left(\sum_{j\in C}\alpha_j - \lambda_f f(C)\right) + P_S}{\sum_{C\in\mathcal{C}} d(C) + P_S} \\ &\le \max_{C\in\mathcal{C}} \frac{\sum_{j\in C}\alpha_j - \lambda_f f(C)}{d(C)}\end{aligned}$$

where the first inequality comes from the fact that $\alpha_j \le p_j$ for any j. The last inequality follows because we can forget about P_S as the numerator is larger than the denominator.

Therefore we can focus on a single star C of the solution. Let $\mathcal{F}$ be the unique facility of this star and let f be the opening cost of this facility. Let also $1, 2, \ldots, k$ be the clients connected in this star to $\mathcal{F}$ and let d_i be the distance between client i and facility $\mathcal{F}$. We also assume that these clients are arranged in the nondecreasing order with respect to t_i. This is the crucial difference with the invalid analysis in [14].

For each $j < i$ we define $r_{j,i}$ as the distance of client j to the closest opened facility in a moment just before t_i. The constraint (3) is valid, because when time increases we may only open new facilities.

To understand constraint (4), i.e., $t_i \le r_{j,i} + d_i + d_j$ for $1 \le j < i \le k$, consider the moment $t = t_i$. Let $\mathcal{F}'$ be the opened facility at distance $r_{j,i}$ from j. By the triangle inequality, the distance from i to $\mathcal{F}'$ is at most $r_{j,i} + d_i + d_j$. The inequality follows from the way we defined t_i (see Fig. 1).

Constraint (5), i.e. $r_{i,l} \le t_i$ for $i < l$ follows, as client i cannot be connected to a facility of distance larger than t_i. Moreover constraint (6) is valid, as otherwise solution S' which does not connect client i to facility $\mathcal{F}$ would be cheaper.

Now we are left with justifying the opening cost constraints (1). Fix $l \in [k]$ and consider the moment just before t_l. We will count the contribution of each client towards opening $\mathcal{F}$. From Algorithm 1 the total contribution cannot exceed f. First, consider $i < l$. It is easy to see that if i was already connected to some facility, then its offer is equal to $[r_{i,l} - d_i]^+$. Otherwise, as $t_i \leq t_l$, we know that i already exhausted its potential, hence its offer is equal to $[p_i - d_i]^+$. Consider now $i \geq l$. From the description of Algorithm 1 and definition of t_i its budget at this point is equal to $\min\{t_i, p_i\} \geq \min\{t_l, p_i\}$. □

Observe that $P(k)$ resembles the factor revealing program used in the analysis of the JMS algorithm for version without penalties [7]. However $P(k)$ has additional variables p and minimas. Consider now the following program $\hat{P}(k,m)$

$$\max \frac{\sum_i m_i t_i - \lambda_f f}{\sum_i m_i d_i} \qquad (\hat{P}(k,m))$$

s.t.

$$\sum_{i=1}^{l-1} m_i \left[r_{i,l} - d_i\right]^+ + \sum_{i=l}^{k} m_i \left[t_l - d_i\right]^+ \leq f \qquad l \in [k] \qquad (9)$$

$$(2), (3), (4), (7)$$

where m is the vector of m_i's. We claim that by losing arbitrarily small ϵ, we can bound the value of $P(k)$ by the $\hat{P}(k,m)$ for some vector m of natural numbers. This is captured by the following theorem, where $v(\text{SOL}, P)$ denotes the value of the solution SOL to program P.

Theorem 2. *For any $\epsilon > 0$ and any feasible solution* SOL *of value $v(\text{SOL}, P(k))$ to the program $P(k)$, there exists a vector m of natural numbers and a feasible solution $\overline{\text{SOL}}_{k,m}$ to program $\hat{P}(k,m)$ such that $v(\text{SOL}, P(k)) \leq v(\overline{\text{SOL}}_{k,m}, \hat{P}(k,m)) + \epsilon$.*

Before we prove the above theorem, we show that it implies the desired bound on the approximation ratio. Note that the program $\hat{P}(k,m)$ is similar to the factor revealing program in the statement of Theorem 6.1 in [7] where their $k = \sum_{i=1}^{k} m_i$. The only difference is that it has additional constraints imposing that some clients are the same (we have m_i copies of each client). However, this cannot increase the value of the program. Therefore, we obtain the same bi-factor approximation as the JMS algorithm [7].

Corollary 1. *Algorithm 1 is a $(1.11, 1.78)$-approximation algorithm*[3] *for the* UFL *problem with penalties, i.e., it produces solutions whose cost can be bounded by* 1.11 *times the optimal facility opening cost plus* 1.78 *times the sum of the optimal connection cost and penalties.*

We are left with the proof of Theorem 2 which we give in the subsection below.

---

[3] see e.g., Lemma 2 in [12] for a proof of these concrete values of bi-factor approximation.

2.2 Reducing the Factor Revealing Programs

Proof (Proof of Theorem 2). First, we give the overview of the proof and the intuition behind it. We have two main steps:

- **Step 1—Getting rid of p and minimas**
  We would like to get a rid of the variables p_i and the minimas. To achieve this, we replace them with appropriate ratios z_i.
- **Step 2—Discretization**
  We then make multiple copies of each client. For some of them, we assign penalty equal to its t_i and for others—d_i. The portion of copies with positive penalty is equal to z_i.

For each step, we construct optimization programs and corresponding feasible solutions. The goal is to show, that in the resulting chain of feasible solutions and programs, the value of each solution can be upper-bounded by the value of the next solution. Formally, take any feasible solution $\text{SOL} = (t^*, d^*, r^*, p^*, f^*)$ to the program $P(k)$. In the following, we will construct solutions $\text{SOL}_1, \text{SOL}_2$ and programs P_1 and P_2 such that $v(\text{SOL}, P(k)) \leq v(\text{SOL}_1, P_1) \leq v(\text{SOL}_2, P_2) + \epsilon \leq v(\overline{\text{SOL}}_{k,m}, \hat{P}(k,m)) + \epsilon$

Step 1—Getting rid of p and Minimas

Define $z_i^* = \frac{\min\{t_i^*, p_i^*\} - d_i^*}{t_i^* - d_i^*}$ if $t_i^* - d_i^* > 0$, and $z_i^* = 0$ otherwise. Observe that $z_i^* \in [0, 1]$ as $p_i^* \geq d_i^*$ by constraint (6). We claim that $\text{SOL}_1 = (t^*, d^*, r^*, z^*, f^*)$ is a feasible solution to the following program $P_1(k)$

$$\max \frac{\sum_i d_i + z_i(t_i - d_i) - \lambda_f f}{\sum_i d_i} \qquad (P_1(k))$$

s.t.

$$\sum_{i=1}^{l-1} z_i \left[r_{i,l} - d_i\right]^+ + \sum_{i=l}^{k} z_i \left[t_l - d_i\right]^+ \leq f \qquad l \in [k] \qquad (10)$$

$$(2), (3)(4), (5), (7)$$

$$0 \leq z_i \leq 1 \qquad (11)$$

and that the value of SOL in $P(k)$ is the same as the value of SOL_1 in $P_1(k)$. The latter property follows trivially as $\sum_i d_i^* + z_i^*(t_i^* - d_i^*) = \sum_i \min\{t_i^*, p_i^*\}$. To show feasibility, we have to argue that (10) is a valid constraint for SOL_1. To this end we will use the following claim. □

Claim. For any $x \leq t_i^*$, we have that

$$\left[\min\{x, p_i^*\} - d_i^*\right]^+ \geq z_i^* \left[x - d_i^*\right]^+$$

Proof. Consider two cases:

1. $p_i^* \leq x$. In this case the left hand side is equal to $[p_i^* - d_i^*]^+$, while the right hand side is equal to $\frac{p_i^* - d_i^*}{t_i - d_i} \cdot [x - d_i^*]^+$. As $x \leq t_i^*$, the claim follows.

2. $p_i^* > x$. In this case the left hand side is equal to $[x - d_i^*]^+$, while the right hand side is equal to $z_i^* \cdot [x - d_i^*]^+$. As $z_i^* \leq 1$, the claim follows. □

Claim 2.2 together with the fact that $r_{i,l}^* \leq t_i^*$ for $i < l$ (constraint (5)) and $t_l^* \leq t_i^*$ for $l \leq i$ (constraint (2)) implies that

$$\sum_{i=1}^{l-1} \left[\min\{r_{i,l}^*, p_i^*\} - d_i^*\right]^+ + \sum_{i=l}^{k} [\min\{t_l^*, p_i^*\} - d_i^*]^+ \geq \sum_{i=1}^{l-1} z_i^* \left[r_{i,l}^* - d_i^*\right]^+ + \sum_{i=l}^{k} z_i^* [t_l^* - d_i^*]^+$$

which shows feasibility of SOL_1.

Step 2—Discretization

Take $M = N \cdot \lceil \max_{i:z_i^*>0} \frac{1}{z_i^*} \rceil$, where $N = \lceil \frac{\lambda_f f^*}{\epsilon} \rceil$. Note that the value of M depends on SOL_1.

Define now program $P_2(k, M)$, by adding to the program $P_1(k)$ the constraints $z_i \in \{0, \frac{1}{M}, \frac{2}{M}, \ldots, \frac{M}{M}\}$ for each i. We construct a solution to this program in the following way. Take $z_i' = \frac{\lceil z_i^* \cdot M \rceil}{M}$ and $f' = \frac{N+1}{N} f^*$. Let now $\mathrm{SOL}_2 = (t^*, d^*, r^*, z', f')$.

First, we claim that SOL_2 is feasible to program $P_2(k, M)$. To see this, observe that $z_i^* \leq z_i' \leq \frac{N+1}{N} z_i^*$ and $z' \in \{0, \frac{1}{M}, \frac{2}{M}, \ldots, \frac{M}{M}\}$. The feasibility follows from multiplying both sides of the constraint (10) by $\frac{N+1}{N}$.

Second, we claim that $v(\mathrm{SOL}_1, P_1(k)) \leq v(\mathrm{SOL}_2, P_2(k, M)) + \epsilon$. We have the following:

$$\begin{aligned}
v(\mathrm{SOL}_1, P_1(k)) &= \frac{\sum_i d_i^* + z_i^*(t_i^* - d_i^*) - \lambda_f f^*}{\sum_i d_i^*} \\
&\leq \frac{\sum_i d_i^* + z_i'(t_i^* - d_i^*) - \frac{N+1}{N}\lambda_f f^* + \frac{1}{N}\lambda_f f^*}{\sum_i d_i^*} \\
&\leq \frac{\sum_i d_i^* + z_i'(t_i^* - d_i^*) - \lambda_f f' + \epsilon}{\sum_i d_i^*} \\
&\leq v(\mathrm{SOL}_2, P_2(k, M)) + \epsilon
\end{aligned}$$

where in the last line we use the fact that for the normalization, the denominator $\sum_i d_i$ can be fixed to be equal 1.

Finishing the Proof

Define now $m_i = M \cdot z_i'$ and consider program $\hat{P}(k, m)$. Note, that all the m_i variables are natural numbers as required. It remains to construct the solution $\overline{\mathrm{SOL}}_{k,m}$ for $\hat{P}(k, m)$. Let $\hat{f} = M \cdot f'$ and $\overline{\mathrm{SOL}}_{k,m} = (t^*, d^*, r^*, \hat{f})$. To see that the constraint (9) is satisfied, multiply by M both sides of valid constraint (10) for SOL_2 (i.e. $\sum_{i=1}^{l-1} z_i' \left[r_{i,l}^* - d_i^*\right]^+ + \sum_{i=l}^{k} z_i' [t_l^* - d_i^*]^+ \leq f'$).

It remains to bound the value of SOL_2 with the value of $\overline{\text{SOL}}_{k,m}$:

$$\begin{aligned}
v(\text{SOL}_2, P_2(k,m)) &= \frac{\sum_i d_i^* + z_i'(t_i^* - d_i^*) - \lambda_f f'}{\sum_i d_i^*} \\
&= \frac{\sum_i d_i^* M + z_i' M(t_i^* - d_i^*) - \lambda_f M f'}{\sum_i M d_i^*} \\
&= \frac{\sum_i d_i^* M + m_i(t_i^* - d_i^*) - \lambda_f M f'}{\sum_i M d_i^*} \\
&= \frac{\sum_i m_i t_i^* - \lambda_f M f' + \sum_i (M - m_i) d_i^*}{\sum_i m_i d_i^* + \sum_i (M - m_i) d_i^*} \\
&\leq \frac{\sum_i m_i t_i^* - \lambda_f M f'}{\sum_i m_i d_i^*} \\
&= v(\overline{\text{SOL}}_{k,m}, \hat{P}(k,m))
\end{aligned}$$

where the last inequality follows from the fact that the nominator is larger than the denominator (as this fraction gives an upper bound on approximation factor which must be greater than 1). □

3 Combining Algorithms for FLP

By Corollary 1, the adapted JMS algorithm is a $(1.11, 1.78)$-approximation algorithm for the UFL problem with penalties.

It remains to note that the applicability of the LP-rounding algorithms for UFL to the FLP problem has already been studied. In particular the algorithm 4.2 of [11] is an adaptation of the LP rounding algorithms for UFL by Byrka and Aardal [1] and Li [10] to the setting with penalties.

Note also that Qiu and Kern [14] made an attempt on finalising the work on UFL with penalties and analysing the adapted JMS algorithm, and correctly argued that once the analogue of the JMS algorithm for the penalty version of the problem is known the improved approximation ratio for the variant with penalties will follow. Their analysis of the adapted JMS was incorrect (and the paper withdrawn from arxiv). By providing the missing analysis of the adapted JMS, we fill in the gap and obtain:

Corollary 2. *There exists a* 1.488*-approximation algorithm for the* FLP.

Corollary 3. *For any* $\lambda_f \geq 1.6774$ *and* $\lambda_{c+p} = 1 + \frac{2}{e^{\lambda_f}}$ *there exists a bifactor* $(\lambda_f, \lambda_{c+p})$*-approximation algorithm for* FLP.

4 Solving NCC-FL with Algorithms for FLP

We will now discuss how to use algorithms for the FLP problem to solve NCC-FL problem. To this end we introduce yet another variant of the problem: facility location with penalties and multiplicities FLPM. In this setting each client j has two additional parameters: penalty p_j and multiplicity m_j, both being nonnegative real numbers. If client j is served by facility i the service cost is $m_j \cdot d_{ij}$ and if it is not served by any facility the penalty cost is $m_j \cdot p_j$.

Lemma 1. *There is an approximation preserving reduction from* NCC-FL *to* FLPM.

Proof. Take an instance $I = (F, D, d, g)$ of the NCC-FL problem with $|F| = n$ facilities. Create the instance $I' = (F, D', d', p, m)$ as follows. The set of the facilities is the same as in original instance. For each client $j \in D$, we will create in I' multiple copies of j.

Fix a single client j. Sort all the facilities by their distance to j and let $d_1^{(j)} \leq d_2^{(j)} \leq \cdots \leq d_n^{(j)}$ be the sorted distances. For every $k \in [n-1]$ define also

$$m_k^{(j)} = \frac{g_j(d_k^{(j)}) - g_j(d_{k-1}^{(j)})}{d_k^{(j)} - d_{k-1}^{(j)}} - \frac{g_j(d_{k+1}^{(j)}) - g_j(d_k^{(j)})}{d_{k+1}^{(j)} - d_k^{(j)}} \tag{12}$$

where for convenience we define $d_0^j = g_j(d_0^{(j)}) = 0$. Let also

$$m_n^{(j)} = \frac{g_j(d_n^{(j)}) - g_j(d_{n-1}^{(j)})}{d_n^{(j)} - d_{n-1}^{(j)}} \tag{13}$$

Observe that concavity of g_j implies that every $m_k^{(j)}$ is nonnegative. Now, for each $k \in [n]$, we create a client j_k in the location of j with penalty set to $p_{j_k} = d_k^{(j)}$ and multiplicity $m_k^{(j)}$. It remains to show that for any subset of facilities $F' \subseteq F$ it holds that $cost_I(F') = cost_{I'}(F')$, where $cost_.(F')$ denotes the cost of a solution obtained by optimally assigning clients to facilities in F'. To see this, consider a client $j \in D$ and its closest facility f in F'. Let k be the index of f in the vector $d^{(j)}$ of sorted distances, i.e. the distance between j and f is $d_k^{(j)}$. Observe that for the solution F' in I' the clients $j_1, \ldots, j_i, \ldots, j_{k-1}$ pay their penalty $d_i^{(j)}$ (as the penalty is at most the distance to the closest facility) and the clients $j_k, \ldots j_n$ can be connected to f. Therefore, we require that for any k,

$$g_j(d_k^{(j)}) = \sum_{i=1}^{k-1} m_i^{(j)} d_i^{(j)} + \sum_{i=k}^{n} m_i^{(j)} d_k^{(j)} \tag{14}$$

It can be observed, that $m_i^{(j)}$ as defined in (12) and (13) satisfy the system of equalities (14). □

It remains to show how to solve FLPM. Recall that our algorithm for FLP is a combination of the JMS algorithm and an LP rounding algorithm. We will now briefly argue that each of the two can be adapted to the case with multiplicities.

To adapt the JMS algorithm, one needs to take multiplicities into account when calculating the contributions of individual clients towards facility opening. Then in the analysis multiplicities can be scaled up and discretized to lose only an epsilon factor. Here we utilize that non-polynomial blowup in the number of clients is not a problem in the analysis.

To adapt the LP rounding algorithm, we first observe that multiplicities can easily be introduced to the LP formulation, and hence solving an LP relaxation of the problem with multiplicities is not a problem. Next, we utilize that the analysis of the expected connection (and penalty) cost of the algorithm is a per-client analysis. Therefore, by the linearity of expectation combined with linearity of the objective function with respect to multiplicities, the original analysis applies to the setting with multiplicities. A similar argument was previously utilized in [2].

5 Approximation Algorithms for SIRPFL

In this section we give the improved algorithms for all the three variants of the Star Inventory Routing Problem with Facility Location that was recently introduced by Jiao and Ravi [9]. First, we recall the definition. We are given the set of facilities F and clients D in the metric space (d), a time horizon $1, \ldots, T$, a set of demand points (j, t) with u_t^j units of demand requested by client j due by day t, facility opening costs f_i, holding costs $h_{s,t}^j$ per unit of demand delivered on day s serving (j, t). The objective is to open a set of facilities and plan the deliveries to each client to minimize the total facility opening costs, client-facility connections and storage costs. Consider first the uncapacitated variant in which every delivery can contain unbounded number of goods.

Observe that once the decision which facilities to open is made, each client can choose the closest open facility and use it for all the deliveries. In that case, we would be left with a single-item lot-sizing problem which can be solved to optimality. The above view is crucial for our approach. Observe that we can precompute all the single-item lot-sizing instances for every pair (i, j). Now we are left with a specific facility location instance that is captured by NCC-FL. The following lemma proves this reduction.

Lemma 2. *There is a 1.488 approximation algorithm for uncapacitated* SIRPFL.

Proof (Proof of Lemma 2). For each $j \in D$ and $i \in F$ solve the instance of a single-item lot-sizing problem with delivery cost $d_{j,i}$ and demands and holding costs for client j to optimality [17]. Now define $g_j(d_{j,i})$ to be the cost of this computed solution and linearly interpolate other values of function g.

It is now easy to see that g_j is an increasing concave function. This follows from the fact, that the optimum solution to the lot-sizing problem with delivery cost x is also a feasible solution to the problem with delivery cost $\alpha \cdot x$ for $\alpha \geq 1$. Moreover the value of this solution for increased delivery cost is at most α times larger.

In this way we obtained the instance of NCC-FL problem which can be solved using the algorithm of Lemma 1. Once we know which facilities to open, we use optimal delivery schedules computed at the beginning. □

Jiao and Ravi studied also capacitated splittable and capacitated unsplittable variants obtaining 24 and 48 approximation respectively [9]. By using corresponding 3 and 6 approximation for the capacitated splittable and unsplittable Inventory Access Problem (IAP) given in [9] (the variants of single-item lot-sizing) and a similar reduction to NCC-FL as above while using a suitable bi-factor algorithm for FLP we are able to give improved approximation algorithms for both capacitated variants of SIRPFL.

Lemma 3. *There is a* 3.236*-approximation algorithm for capacitated splittable* SIRPFL *and* 6.029*-approximation algorithm for capacitated unsplittable* SIRPFL.

Proof (Proof of Lemma 3**).** The approach is the same as in proof of Lemma 2 but with a little twist. We give details only for splittable case as the unsplittable variant follows in the same way.

For each $j \in D$ and $i \in F$ run the 3-approximation algorithm [9] for the instance of a corresponding splittable Inventory Access Problem problem with delivery cost $d_{j,i}$ and demands and holding costs for client j.

Notice that we cannot directly define $g_j(d_{j,i})$ to be the cost of the computed solution as the resulting function would not necessarily be concave (due to using approximate solutions instead of optimal).

Therefore, we construct g_j for each $j \in D$ in a slightly different way. W.l.o.g assume that $d_{j,1} \leq d_{j,2} \cdots \leq d_{j,n}$. Let also $A(x)$ be the computed 3-approximate solution for IAP with delivery cost x and let $V(S, x)$ be the value of solution S for IAP with delivery cost x. Notice that for $x < y$, the solution $A(x)$ is feasible for the same IAP instance but with delivery cost y. In particular, the following bound on cost is true: $V(A(x), y) \leq \frac{y}{x} \cdot V(A(x), x)$.

We now construct a sequence of solutions. Let $S_1 = A(d_{j,1})$. Now, for each $i \in [n-1]$ define:

$$S_{i+1} = \begin{cases} S_i, & \text{if } V(S_i, d_{j,i+1}) < V(A(d_{j,i+1}), d_{j,i+1}) \\ A(d_{j,i+1}), & \text{otherwise} \end{cases}$$

Finally take $g(d_{j,i}) = V(S_i, d_{j,i})$ and linearly interpolate other values. It can be easily observed that g_j is a nondecreasing concave function.

Finally, we are using the bifactor $(\lambda_f, 1 + 2e^{-\lambda_f})$-approximation algorithm to solve the resulting instance of NCC-FL. Because we also lose a factor of 3 for connection cost, the resulting ratio is equal to $\max\{\lambda_f, 3 \cdot (1 + 2e^{-\lambda_f})\}$. The two values are equal for $\lambda_f \approx 3.23594$. □

References

1. Byrka, J., Aardal, K.: An optimal bifactor approximation algorithm for the metric uncapacitated facility location problem. SIAM J. Comput. **39**(6), 2212–2231 (2010)
2. Byrka, J., Skowron, P., Sornat, K.: Proportional approval voting, harmonic k-median, and negative association. In: 45th International Colloquium on Automata, Languages, and Programming (ICALP 2018). Schloss Dagstuhl-Leibniz-Zentrum fuer Informatik (2018)
3. Charikar, M., Khuller, S., Mount, D.M., Narasimhan, G.: Algorithms for facility location problems with outliers. In: Proceedings of the Twelfth Annual Symposium on Discrete Algorithms, 7–9 January 2001, Washington, DC, USA, pp. 642–651 (2001)
4. Chudak, F.A., Shmoys, D.B.: Improved approximation algorithms for the uncapacitated facility location problem. SIAM J. Comput. **33**(1), 1–25 (2003)
5. Geunes, J., Levi, R., Romeijn, H.E., Shmoys, D.B.: Approximation algorithms for supply chain planning and logistics problems with market choice. Math. Program. **130**(1), 85–106 (2011). https://doi.org/10.1007/s10107-009-0310-9
6. Guha, S., Khuller, S.: Greedy strikes back: improved facility location algorithms. J. Algorithms **31**(1), 228–248 (1999)
7. Jain, K., Mahdian, M., Markakis, E., Saberi, A., Vazirani, V.V.: Greedy facility location algorithms analyzed using dual fitting with factor-revealing LP. J. ACM **50**(6), 795–824 (2003)
8. Jain, K., Vazirani, V.V.: Approximation algorithms for metric facility location and k-median problems using the primal-dual schema and lagrangian relaxation. J. ACM **48**(2), 274–296 (2001)
9. Jiao, Y., Ravi, R.: Inventory routing problem with facility location. In: Friggstad, Z., Sack, J.-R., Salavatipour, M.R. (eds.) WADS 2019. LNCS, vol. 11646, pp. 452–465. Springer, Cham (2019). https://doi.org/10.1007/978-3-030-24766-9_33
10. Li, S.: A 1.488 approximation algorithm for the uncapacitated facility location problem. In: Aceto, L., Henzinger, M., Sgall, J. (eds.) ICALP 2011, Part II. LNCS, vol. 6756, pp. 77–88. Springer, Heidelberg (2011). https://doi.org/10.1007/978-3-642-22012-8_5
11. Li, Yu., Du, D., Xiu, N., Xu, D.: Improved approximation algorithms for the facility location problems with linear/submodular penalty. In: Du, D.-Z., Zhang, G. (eds.) COCOON 2013. LNCS, vol. 7936, pp. 292–303. Springer, Heidelberg (2013). https://doi.org/10.1007/978-3-642-38768-5_27
12. Mahdian, M., Ye, Y., Zhang, J.: Improved approximation algorithms for metric facility location problems. In: Jansen, K., Leonardi, S., Vazirani, V. (eds.) APPROX 2002. LNCS, vol. 2462, pp. 229–242. Springer, Heidelberg (2002). https://doi.org/10.1007/3-540-45753-4_20
13. Nagarajan, V., Shi, C.: Approximation algorithms for inventory problems with submodular or routing costs. Math. Program. **160**(1), 225–244 (2016). https://doi.org/10.1007/s10107-016-0981-y
14. Qiu, X., Kern, W.: On the factor revealing lP approach for facility location with penalties. arXiv preprint arXiv:1602.00192 (2016)
15. Sviridenko, M.: Personal communication. Cited in S. Guha, Approximation algorithms for facility location problems, PhD thesis, Stanford (2000)
16. Sviridenko, M.: An improved approximation algorithm for the metric uncapacitated facility location problem. In: Cook, W.J., Schulz, A.S. (eds.) IPCO 2002. LNCS, vol. 2337, pp. 240–257. Springer, Heidelberg (2002). https://doi.org/10.1007/3-540-47867-1_18

17. Wagner, H.M., Whitin, T.M.: Dynamic version of the economic lot size model. Manage. Sci. **5**(1), 89–96 (1958)
18. Xu, G., Xu, J.: An LP rounding algorithm for approximating uncapacitated facility location problem with penalties. Inf. Process. Lett. **94**(3), 119–123 (2005)
19. Xu, G., Xu, J.: An improved approximation algorithm for uncapacitated facility location problem with penalties. J. Comb. Optim. **17**(4), 424–436 (2009)

An Improved Approximation Algorithm for the Uniform Cost-Distance Steiner Tree Problem

Ardalan Khazraei[1(✉)] and Stephan Held[2(✉)]

[1] Linguistics Department, University of Potsdam, Potsdam, Germany
khazraei@uni-potsdam.de
[2] Research Institute for Discrete Mathematics, University of Bonn, Bonn, Germany
held@dm.uni-bonn.de

Abstract. The cost-distance Steiner tree problem asks for a Steiner tree in a graph that minimizes the total cost plus a weighted sum of path delays from the root to the sinks. We present an improved approximation for the uniform cost-distance Steiner tree problem, where the delay of a path corresponds to the sum of edge costs along that path.

Previous constant-factor approximation algorithms deploy well-known bicriteria results for shallow-light Steiner trees. They do not take the actual delay weights into account. Our algorithm modifies a similar algorithm for the single-sink buy-at-bulk problem by Guha et al. [7], allowing a better approximation factor for our problem. In contrast to the bicriteria algorithms it considers delay weights explicitly. Thereby, we achieve an approximation factor of $(1 + \beta)$, where β is the approximation factor for the Steiner tree problem. This improves the previously best known approximation factor for the uniform cost-distance Steiner tree problem from 2.87 to 2.39.

This algorithm can be extended to the problem where the ratio of edge costs to edge delays throughout the graph is bounded from above and below. In particular, this shows that a previous inapproximability result [15] requires large variations between edge delays and costs.

Finally, we present an important application of our new algorithm in chip design. The cost-distance Steiner tree problem occurs as a Lagrangean subproblem when optimizing millions of Steiner trees with mutually depending path length bounds. We show how to quickly approximate a continuous relaxation of this problem with our new algorithm.

1 Introduction

Steiner trees and arborescences that balance the objectives of minimum total length and shortest path lengths in the tree play an important role in the interconnect optimization of computer networks or chips [1,3,4,9,10,12,14]. In fact, the decreasing feature sizes have turned interconnect optimization into a key problem in the physical design of computer chips [1,3].

In this work, we consider a special case of the *cost-distance Steiner tree problem* that was introduced in [14]. In its general form, we are given a graph

C. Kaklamanis and A. Levin (Eds.): WAOA 2020, LNCS 12806, pp. 189–203, 2021.
https://doi.org/10.1007/978-3-030-80879-2_13

G with edge costs $c : E(G) \to \mathbb{R}_{\geq 0}$ and edge delays $d : E(G) \to \mathbb{R}_{\geq 0}$, a root $r \in V(G)$, a terminal set $T \subseteq V(G) \setminus \{r\}$, and delay weights $w : T \to \mathbb{R}_{\geq 0}$. For a tree A rooted at r spanning the terminals $T \cup \{r\} \subseteq V(G)$ we consider the objective function

$$\sum_{e \in E(A)} c(e) + \sum_{t \in T} w(t) \cdot d(E(A_{[r,t]})), \tag{1}$$

where $A_{[r,t]}$ is the unique r-t-path in A and $d(E(A_{[r,t]})) = \sum_{e \in E(A_{[r,t]})} d(e)$. We shall refer to the first term in the objective function as the *connection cost* and second term as the *delay cost.*

In [14] a $\mathcal{O}(\log |T|)$-factor approximation algorithm was presented, and in [15] it was shown that the problem cannot be approximated better than $\Omega(\log \log |T|)$ unless NP $\subseteq$ DTIME$(|T|^{\mathcal{O}(\log \log \log |T|)})$. Here, we consider the *uniform cost-distance Steiner tree problem* that arises if there is a fixed relation $\theta c = d$ for some constant $\theta \in \mathbb{R}_{>0}$ for which our algorithm provides an improved approximation from a previous constant factor. We may assume without loss of generality that the constant θ is equal 1 as it can otherwise be factored into the delay weights.

We later extend this to the *bounded-ratio cost-distance Steiner tree problem* that arises if there are constants $\Theta \geq \theta > 0$ such that $\theta c(e) \leq d(e) \leq \Theta c(e)$ for all $e \in E(G)$ and provide an approximation factor based on the parameter Θ/θ. This in particular shows that the inapproximability result in [15] relies on allowing this parameter to increase indefinitely with the input size. The conditions imply that $c(e) = 0$ if and only if $d(e) = 0$. We assume that edges with $c(e) = 0 = d(e)$ are contracted in a preprocessing step, i.e. $c, d > 0$.

As noted in [14], the bicriteria algorithm balancing minimum spanning trees and shortest path trees [12] can be used to achieve a a constant-factor approximation algorithm for the uniform cost-distance Steiner tree problem, leading to a 3.57-factor approximation. Using a variant that constructs shallow-light Steiner arborescences instead of (terminal) spanning trees [9], the approximation factor improves to 2.87 as shown implicitly in [8] and explicitly in [17].

Many approaches consider a single Steiner tree problem with bounded path lengths [1,3,9,10], and modify the bicriteria approximation algorithm for the shallow-light spanning tree problem by [4], which was later refined by [12]. Note that relaxing the bounds on the path lengths via Lagrangean relaxation results in the uniform cost-distance Steiner arborescence problem.

A related problem is the *single-sink buy-at-bulk problem* where a set of demands need to be routed from a set of sources to a single sink. K different types of pipes are available for installation on any of the edges. Each pipe type has a different cost and capacity. This problem has been extensively studied in the literature with a series of improving constant approximation factors of 292 [7], 216 [18], 153.6 [11], and 40.82 [6]. For the splittable case, the best factor is 20.41 [6]. In [18] the single-sink buy-at-bulk problem was related to the *generalized deep discount problem*, where the pipe capacities are replaced by *'incremental'* costs per unit of flow. With only a single pipe type it reduces to the uniform cost-distance Steiner tree problem. The approximation factor of the generalized deep discount problem is within a factor of two of the approximation factor for the single-sink buy-at-bulk problem [18].

1.1 Our Contribution

Here, we propose a new algorithm that achieves an approximation factor of $(1+\beta)$, where β is the approximation factor for the Steiner tree problem. Using $\beta = \ln(4)+\epsilon$ [2], this improves the best known approximation factor from 2.87 to 2.39. In contrast to the employment of the bicriteria approximation algorithm for shallow-light arborescences [8,17], our algorithm uses the delay weights explicitly.

We also show NP hardness of the uniform cost-distance Steiner tree problem for the special case $V(G) = \{r\} \cup T$, $c \equiv d : E(G) \to \{1,2\}$. This demonstrates that the hardness stems not only from the Steiner tree problem.

The bounded-ratio and uniform cost-distance Steiner tree problems have important applications. We show how they arise as Lagrangean subproblems in a fast approximation algorithm [10] for computing (fractional) Steiner trees for all nets in a VLSI netlist with mutual delay constraints minimizing the total length of all trees.

The remainder of this paper is organized as follows. In Sect. 2, we show that the uniform cost-distance Steiner tree problem is NP-hard even if $V(G) = T \cup \{r\}$, $c : E(G) \to \{1,2\}$, and $w(t) = w > 0$ $(t \in T)$. Then, in Sect. 3, we briefly summarize how to obtain a 2.87-factor approximation algorithm from bicriteria algorithms for shallow-light Steiner trees. Our main result, a $(1+\beta)$-factor approximation algorithm for the uniform cost-distance Steiner tree problem is presented in Sect. 4. Finally, a significant application in chip design is given in Sect. 6, followed by Conclusions.

2 Hardness of the Spanning Tree Case

The NP-hardness of the uniform cost-distance Steiner tree problem follows directly from the NP-hardness of the ordinary Steiner tree problem, when setting all delay weights to zero. Here, we show that the special case $V(G) = \{r\} \cup T$ and $c : E(G) \to \{1,2\}$ is already NP-hard. Thus, the difficulty arises from the two-folded objective function and not only from the Steiner tree problem.

Note that hardness of the buy-at-bulk problems can still stem from the Steiner tree problem when all vertices are demand vertices, because demands can be chosen as zero, leaving vertices unconnected. In contrast, in the cost-distance Steiner tree problem all terminals have to be connected to r regardless of their delay weight.

Theorem 1. *The cost-distance Steiner tree problem is NP-hard even if* $c : E(G) \to \{1,2\}$, $d(e) = c(e)$ $(e \in E(G))$, *and* $V(G) = \{r\} \cup T$.

Proof. We prove the statement by reduction from 3-SAT. Let $\phi = C_1 \wedge C_2 \wedge \ldots \wedge C_m$ be a 3-SAT formula on variables $x_1, x_2, \ldots, x_n$. We construct a graph G as follows:

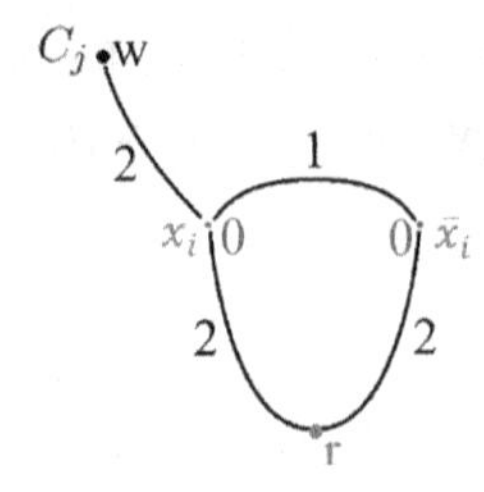

Fig. 1. Graph construction used to reduce 3SAT to uniform cost-distance spanning tree

$$V(G) := \{C_1, \ldots, C_m, x_1, \ldots, x_n, \bar{x}_1, \ldots, \bar{x}_n, r\}, \text{ and} \tag{2}$$

$$E(G) := \{\{r, x_i\}, \{r, \bar{x}_i\}, \{x_i, \bar{x}_i\} : i = 1, \ldots, n\} \cup \{\{C_j, \alpha\} : \alpha \text{ is a literal in } C_j\} \tag{3}$$

Edge costs are chosen as $c(\{x_i, \bar{x}_i\}) = 1$ for $i \in \{1, \ldots, n\}$, and $c(e) = 2$ for all other edges. The delay weights are chosen uniformly as $w(C_j) = w$ ($j \in \{1, \ldots, m\}$) for an arbitrary constant $w > 0$ and zero for all other nodes. Here, any arbitrary choice of w works as long as it is positive. Figure 1 shows exemplified how the root, one pair of literals, and a clause are connected and the edge costs and weights are chosen. We claim that ϕ is satisfiable, if and only if G has a spanning tree A with objective value at most $3n + 2m + 4mw$.

To prove the claim, assume first that we are given a satisfying truth assignment for ϕ. We will construct a desired spanning tree A from the truth assignment. Add all edges between a true literal and the root to $E(A)$ and add all edges of length one. Finally, connect each clause node to one of its true literals in the truth assignment. Then, the connection cost of the tree is $3n + 2m$. The path from each clause to the root consists of an edge of length two to the literal and another edge of length two to the root, so it has length four, resulting in the objective value of $3n + 2m + 4mw$.

For the reverse direction, observe that a connection cost of $3n + 2m$ is the cost of a minimum spanning tree and, thus, the least possible connection cost. Also, the distance of the clause nodes to the root in the graph is four, therefore the above objective value is simultaneously minimum in both factors. Thus, a tree with this value must be a minimum spanning tree and it selects all edges of length one. So at most one literal per variable is adjacent to the root. The connected literals give us a truth assignment. Now the spanning tree must also contain a shortest r-C_i-path for all $i = 1, \ldots, m$. Consequently, each clause node must be connected to a literal adjacent to the root, i.e. a true literal. This means each clause evaluates to true and ϕ is satisfiable. □

By Theorem 1, the uniform cost-distance Steiner tree problem in graphs is NP-hard even when restricting to the case where $T = V(G) \setminus \{r\}$ rendering the Steiner tree to become a spanning tree. We therefore turn our attention to approximation algorithms.

3 Approximation Factors from Shallow-Light Steiner Trees

Previous constant-factor approximation algorithms for the uniform cost-distance Steiner tree problem are based on bicriteria approximation algorithms for shallow-light Steiner trees. For a rooted Steiner tree, being shallow refers to bounding path lengths in the tree between terminals and the root, and being light refers to bounding the sum of edge lengths of the tree. Instead of delay weights w, they are given a delay bound $\bar{d}(t)$ for every $t \in T$. Sometimes this is a uniform deadline $\bar{d}(t) = \Delta \in \mathbb{R}_{\geq 0}$ [4], or the distance from the root $\bar{d}(t) = \text{dist}_{(G,c)}(r,t)$ [12].

These algorithms [3,4,8,9,12] take as input an initial approximate minimum length Steiner tree A_0 and a parameter $\delta > 0$. Then, they proceed roughly as follows. They traverse A_0 along an Eulerian walk in the directed graph $\overleftrightarrow{A_0}$ which has each edge of A_0 replaced by two opposite arcs, starting at r and build an auxiliary graph H that is initialized with A_0. Whenever they enter a terminal $t \in T$ for which the current r-t-distance in H w.r.t. d is greater than $(1+\delta)\bar{d}(t)$, they add a shortest path in (G,d) to H. Finally, they return a shortest path tree in H. The length of the resulting tree is at most $\left(1+\frac{2}{\delta}\right)$ times the length of the input tree, and it violates the delay bounds by at most a factor $(1+\delta)$ (see [8] for more details).

We can employ a simple lower bound for the minimum total cost of a cost-distance Steiner tree that is also valid in the non-uniform case. The connection cost is minimized for a Steiner minimum tree T^{SMT} in (G,c) and the delay cost is minimized for a shortest path tree T^{SP} in (G,d). Thus, with $C_{SMT} := \sum_{e \in E(T^{SMT})} c(e)$ and $D_{SP} := \sum_{t \in T} w_t \text{dist}_{(G,d)}(r,t)$ a lower bound for the objective value is

$$C_{SMT} + D_{SP}. \tag{4}$$

As proposed but not explicitly conducted in [14], we can use a bicriteria algorithm to derive a constant factor approximation for the uniform cost-distance problem. The bicriteria algorithm returns a solution with total cost $(1+\frac{2}{\delta})\beta C_{SMT} + (1+\delta)D_{SP}$. Now choose δ such that $(1+\delta) = (1+\frac{2}{\delta})\beta$ (see also [8,17]), equalizing the approximation factors for the delay and for the total length. Using a spanning tree for A_0 [12], where $\beta = 2$, this yields an approximation factor of 3.57. With the currently best known approximation factor $\beta = \ln(4) + \epsilon$ [2], this gives an approximation factor of 2.87.

Note that the bicriteria algorithm does not take the actual delay weights w into account. It returns the same solutions for $w \equiv 0$ or $w \nearrow \infty$, for which the optimum solutions would be T^{SMT} or approach T^{SP}.

Algorithm 1: COST-DISTANCE STEINER TREE APPROXIMATION

Input : A graph G, $c, d : E(G) \to \mathbb{R}_{\geq 0}$, $T \subset V(G)$, $r \in V(G)$, a weight threshold $\mu > 0$.

Output: A Steiner arborescence on the terminal set T rooted at r.

1 $A_0 :=$ β-approximate minimum cost Steiner arborescence for $T \cup \{r\}$ in $(\overleftrightarrow{G}, c)$ rooted at r, satisfying $\delta^+(t) = \emptyset$ for all $t \in T$, $|\delta^+_{A_0}(v)| = 2$ for all $v \in V(A_0) \setminus T$.

2 $H := A_0$; $\mathcal{T} = \emptyset$.

3 **for** $v \in V(A_0)$ *in reverse topological order* **do**

4 Let H_v be the sub-arborescence of H rooted at v.

5 **if** $w(V(H_v) \cap T) > \mu$

6 $T' := V(H_v) \cap T$; $A' := H_v$.

7 $E(H) := E(H) \setminus ((E(A') \cup \delta^-(v))$.

8 $\mathcal{T} = \mathcal{T} \cup \{T'\}$.

9 **for** $T' \in \mathcal{T}$ **do**

10 Let A' be the Steiner tree connecting T' according to Step 6 or $(T', \emptyset)$ if $|T'| = 1$.

11 Select a "port" vertex $t \in T'$ and a shortest r-t-path P in (G, d) such that the arborescence A'' rooted at r, which results from re-orienting $P + A'$, minimizes

$$\sum_{e \in E(A'')} c(e) + \sum_{t \in T'} w(t) \cdot d(E(A''_{[r,t]})). \tag{5}$$

$E(H) := E(H) \cup E(A'')$.

12 Return shortest path arborescence in H w.r.t. d rooted at r and covering T.

4 Improved Approximation for the Uniform Cost-Distance Steiner Tree Problem

We show how to improve the approximation factor for the uniform cost-distance Steiner tree problem by employing the delay weights algorithmically.

Algorithm 1 shows the overall algorithm, which we will now describe in detail. The solution to the cost-distance Steiner tree problem is a tree. Our algorithm formally computes a Steiner arborescence in $\overleftrightarrow{G}$ rooted at r, where $\overleftrightarrow{G}$ is the directed graph G with each edge replaced by two oppositely oriented arcs of cost $c(\overleftarrow{e}) := c(\overrightarrow{e}) := c(e)$ and delay $d(\overleftarrow{e}) := d(\overrightarrow{e}) := d(e)$ for $e \in E(G)$. It can easily be turned into the desired Steiner tree by neglecting its orientation. Note that the description of the algorithm uses separate functions for cost and delay that we use later for the bounded-ratio case in Sect. 5. However, the analysis of the uniform case in this section assumes $c \equiv d$.

We partition the terminal sets into clusters of bounded delay weight based on a threshold parameter $\mu > 0$. To this end, we first compute a light Steiner tree for $T \cup \{r\}$ in (G, c) by a β-factor approximation algorithm and turn it into an arborescence A_0 rooted at r. By inserting dummy vertices and edges of zero cost and zero delay, as well as short-cutting Steiner vertices with out-degree one, we

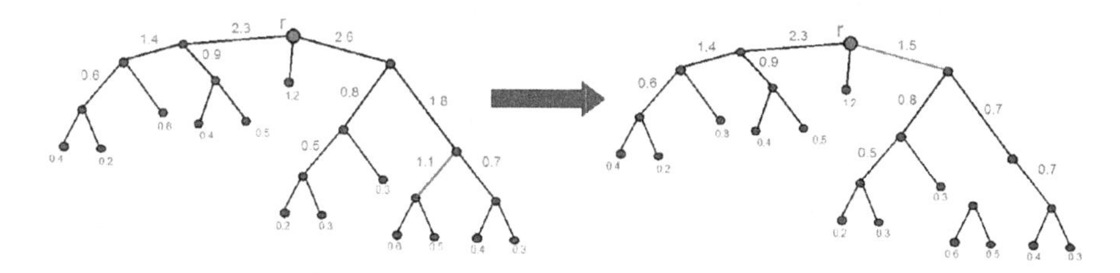

Fig. 2. Splitting off sub-arborescences of aggregate delay weight higher than the threshold μ, here with $\mu = 1$. Traversing in reverse topological order we remove the red edge into a sub-arborescence with delay weight $1.1 > \mu$. (Color figure online)

can assume that A_0 is a full binary tree, i.e. $|\delta^+_{A_0}(v)| = 2$ $(v \in V(A_0) \setminus T)$, with leave set T. Here, we use $\delta^+_{A_0}(v)$ and $\delta^-_{A_0}(v)$ to denote the set of edges leaving vertex v and the set of edges entering vertex v in the graph A_0, respectively. Note that formally A_0 is not a subgraph of $\overleftrightarrow{G}$ any more.

Then, we traverse the vertices of A_0 in reverse topological order. Whenever the delay weight of the terminals spanned by the sub-arborescence H_v rooted at the currently visited vertex $v \in V(A_0)$ is larger than μ, we split off the sub-arborescence rooted at v as in Fig. 2. The sub-arborescences are stored for later reuse.

This clustering results in a partition of $T = T_r \dot{\cup} T_1 \dot{\cup} \ldots \dot{\cup} T_k = T_r \dot{\cup} \dot{\bigcup}_{T' \in \mathcal{T}} T'$ and arborescences $A_1, \ldots, A_k$, where A_i connects the terminals in T_i $(i = 1, \ldots, k)$ rooted at some vertex r_i where the sub-arborescence was cut off such as in Fig. 3. i.e., r_i corresponds to some vertex v and T_i to some set T' according to Step 6. T_r consists of the terminals that are still residing in the component H_r when the for-loop in Step 3 terminates.

The following proposition provides a key characteristic of the clusters.

Proposition 1. *Each cluster $T' \in \mathcal{T}$ generated in Step 6 is either a heavy singleton $\{t\}$ $(t \in T)$ with delay weight $w(t) > \mu$ or a cluster $T' \in \mathcal{T}$ with delay weight $\mu < w(T') \leq 2\mu$.*

Proof. If a visited vertex is not a heavy singleton, then its at most two sub-arborescences have weight at most μ. Otherwise, they would have been split off. In turn, the entering edge is removed only if $w(T') > \mu$. □

After the clusters are determined, we connect each cluster $T' \in \mathcal{T}$ to the root r in Step 11. To this end we consider each $t \in T'$ as a possible "port" vertex for which we select an r-t-path P minimizing $d(E(P))$. The other vertices in T' are connected through the initial sub-arborescence A' after re-orienting arcs if necessary. For each cluster, we pick a port vertex minimizing the total cost ((5)) for the cluster. Finally, we return a shortest path tree in H w.r.t. the edge delays rooted in r and containing all terminals T.

Next, we analyze the cost caused by a cluster after choosing the terminal and shortest path through which we connect it in Step 11.

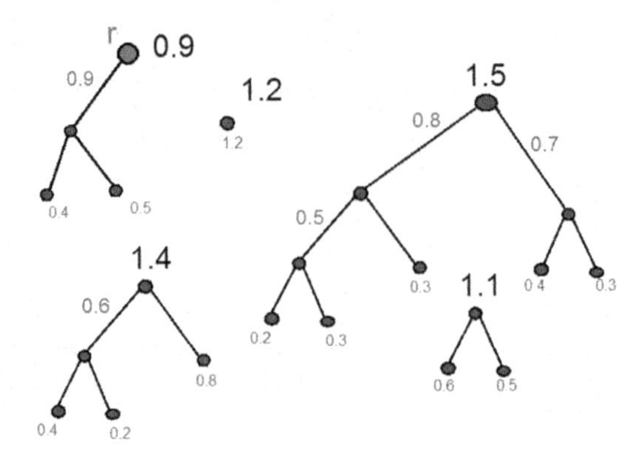

Fig. 3. The clusters at the end of the splitting-off process with $\mu = 1$.

Lemma 1. *Let $T' \in \mathcal{T}$, and let A' be the sub-arborescence stored with T'. In the uniform case $c \equiv d$, the terminal $t \in T'$ and the path P chosen in Step 11 provide a shortest path tree in $P \cup A'$ that has a total cost of at most*

$$(1 + \frac{1}{\mu}) \sum_{t \in T'} w(t) \mathrm{dist}_{(G,d)}(r, t) + (1 + \mu) c(E(A')).$$

Proof. If T' is a singleton $\{t\}$, the statement becomes evident since $c(E(A')) = 0$, $\frac{w(t)}{\mu} > 1$ and $c \equiv d$. Otherwise, let us randomly choose a "port" terminal $t \in T'$ together with a shortest r-t-path P in (G, d). We will show that the expected value of this choice satisfies the desired cost bound. Thus, the deterministic best choice of terminal and path in the algorithm must also satisfy this bound.

Let $W := w(T')$. The terminal $t \in T'$ will be selected with probability p_t proportional to its delay weight:

$$p_t := \frac{w(t)}{W}.$$

The expected value of the connection cost can be upper bounded by the expected value of $c(E(P))$ plus $c(E(A'))$. The expected value of the delay cost can be upper bounded by the expected value of $W \cdot d(E(P))$ plus the expected delay cost for serving $T' \setminus \{t\}$ through A'.

The delay cost on P has the expected value

$$\mathbb{E}\left(W \cdot d(E(P))\right) = W \sum_{t \in T'} p_t \cdot \mathrm{dist}_{(G,d)}(r, t) = \sum_{t \in T'} w(t) \cdot \mathrm{dist}_{(G,d)}(r, t).$$

This is the delay cost of a shortest path tree connecting T' to the root, i.e. the minimum possible delay cost for T'.

For the delay cost through A', consider an edge $(u, v) \in E(A')$ in the sub-arborescence A' before re-orienting it. The edge separates the sub-arborescence A'_v of $A' - (u, v)$ containing v from the rest of A'. We compute the expected

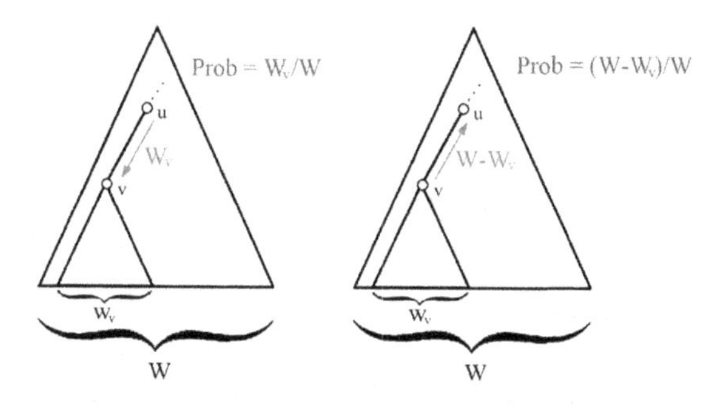

Fig. 4. The delay weight on the edge (u, v) is W_v with probability (W_v/W) if the port terminal is chosen from A'_v. It is $W - W_v$ with probability $(W - W_v)/W$ if the port is chosen from $A' \setminus A'_v$.

delay cost contribution of (u, v). It is the product of $d(u, v)$ and the delay weight served through (u, v) or (v, u) if it is re-oriented. Let $T'_v := V(A'_v) \cap T'$ and $W_v := w(T'_v)$.

With probability $\frac{W_v}{W}$ one of the terminals in T'_v is chosen by the random selection as the port t. In this case, the delay weight served by (u, v) equals the sum of the weights of all other terminals, $W - W_v$. Otherwise, the port is chosen from $T' \setminus T'_v$ and the delay weight served by (u, v) (after reverting it) is W_v (see Fig. 4). Therefore, the expected delay weight served by (u, v) is

$$\frac{W_v}{W} \cdot (W - W_v) + \frac{W - W_v}{W} \cdot W_v = \frac{2W_v(W - W_v)}{W} \leq \frac{W}{2} \leq \mu,$$

where the last inequality follows from Proposition 1. This means that the delay cost in A' has expected value at most $\mu \sum_{e \in E(A')} d(e) = \mu d(E(A'))$.

The connection cost of P has expected value

$$\begin{aligned}\mathbb{E}\left(c(E(P))\right) &= \sum_{t \in T'} \frac{w(t)}{W} d(E(P_t)) = \sum_{t \in T'} \frac{w(t)}{W} \mathrm{dist}_{(G,d)}(r, t) \\ &\leq \frac{1}{\mu} \sum_{t \in T'} w(t) \mathrm{dist}_{(G,d)}(r, t),\end{aligned}$$

where P_t denotes a shortest r-t-path in (G, d), and we are using that $W \geq \mu$ and $c(E(P) = d(E(P))$. As the expected cost of serving T' through a randomly chosen t is at most

$$\sum_{t \in T'} w(t) \mathrm{dist}_{(G,d)}(r, t) + \mu c(E(A')) + \frac{1}{\mu} \sum_{t \in T'} w(t) \mathrm{dist}_{(G,d)}(r, t) + c(E(A'))$$

$$= \left(1 + \frac{1}{\mu}\right) \sum_{t \in T'} w(t) \mathrm{dist}_{(G,d)}(r, t) + (1 + \mu) c(E(A')),$$

this cannot be exceeded by the cheapest choice for t in Step 11. $\square$

Theorem 2. *Given an instance $(G, c \equiv d, w)$ of the uniform cost-distance Steiner tree problem and a threshold parameter $\mu > 0$, a Steiner tree A, whose objective value (1) is at most*

$$(1+\mu)\beta C_{SMT} + (1+\frac{1}{\mu})D_{SP},$$

where $C_{SMT} := \sum_{e \in E(T^{SMT})} c(e)$ and $D_{SP} := \sum_{t \in T} w_t \text{dist}_{(G,d)}(r,t)$, can be computed in $\mathcal{O}(\Lambda + |T|^2 + m + n \log n)$ time. Here, Λ is the running time for computing a β-approximate minimum Steiner tree for $T \cup \{r\}$, $n = |V(G)|$, and $m = |E(G)|$.

Proof. We run Algorithm 1. The resulting objective value is given by the cost of the root component H_r, just before the for loop in Step 9 starts adding clusters in $\mathcal{T}$, plus the cost of all added clusters. As the total delay weight in any sub-arborescence of $H_r - r$ is at most μ, the total cost of H_r is at most $\mu d(E(H_r)) + c(E(H_r)) = (\mu + 1)c(E(H_r))$.

As we use edges from the initial approximate minimum Steiner tree at most once in either $E(H_r)$ or one of the edge sets $E(A')$ of the stored sub-arborescences, we can simply add the above cost of H_r and the cluster costs according to Lemma 1. This gives the desired solution cost of at most

$$(1+\mu)\beta C_{SMT} + (1+\frac{1}{\mu})D_{SP}.$$

The running time of the algorithm is given by the running time $\Lambda + \mathcal{O}(n)$ for computing the β-approximate minimum Steiner tree and transforming it into the arborescence A_0 plus the running time for clustering and selecting the port terminal and shortest path for all clusters. The clustering can be accomplished in time $\mathcal{O}(|E(A_0)|) \leq \mathcal{O}(|T|)$.

Rerouting a cluster tentatively connects a cluster through a terminal in the cluster. To this end we once compute a shortest path tree rooted at r covering all terminals using Dijkstra's algorithm with Fibonacci heaps in time $\mathcal{O}(m+n\log n)$ [5]. As the arborescence stored for the cluster has size $\mathcal{O}(|T|)$, the total runtime of all the rerouting steps is $\mathcal{O}(|T|^2 + m + n\log n)$. Thus, the overall running time is bounded by $\mathcal{O}(\Lambda + |T|^2 + m + n\log n)$. □

Corollary 1. *The uniform cost-distance Steiner tree problem can be approximated in polynomial time within a factor $(1+\beta)$, where β is the Steiner approximation ratio.*

Proof. Let OPT be the total cost of an optimum solution. Setting $\mu = \frac{1}{\beta}$ yields a solution with total cost at most

$$(1+\mu)\beta C_{SMT} + (1+\frac{1}{\mu})D_{SP} = (1+\beta)(C_{SMT} + D_{SP}) \leq (1+\beta)OPT.$$

□

Using [2] to compute A_0 with $\beta = \ln(4)+\epsilon \leq 1.39$ we get the following result.

Corollary 2. *The uniform cost-distance Steiner tree problem can be approximated within a factor 2.39 in polynomial time.*

Many practical applications [1,3,10] require a faster running time than provided by the Steiner tree approximation algorithm in [2]. We can lower the quality of the initial Steiner arborescence to get the following result.

Corollary 3. *Given an instance (G, c, d, w) of the uniform cost-distance Steiner tree problem, a Steiner tree A, whose cost is at most 3 times the optimum cost can be computed in $\mathcal{O}(|T|^2 + m + n\log n)$, where $n = |V(G)|$ and $m = |E(G)|$.*

Proof. Compute A_0 by a terminal spanning tree algorithm in time $\mathcal{O}(m+n\log n)$ [13]. Thereby, $\beta = 2$ and the approximation factor is 3. □

Algorithm 1 is similar to [7] for the single-sink buy-at-bulk problem with a single pipe type ($K = 1$), we would like to point out a few key differences. Both start from a light arborescence and split off sub-arborescences that are heavy w.r.t. the delay weight and demand respectively. However, we reuse the initial sub-arborescence to connect terminals within a cluster. To this end, we introduce a novel computation of the expected delay cost across the edges of the stored sub-arborescences. In contrast, new pipes between the initial root of the sub-arborescence and the port vertex are inserted in [7]. Furthermore, our clustering enforces an upper bound of 2μ on the delay weights (except for heavy singletons), which is crucial to obtain the desired approximation factor.

5 Bounded-Ratio Cost-Distance Steiner Trees

The approximation algorithms in Sects. 3 and 4 for the uniform cost-distance Steiner can be extended to the bounded-ratio cost-distance Steiner tree problem, where $\theta c(e) \leq d(e) \leq \Theta c(e)$ for two constants $\theta \leq \Theta$. For the bicriteria algorithm (and Algorithm 1), we compute the initial light Steiner tree w.r.t. (G, c), i.e. aiming for minimum connection cost. Then, for the bicriteria algorithm in Sects. 3, the splitting into sub-instances is then conducted based on delays in (G, d).

Similarly, the choice of the port vertex $t \in T'$ in Step 11 of Algorithm 1 is done based on the combined cost function (4). With a β-approximate Steiner tree A_0 and a delay violation tolerance $\delta > 0$, the bicriteria algorithm computes a solution of objective value at most

$$\left(1 + \frac{2}{\delta}\frac{\Theta}{\theta}\right)\beta C_{SMT} + (1+\delta)D_{SP}. \tag{6}$$

For given C_{SMT} and D_{SP}, we can choose δ such that the ratio of the final cost (6) to the trivial lower bound (4) is minimized. This leads to a choice of $\delta = \sqrt{2\beta\frac{C_{SMT}}{D_{SP}}\frac{\Theta}{\theta}}$ and to the approximation factor

$$\frac{D_{SP} + \beta C_{SMT} + 2\sqrt{2\beta C_{SMT} D_{SP} \frac{\Theta}{\theta}}}{D_{SP} + C_{SMT}}. \tag{7}$$

It turns into an approximation factor of $\frac{1}{2}\left(\beta + 1 + \sqrt{(\beta-1)^2 + 8\beta\frac{\Theta}{\theta}}\right)$, independent of C_{SMT} and D_{SP}. The same value arises when choosing δ such that $(1 + \frac{2}{\delta}\frac{\Theta}{\theta})\beta = 1 + \delta$.

Similarly, following the proof from Sect. 4, Algorithm 1 achieves a solution of cost at most

$$(1 + \mu\Theta)\beta C_{SMT} + (1 + \frac{1}{\mu\theta})D_{SP}.$$

Setting $\mu = \frac{1}{\sqrt{\beta \frac{C_{SMT}}{D_{SP}} \frac{\Theta}{\theta}}}$, results in an approximation factor of

$$\frac{D_{SP} + \beta C_{SMT} + 2\sqrt{\beta C_{SMT} D_{SP} \frac{\Theta}{\theta}}}{D_{SP} + C_{SMT}},$$

which compared to (7) avoids a factor 2 under the square root. We can state the following generalization of Theorem 2.

Theorem 3. *Given an instance (G, c, d, w) of the bounded-ratio cost-distance Steiner tree problem, a Steiner tree A, whose objective value (1) is at most $\frac{\beta+1+\sqrt{(\beta-1)^2+4\beta\frac{\Theta}{\theta}}}{2}$ times the optimum, can be computed in $\mathcal{O}(\Lambda + |T|^2 + m + n\log n)$ time, where Λ is the running time for computing a β-approximate minimum Steiner tree for $T \cup \{r\}$, $n = |V(G)|$, and $m = |E(G)|$.*

For $\theta = \Theta$, it matches the previous ratio $(1 + \beta)$.

6 Application: Delay Constrained Steiner Netlists

We now present an interesting application of (uniform) cost-distance Steiner trees in chip design. Here, Steiner trees need to be chosen for millions of interconnects subject to mutual signal delay constraints. The graph G can be considered a (possibly 3-dimensional) grid graph with holes. Holes arise from blockages that prevent interconnect routing.

The Steiner tree instances and their mutual delay dependencies are defined through a hypergraph D with $V(D) \subseteq V(G)$, which is often called *netlist*. Vertices in $V(D)$ represent start and end points of signal paths or small logic gates such as two-way AND or OR functions. Edges $E(D)$ represent interconnect nets. Each edge $e \in E(D)$ has a designated root $r_e \in e$ $(\subseteq V(G))$, while $T_e := e \setminus \{r_e\}$ is its sink set. This implies an orientation of the hyperedges from the root to the sinks. A path is an alternating sequence of vertices and hyperedges $v_1, e_1, v_2, e_2, \ldots, e_k, v_{k+1}$ such that v_i is the root of e_i and v_{i+1} is a sink of e_i. As signal paths are intercepted by registers where signal paths start and end, the hypergraph is acyclic, i.e. there are no cycles $v_1, e_1, v_2, e_2, \ldots, e_k, v_{k+1} = v_1$.

Each vertex $v \in V(D)$ has a fixed delay $d(v) \in \mathbb{R}_{\geq 0}$, where d is extended from $E(G)$ to $d : E(G) \cup V(D) \rightarrow \mathbb{R}_{\geq 0}$.

Given a Steiner tree $A \subseteq G$ connecting the root r_e of a hyperedge $e \in E(D)$ with its sinks T_e, the delay from r_e to a sink $v \in T_e$ is defined as $d_A(r_e, v) := d(r_e) + \sum_{f \in E(A_{[r_e,v]})} d(f)$.

The *delay-constrained Steiner netlist problem* is to compute a Steiner tree A_e for each $e \in E(D)$, minimizing the total connection cost $\sum_{e \in E(D)} c(E(A_e))$ such that all (maximal) paths $v_1, e_1, v_2, e_2, \ldots, e_k, v_{k+1}$ of D satisfy

$$\sum_{i=1}^{k} d_{A_{e_i}}(v_i, v_{i+1}) \leq \Delta,$$

where Δ is the given cycle time of the chip.

Using vertex potentials $a : V(D) \rightarrow \mathbb{R}$ for longest paths, it is easy to see that the path delay constraints can be represented in a compact form using $|V(D)|$ variables and $2|V(D)| + \sum_{e \in E(D)} |e|$ constraints:

$$\begin{aligned} a(v) &\geq 0 && (v \in V(D) : \delta^-(v) = \emptyset) \\ a(v) &\leq \Delta && (v \in V(D) : \delta^+(v) = \emptyset) \\ a(r_e) + d_{A_e}(r_e, v) &\leq a(v) && (e \in E(D), v \in e \setminus \{r_e\}). \end{aligned} \tag{8}$$

The delay-constrained Steiner netlist problem is a special case of the timing-constrained global routing problem, where G is additionally equipped with edge capacities $u : E(G) \rightarrow \mathbb{R}_{\geq 0}$. Now, the Steiner trees also have to be packed w.r.t. the edge capacities. Recently, a global resource sharing model for timing-constrained global routing was proposed by [10]. It considers the feasibility problem for the constraints (8), the global routing capacities, and a guessed upper bound on the total wire length. In this model $C := |V(D)| + |E(D)|$ customers/variables consume from $R := \sum_{e \in E(D)} |e| + |E(G)| + 1$ resources/inequality constraints. The algorithm by [16] efficiently approximates an optimum solution to the fractional relaxation via Lagrangean relaxation and multiplicative-weight updates. The output is a convex combination of Steiner trees for each $e \in E(D)$. The fractionality arises as the average of finitely many discrete solutions.

However, it requires an approximate oracle for the cost-distance Steiner tree problem that occurs as a Lagrangean subproblem, where prices $c : E(G) \rightarrow \mathbb{R}_{\geq 0}$, penalizing routing congestion and the total tree length, and prices $w : \cup_{e \in E(D)}(e \setminus r_e) \rightarrow \mathbb{R}_{\geq 0}$, penalizing signal delays, are maintained by the resource sharing algorithm. Details can be found in [10,16]. In this general setting, [14] still provides the best known approximation factor $\mathcal{O}(\log |T|)$, while only oracles for constantly bounded terminal sizes or heuristics were presented by [10].

The special case of neglecting the packing aspect, i.e. $u \equiv \infty$, is still relevant. Several recent practical algorithms for delay-bounded Steiner trees in chip design [1,3] compute delay-constrained Steiner trees on a net-by-net basis without considering mutual delay dependencies or routing congestion. We can improve over such approaches by directly considering delay constraints on signal paths.

In this special case, the resource sharing algorithm [10,16] maintains a single price penalizing the total length of all trees and delay prices $w : \cup_{e \in E(D)}(e \setminus r_e) \rightarrow \mathbb{R}_{\geq 0}$ so that the required oracle function corresponds to a uniform cost-distance Steiner tree problem with $\theta c(e) = d(e)$ $(e \in E(G))$ for some θ depending on the current resource price for the total length. Using Theorem 2 as an oracle for [10,16] yields the following results.

Theorem 4. *Given $\omega > 0$, we can compute a fractional solution to the delay-bounded Steiner netlist problem that violates the path delay constraints by at most a factor $(1 + \omega)2.39$ and deviates from the length of an optimum solution by at most a factor $(1 + \omega)2.39$. The running time of the algorithm is*

$$\mathcal{O}\left(|E(D)|(m + n \log n) + (\Lambda + |e_{\max}|n) \log R \left((C + R) \log \log R + \frac{(C + R)}{\omega}\right)\right),$$

where $m := |E(G)|$ and $n := |V(G)|$.

Proof. The approximation factor follows immediately from Theorem 8 in [16] and Theorem 2. The resource sharing algorithm [16] makes $\mathcal{O}\left(\log R \left((C + R) \log \log R + (C + R)\omega^{-1}\right)\right)$ calls to the oracle. For each hyperedge $e \in E(D)$, the shortest path trees w.r.t. (G, c) needs to be computed only once in the beginning so that the running time (taking $\mathcal{O}(|E(D)|(m + n \log n))$ time in total) per oracle call is only $(\Lambda + |e_{\max}|n)$. □

Note that the number of oracle calls is near-linear in the instance size, leading to a fast algorithm in practice. While the solution is only fractional, it can be discretized using randomized rounding as in [16]. The models and algorithms in [10,16] constitute state-of-the-art algorithms for global routing on industrial instances. Thus, our improved approximation ratio for uniform cost-distance Steiner trees is of theoretical relevance and likewise also of practical interest.

As modern metal stacks exhibit routing layers with different metal widths and spacings, it would be reasonable to have different cost and delay parameters on different layers of the 3d grid graph G. As the number of layers is typically bounded by a small constant 10–16, the oracle to solve the Lagrangean subproblem would correspond to the bounded-ratio cost-distance Steiner tree problem, for which Algorithm 1 also provides a constant approximation factor.

7 Conclusion

We presented an improved approximation factor $(1 + \beta)$ for the uniform cost-distance Steiner tree problem, which for $\beta = \ln(4) + \epsilon$ [2] results in a factor of 2.39. The best previous factor based on bicriteria algorithms for shallow-light Steiner arborescences was 2.87. We achieve this by using delay weights explicitly.

The practical importance of the bounded-ratio and uniform cost-distance Steiner tree problems is demonstrated applications from chip design, where these problems occur as Lagrangean subproblems. In the application Steiner trees for all nets of minimum total length have to be computed subject to mutual delay dependencies.

References

1. Alpert, C.J., et al.: Prim-Dijkstra revisited: achieving superior timing-driven routing trees. In: International Symposium on Physical Design (ISPD 2018), pp. 10–17 (2018)
2. Byrka, J., Grandoni, F., Rothvoß, T., Sanità, L.: Steiner tree approximation via iterative randomized rounding. J. ACM **60**(6), 1–33 (2013)
3. Chen, G., Young, E.F.: SALT: provably good routing topology by a novel Steiner shallow-light tree algorithm. IEEE Trans. Comput. Aided Des. Integr. Circ. Syst. **39**(6), 1217–1230 (2019)
4. Cong, J., Kahng, A.B., Robins, G., Sarrafzadeh, M., Wong, C.K.: Provably good performance-driven global routing. IEEE Trans. Comput. Aided Des. Integr. Circ. Syst. **11**(6), 739–752 (1992)
5. Fredman, M.L., Tarjan, R.E.: Fibonacci heaps and their uses in improved network optimization problems. J. ACM **34**, 596–615 (1987)
6. Grandoni, F., Rothvoß, T.: Network design via core detouring for problems without a core. In: International Colloquium on Automata, Languages, and Programming (ICALP 2010), pp. 490–502 (2010)
7. Guha, S., Meyerson, A., Mungala, K.: A constant factor approximation for the single sink edge installation problem. SIAM J. Comput. **38**(6), 2426–2442 (2009)
8. Guo, L., Zou, N., Li, Y.: Approximating the shallow-light Steiner tree problem when cost and delay are linearly dependent. In: Symposium on Parallel Architectures, Algorithms and Programmng (PAAP 2014), pp. 99–103 (2014)
9. Held, S., Rotter, D.: Shallow-light Steiner arborescences with vertex delays. In: Goemans, M., Correa, J. (eds.) IPCO 2013. LNCS, vol. 7801, pp. 229–241. Springer, Heidelberg (2013). https://doi.org/10.1007/978-3-642-36694-9_20
10. Held, S., Müller, D., Rotter, D., Scheifele, R., Traub, V., Vygen, J.: Global routing with timing constraints. IEEE Trans. Comput.-Aided Des. Integr. Circ. Syst. **37**(2), 406–419 (2018)
11. Jothi, R., Raghavachari, B.: Improved approximation algorithms for the single-sink buy-at-bulk network design problems. J. Discrete Algorithms **7**(2), 249–55 (2009)
12. Khuller, S., Raghavachari, B., Young, N.: Balancing minimum spanning and shortest path trees. Algorithmica **14**(4), 305–321 (1995)
13. Mehlhorn, K.: A faster approximation algorithm for the Steiner problem in graphs. Inf. Process. Lett. **27**, 125–128 (1988)
14. Meyerson, A., Munagala, K., Plotkin, S.: Cost-distance: two metric network design. SIAM J. Comput. **38**(4), 1648–1659 (2008)
15. Chuzhoy, J., Gupta, A., Naor, J., Sinha, A.: On the approximability of some network design problems. ACM Trans. Algorithms (TALG) **4**(2), 1–17 (2008)
16. Müller, D., Radke, K., Vygen, J.: Faster min-max resource sharing in theory and practice. Math. Prog. Comput. **3**, 1–35 (2011)
17. Rotter, D.: Timing-constrained global routing with buffered Steiner trees. Dissertation thesis. University of Bonn (2017)
18. Talwar, K.: The single-sink buy-at-bulk LP has constant integrality gap. In: International Conference on Integer Programming and Combinatorial Optimization (IPCO 2002), pp. 475–486 (2002)

A Constant-Factor Approximation Algorithm for Red-Blue Set Cover with Unit Disks

Raghunath Reddy Madireddy(✉) and Apurva Mudgal

Department of Computer Science and Engineering,
Indian Institute of Technology Ropar, Rupnagar, Punjab, India
{raghunath.reddy,apurva}@iitrpr.ac.in

Abstract. The main contribution of this paper is the first constant factor approximation algorithm for red-blue set cover problem with unit disks. To achieve this, we first give a polynomial time algorithm for line-separable red-blue set cover problem with unit disks. We next obtain a factor 2 approximation algorithm for strip-separable red-blue set cover problem with unit disks. Finally, we obtain a constant factor approximation algorithm for red-blue set cover problem with unit disks by combining our algorithm for the strip-separable problem with the results of Ambühl et al. [1].

Our methods involve a novel decomposition of the optimal solution to line-separable problem into blocks with special structure and extensions of the sweep-line technique of Erlebach and van Leeuwen [9].

Keywords: Red-blue set cover · Unit disks · Line-separable · Strip-separable · Sweep-line method

1 Introduction

In this paper, we consider the geometric red-blue set cover problem defined below:

Problem 1. ***Red-Blue Set Cover with Unit Disks.*** *Let $\mathcal{R}$ and $\mathcal{B}$ be sets of red and blue points, respectively, in the plane. Further, let $\mathcal{D}$ be a set of unit disks which cover all points in $\mathcal{B}$. Find a subset of disks $\mathcal{D}' \subseteq \mathcal{D}$ such that disks in $\mathcal{D}'$ cover all points in $\mathcal{B}$ and cover the minimum number of points in $\mathcal{R}$ (see an instance in Fig. 1(a)). We denote the instance with $(\mathcal{R}, \mathcal{B}, \mathcal{D})$.*

We also consider two special cases of the above problem (both with unit disks): line-separable red-blue set cover and strip-separable red-blue set cover. These problems restrict the locations of disk centers and blue points with respect

A. Mudgal—Partially supported by grant No. SB/FTP/ETA-434/2012 under DST-SERB Fast Track Scheme for Young Scientist.

C. Kaklamanis and A. Levin (Eds.): WAOA 2020, LNCS 12806, pp. 204–219, 2021.
https://doi.org/10.1007/978-3-030-80879-2_14

to a horizontal line or a horizontal strip. These problems are motivated by similar special cases of the discrete unit disk cover problem which are solvable in polynomial time [1,7].

Problem 2. ***Line-Separable Red-Blue Set Cover with Unit Disks.*** *This is a special case of Problem 1 where all blue points in $\mathcal{B}$ are below a horizontal line l and all disk centers are above line l. There are no restrictions on the positions of the red points. See Fig. 1(b).*

Problem 3. ***Strip-Separable Red-Blue Set Cover with Unit Disks.*** *This is a special case of Problem 1 where all blue points in $\mathcal{B}$ are inside a horizontal strip $\mathcal{T}$ and all disk centers are outside strip $\mathcal{T}$. Further, red points in $\mathcal{R}$ can lie anywhere on the plane. See Fig. 1(c).*

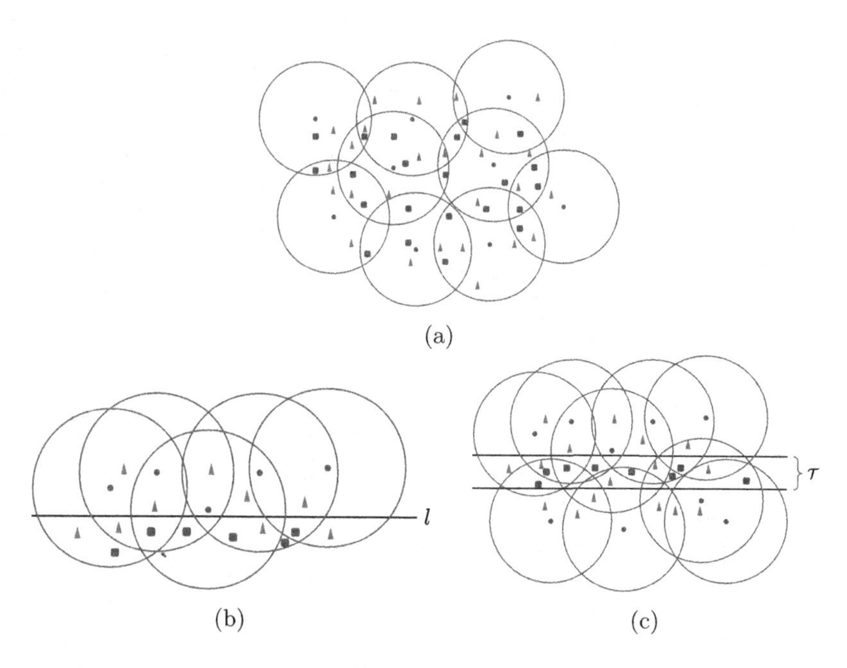

Fig. 1. Blue points are represented by squares, red points by triangles, and disk centers by filled circles. (a) Instance of red-blue set cover problem. (b) Instance of line-separable red-blue set cover problem. (c) Instance of strip-separable red-blue set cover problem. (Color figure online)

In the following, we assume that m denotes the number of points in $\mathcal{R} \cup \mathcal{B}$ and n denotes the number of disks in $\mathcal{D}$.

Applications. Several real-world scenarios can be modeled as the red-blue set cover with unit disks problem. One such case can occur in wireless sensor networks used for security purposes. Suppose we are given a set of potential locations to place sensors of uniform circular transmission range to regularly monitor a set of suspicious users in the plane. However, these users are surrounded by some fair users. Monitoring a fair user is associated with some penalty as it damages the privacy of the fair user. Hence, we need to place the sensors at a subset of given potential locations such that all suspicious users are within the transmission range of at least one sensor and the minimum number of fair users are within the transmission range of any sensor that is placed. In this model, the cost of establishing the network is assumed to be negligible as compared with the penalty for breaking the privacy of fair users. This problem can be modeled as red-blue set cover problem with disks as range of sensors, red points are fair users, and blue points are suspicious users.

Previous Work: Red-blue set cover problem (with arbitrary sets) was introduced by Carr et al. [5]. In this problem, we are given two sets of elements $\mathcal{R}$ (red) and $\mathcal{B}$ (blue). Further, we are given a family of subsets $\mathcal{A}$ of $\mathcal{R} \cup \mathcal{B}$. The goal is to find a subset $\mathcal{A}' \subseteq \mathcal{A}$ such that (1) all elements in $\mathcal{B}$ are covered by $\mathcal{A}'$ and (2) the minimum number of elements in $\mathcal{R}$ are covered by $\mathcal{A}'$. Carr et al. [5] proved that unless $P = NP$, the problem (even when every set in $\mathcal{A}$ contains three elements, one from $\mathcal{R}$ and two from $\mathcal{B}$) cannot be approximated within $2^{\log^{1-\delta} n}$ factor where $\delta = 1/\log^c \log n$ and $c < 1/2$. Note that Problem 1 is a geometric special case of the above problem.

We now describe the related previous work on dynamic programming methods using sweep-line for the following problems: (a) geometric set cover problem with unit disks and (b) geometric set cover and red-blue set cover problems with unit squares.

(a) *Geometric set cover with unit disks* (also known as discrete unit disk cover problem). In this problem, we are given an instance $(\mathcal{P}, \mathcal{D})$ where $\mathcal{P}$ is a set of points in the plane and $\mathcal{D}$ is a set of unit disks which covers all points in $\mathcal{P}$. One has to find a minimum size subset $\mathcal{D}' \subseteq \mathcal{D}$ which covers all points in $\mathcal{P}$. This problem is known to be NP-hard [10]. Further, several dynamic programming based algorithms with approximation factor 108 [3], 72 [1], 38 [4], 22 [7], 18 [8], 15 [11], and $(9 + \epsilon)$ [2] have been proposed. Line-separable and strip-separable versions of geometric set cover with unit disks were defined during this line of work and they have polynomial-time algorithms [1,7]. Recently, Li and Jin [12] gave a polynomial time approximation scheme (PTAS) for weighted version of geometric set cover with unit disks.

(b) *Geometric set cover and red-blue set cover problems with unit squares.* These problems are variations of geometric set cover and red-blue set cover with unit disks where unit disks are replaced with unit squares. Both problems are known to be NP-hard [6,10]. First, Erlebach and van Leeuwen [9] gave a PTAS for the weighted version of the geometric set cover with unit squares using a sweep-line method. Later, Chan and Hu [6] came up with a PTAS for red-blue set cover with unit squares using mod-one technique and sweep-line method. We note that, one can prove that Problem 1 is NP-hard by doing necessary modifications to the redcution of Chan and Hu [6].

On a related note, PTAS-es can be obtained by local search for both geometric set cover with unit disks and unit squares [14].

Our Results: The results proved in this paper are listed below:

1. A polynomial time algorithm for Problem 2 (see Sect. 2).
2. A 2-factor approximation algorithm for Problem 3 (see Sect. 3).
3. A 368-factor approximation algorithm for Problem 1 (see Sect. 4).

Although our constant factor for red-blue set cover with unit disks is high but we note that this is the first constant factor algorithm for this important problem. Further we hope that our work opens up a line of research which leads to better and better approximations for red-blue set cover with unit disks problem.

The following questions are left open by our work:

1. Is there a polynomial-time algorithm for strip-separable red-blue set cover with unit disks ? A polynomial-time algorithm is known for strip-separable discrete unit disk cover problem [1].
2. Is there a PTAS for red-blue set cover with unit disks ?

The rest of this paper is organized as follows. In Sect. 2, we give a $O(n^{14}m)$-time exact algorithm for Problem 2 (Theorem 3). In Sect. 3, we give a factor 2 approximation algorithm for Problem 3 (Theorem 4). Finally, in Sect. 4, we describe the constant factor algorithm for Problem 1 (Theorem 5).

2 Polynomial-Time Algorithm for Line-Separable Red-Blue Set Cover

In this section we give a $O(n^{14}m)$-time algorithm for Problem 2. We are given an instance $(\mathcal{R}, \mathcal{B}, \mathcal{D})$ of red-blue set cover problem and a horizontal line l such that the centers of disks in $\mathcal{D}$ are above line l and points in $\mathcal{B}$ are below line l. Further, points in $\mathcal{R}$ can lie on the both sides of line l. Let l^+ and l^- be the regions above and below the line l respectively. We say that two disks D and D' intersect at point p *iff* the boundaries of both disks D and D' pass through p. Similarly, by the arc of a disk D, we mean an arc of the boundary of D. Let g and h be two points on the boundary of a disk D. We denote the arc of D from g to h in clockwise direction by $arc(g, h, D)$.

Let $\mathcal{O} \subseteq \mathcal{D}$ be an optimal solution for the considered problem. We say that a point q is *reachable* from a point p if there exists a path (curve) P between p and q such that P does not intersect any disk in $\mathcal{O}$ or line l. Let p_0 be a point in region l^- such that the vertical distance from p_0 to line l is at least one. Let $\mathcal{X}_{lower}$ be the open set of points which are reachable from p_0. We call the boundary of region $\mathcal{X}_{lower}$ the lower envelope $\mathcal{E}_{\mathcal{O}}^-$ of disks in $\mathcal{O}$. The lower envelope consists of a sequence of arcs of disk boundaries and segments of line l

(see Fig. 2). Further, exactly one continuous arc of every disk in $\mathcal{O}$ contributes to lower envelope $\mathcal{E}_{\mathcal{O}}^-$ and we call the arc as the appearance of the disk on $\mathcal{E}_{\mathcal{O}}^-$. We denote the smallest and largest x-coordinates of the appearance of a disk D on envelope $\mathcal{E}_{\mathcal{O}}^-$ by $start(D)$ and $end(D)$ respectively.

Let $D_1, D_2, \ldots, D_{|\mathcal{O}|}$ be the disks in $\mathcal{O}$ ordered according to their appearances on the lower envelope $\mathcal{E}_{\mathcal{O}}^-$ from left to right i.e., $end(D_i) \leq start(D_{i+1})$ for $i = 1, 2, \ldots, |\mathcal{O}| - 1$. We note that $(D_{i-1} \cap D_{i+1} \cap l^-) \subseteq (D_i \cap l^-)$. We denote the x-coordinate of left intersection point of D and line l by $L(D)$ and the x-coordinate of right intersection point of D and line l by $R(D)$. One can note that $L(D_1) < L(D_2) < \cdots < L(D_{|\mathcal{O}|})$ and $R(D_1) < R(D_2) < \cdots < R(D_{|\mathcal{O}|})$.

Let q_0 be a point in region l^+ such that the vertical distance from q_0 to line l is at least two. Let $\mathcal{X}_{upper}$ be the open set of points which are reachable from q_0. We call the boundary of region $\mathcal{X}_{upper}$ the upper envelope $\mathcal{E}_{\mathcal{O}}^+$ of disks in $\mathcal{O}$. Not all disks from $\mathcal{O}$ appear on $\mathcal{E}_{\mathcal{O}}^+$ (see Fig. 2).

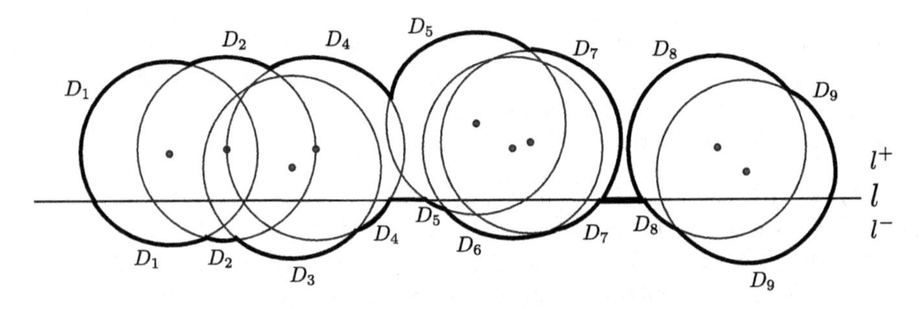

Fig. 2. The union of the thick arcs and line segment (in black color), in region l^-, defines the lower envelope of the disks. $D_1, D_2, D_3, D_4, D_5, D_6, D_7, D_8$, and D_9 be the order of disks on the lower envelope according to their appearance from left to right. Further, the upper envelope is defined by thick arcs of disks and line segments in region l^+. The sub-sequence $D_1, D_2, D_4, D_5, D_7, D_8$, and D_9 forms the upper envelope. (Color figure online)

For any two intersecting disks D_a and D_b in $\mathcal{O}$ $(1 \leq a < b \leq |\mathcal{O}|)$, we denote the intersection points of D_a and D_b with $p^1_{a,b}$ and $p^2_{a,b}$ where the y-coordinate of $p^1_{a,b}$ is greater than the y-coordinate of $p^2_{a,b}$. For any two disks D_a and D_b in $\mathcal{O}$ which intersect twice in region l^+, let H_{ab} denote the closed region bounded by $arc(p^2_{a,b}, R(D_a), D_a)$, $arc(L(D_b), p^2_{a,b}, D_b)$, and the horizontal line segment $\overline{R(D_a)L(D_b)}$. We call H_{ab} the *hole* induced by disks D_a and D_b (Fig. 3).

We note that the existing polynomial-time algorithm for line-separable discrete unit disk cover with unit disks [7] cannot be extended to our line-separable red-blue set cover with unit disks. This is due to, in the case of our problem, red-points can lie on both sides of line l. Further, the behavior of the upper envelope is not the same as the lower envelope i.e., every disk D in an optimal solution covers a unique interval $[start(D), end(D)]$ in region l^-, but the same

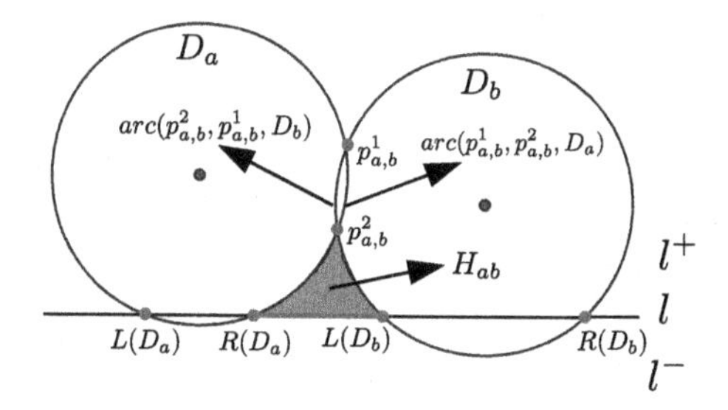

Fig. 3. The shaded region denotes the hole H_{ab} induced by D_a and D_b. (Color figure online)

is not true for the disks which appear on the upper envelope. Furthermore, two consecutive disks on the upper envelope may create a hole in region l^+, and it can contain some red points. Further, the red points in the hole can be covered by disks which are non-adjacent in the order $D_1, D_2, \ldots, D_{|\mathcal{O}|}$. The main goal is to ensure that no red point covered above line l is either missed or double counted.

Our algorithm for the line-separable problem is based on a new theorem about the structure of an optimal solution which we call the *block decomposition theorem*. The theorem essentially states that the disks in an optimal solution to line-separable red-blue set cover with unit disks problem can be partitioned into a sequence of special structures which we call blocks of types $\mathscr{A}$, $\mathscr{B}$, $\mathscr{C}$, and $\mathscr{D}$ (see Sect. 2.1 for the exact statement). The blocks divide the optimal solution into non-interacting pieces: the red points covered by disks in a block except the first and last disk are not covered by disks in any other block.

2.1 Statement of Block Decomposition Theorem

We start this section with definition of a block and possible special structures of disks in a block. Then the statement of block decomposition theorem follows.

Definition 1 ***(Block).*** *A block of the optimal solution $\mathcal{O}$ is a contiguous subsequence $D_i, D_{i+1}, \ldots, D_j$ of disks from $\mathcal{O}$ such that the region in l^+ covered by disks $D_{i+1}, D_{i+2}, \ldots, D_{j-1}$ lies completely inside $(D_i \cup D_j \cup H_{ij}) \cap l^+$. In region l^-, the block spans the region between $x = end(D_i)$ and $x = start(D_j)$.*

An example is illustrated in Fig. 4.

We now describe four types of blocks $\mathscr{A}$, $\mathscr{B}$, $\mathscr{C}$, and $\mathscr{D}$. These block types have additional structure which is used by the sweep-line method in Sect. 2.2.

Block of type $\mathscr{A}$. Disks D_i and D_j do not intersect and hence $j = i + 1$.
Block of type $\mathscr{B}$. Disks D_i and D_j intersect exactly once in region l^+.
Block of type $\mathscr{C}$. Disks D_i and D_j intersect twice in region l^+. Further, there exists a disk $D' \in \{D_{i+1}, \ldots, D_{j-1}\}$ such that D' completely covers the hole H_{ij} induced by D_i and D_j.

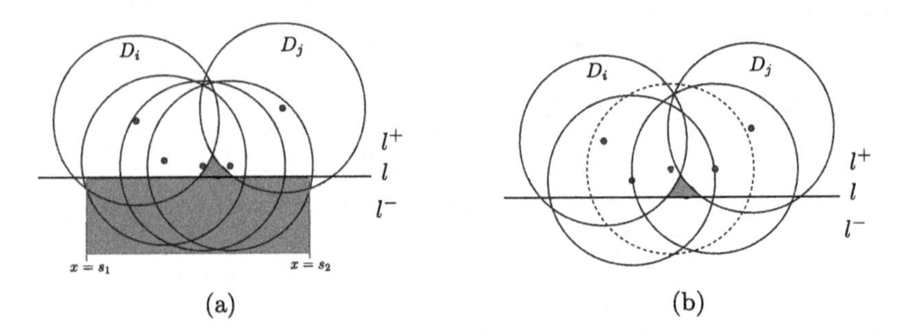

Fig. 4. (a) An example of a block and the span of block in region l^- between $x = s_1$ and $x = s_2$. (b) An example of a set of disks which do not form a block since the disk with dotted boundary covers region outside $D_i \cup D_j \cup H_{ij}$. (Color figure online)

Block of type $\mathscr{D}$. As in type $\mathscr{C}$ block, disks D_i and D_j intersect twice in region l^+. Unlike type $\mathscr{C}$ block, there exists no disk $D' \in \{D_{i+1}, \ldots, D_{j-1}\}$ such that D' completely covers hole H_{ij}.

In this case, no disk in $\{D_{i+1}, \ldots, D_{j-1}\}$ intersect $\overline{R(D_i)L(D_j)}$ twice. Then we can partition set $\{D_{i+1}, \ldots, D_{j-1}\}$ into two sets $\mathcal{O}_1 = \{D_{i+1}, \ldots, D_{k'}\}$ and $\mathcal{O}_2 = \{D_{k'+1}, \ldots, D_{j-1}\}$ where $D_{k'}$ is the rightmost disk with $R(D_{k'})$ on segment $\overline{R(D_i)L(D_j)}$. First, we note that there can exist disks in $\mathcal{O}_1$ (and also in $\mathcal{O}_2$) which do not cover any intersection point of D_i and D_j. Further, any two such disks can cover the red points inside the hole formed by D_i and D_j in region l^+. In the following, based on the behaviour of disks in $\mathcal{O}_1$ and $\mathcal{O}_2$, we categorize type $\mathscr{D}$ block into $\mathscr{D}_1$ and $\mathscr{D}_2$ blocks as follows:

Type $\mathscr{D}_1$. The disks in $\mathcal{O}_1$ and $\mathcal{O}_2$ satisfy the following properties:

1. Disk $D_{k'+1}$ completely covers the region $D_t \cap H_{ij}$ for all disks $D_t \in \mathcal{O}_2$.
2. There exists a subsequence $D_{x_1}, D_{x_2}, \ldots, D_{x_\mu}$ $(i+1 \leq x_1 < x_2 < \cdots < x_\mu \leq k')$ such that every set $\mathcal{O}_{x_t x_{t+1}} = \{D_{x_t}, D_{x_t+1}, \ldots, D_{x_{t+1}}\}$ forms a block of type $\mathscr{B}$. Further, no point in $D_{x_t} \setminus D_{x_{t-1}}$ is covered by $D_{x_1} \cup D_{x_2} \cup \cdots \cup D_{x_{t-2}}$.

Type $\mathscr{D}_2$ block is symmetric to type $\mathscr{D}_1$ block. Disk $D_{k'}$ covers the region $D_t \cap H_{ij}$ for all $i < t \leq k'$ and the disks in $\mathcal{O}_2$ can be partitioned into type $\mathscr{B}$ blocks.

The following theorem gives the statement of the block decomposition theorem. It essentially proves two things. (i) For any three disks on upper envelope $\mathcal{E}_{\mathcal{O}}^+$, the region in l^+ which is common to the both the first and last disk is completely contained in the middle disk. The correctness of the sweep-line method for line-separable problem depends on this fact. (ii) The optimal solution $\mathcal{O}$ can be partitioned into a sequence of blocks such that the last disk of the current block is the same as the first disk of next block. Further, each block is one of the four types $\mathscr{A}$, $\mathscr{B}$, $\mathscr{C}$, and $\mathscr{D}$. The proof of the following theorem can be found in [13].

Theorem 1 (***Block decomposition theorem***). *Let $D_{\sigma_1}, D_{\sigma_2}, \ldots, D_{\sigma_k}$ $(1 \leq \sigma_1 < \sigma_2 < \cdots < \sigma_k \leq |\mathcal{O}|)$ be the disks which appear on upper envelope $\mathcal{E}_{\mathcal{O}}^+$ ordered from left to right. Then the following are true.*

1. *For* $1 \leq x < y < z \leq k$, $(D_{\sigma_x} \cap D_{\sigma_z} \cap l^+) \subseteq (D_{\sigma_y} \cap l^+)$.
2. *For* $1 \leq \kappa \leq k-1$, *the set* $\mathcal{O}_{\sigma_\kappa \sigma_{\kappa+1}} = \{D_{\sigma_\kappa}, D_{\sigma_\kappa+1}, D_{\sigma_\kappa+2}, \ldots, D_{\sigma_{\kappa+1}}\}$ *is a block of one of the four types* $\mathscr{A}, \mathscr{B}, \mathscr{C}$, *and* $\mathscr{D}$.

2.2 Sweep-Line Method

We now outline our sweep-line based dynamic programming method which returns an optimal solution $\mathcal{O}$ in $O(n^{14}m)$ time for Problem 2. We first note that, by block decomposition theorem (Theorem 1), once we know the sub-sequence of disks forming the upper envelope, the hardness lies in finding optimal blocks of type $\mathscr{A}, \mathscr{B}, \mathscr{C}$, and $\mathscr{D}$ between every pair of consecutive disks in the sub-sequence. We call this optimal finding blocks as *block red-blue set cover* (see Definition 2). From the first property given in Theorem 1 and by the properties of disks in the optimal solution (about the coverage on the lower envelope), a red point covered by an optimal solution of a block cannot be covered by a disk in an optimal solution of another block except the points covered by the left and right disks.

Before giving a polynomial-time algorithm for the line-separable red-blue set cover with unit disks problem in Sect. 2.2.2, we consider a special case of the problem, called *block red-blue set cover* problem in Sect. 2.2.1 and we give a polynomial-time algorithm for this special problem. We use this polynomial-time algorithm to give a polynomial-time algorithm for the line-separable red-blue set cover problem.

2.2.1 Block Red-Blue Set Cover Problem

The block red-blue set cover problem is defined as follows.

Definition 2. *We are given two disks* D_{left} *and* D_{right} *such that* $L(D_{left}) < L(D_{right})$, *a set of disks* $\mathcal{D}_B$, *and two x-coordinates* s_1 *and* s_2 *where* $s_1 < s_2$. *A subset* $\mathcal{D}'_B$ *of* $\mathcal{D}_B$ *is a valid solution if it satisfies the following conditions:*

1. $\mathcal{D}'_B$ *covers all the blue points between* $x = s_1$ *and* $x = s_2$ *which are not covered by* D_{left} *and* D_{right}.
2. $\{D_{left}, D_{right}\} \cup \mathcal{D}'_B$ *forms a block such that* $L(D_{left}) < L(D) < L(D_{right})$ *for all* $D \in \mathcal{D}'_B$.
3. *Appearance of every disk in* $\mathcal{D}'_B$ *on the lower envelope formed by disks in* $\{D_{left}, D_{right}\} \cup \mathcal{D}'_B$ *is between* $x = s_1$ *and* $x = s_2$.

We denote this instance with $Block(D_{left}, D_{right}, s_1, s_2, \mathcal{D}_B)$. *The cost of the solution* $\mathcal{D}'_B$ *is equal to the number of red points covered by disks in* $\mathcal{D}'_B$ *which are not already covered by* D_{left} *and* D_{right}. *Our objective is to find a valid solution* $\mathcal{D}^*_B$ *with minimum cost. The optimal cost is* ∞ *if no valid solution exists.*

Since any valid solution $\mathcal{D}'_B$ forms a block with D_{left} and D_{right} as boundary disks, for all disks $D \in \mathcal{D}'_B$: (*i*) $(L(D), R(D)) \subseteq (L(D_{left}), R(D_{right}))$ and (*ii*) the portion of D in region l^+ is completely contained in $D_{left} \cup D_{right} \cup H_{lr}$ where H_{lr} is the hole formed by D_{left} and D_{right}. Note that H_{lr} can be empty. We can therefore remove all disks from $\mathcal{D}_B$ which do not satisfy the above two conditions. For any two disks D_y and D_z such that $L(D_y) < L(D_z)$, let $q_{y,z}$ be the x-coordinate of the intersection point of D_y and D_z in region l^- if both disks D_y and D_z intersect in region l^- otherwise $q_{y,z} = L(D_z)$.

In the following, we show that block red-blue set cover problem can be solved in polynomial time by using the sweep-line method similar to that of Erlebach and van Leeuwen [9]. We maintain a vertical sweep-line in region l^- which moves from $x = s_1$ to $x = s_2$. At any point of the time, the sweep-line passes through $x = q_{y,z}$ for some disks D_y and D_z in $\mathcal{D}_B$ such that $L(D_y) < L(D_z)$. The snapshot of the sweep-line position is captured in a 5-tuple $(D_{prev}, D_{curr}, D_{upper\_prev}, D_{upper\_curr}, D_{extra})$.

Definition 3. *We say a 5-tuple is valid if the following conditions hold true:*

1. $L(D_{upper\_prev}) \leq L(D_{prev}) < L(D_{curr}) \leq L(D_{upper\_curr})$ *and* $L(D_{left}) \leq L(D_{extra}) \leq L(D_{right})$.
2. $(D_{prev} \cup D_{curr}) \cap l^+ \subseteq (D_{upper\_prev} \cup D_{upper\_curr} \cup D_{extra} \cup D_{left} \cup D_{right}) \cap l^+$.

For a valid 5-tuple $(D_{prev}, D_{curr}, D_{upper\_prev}, D_{upper\_curr}, D_{extra})$, the sweep-line is at $x = q_{prev,curr}$. We note that the disks in the 5-tuple need not be distinct and further some of the disks can be empty. In the above definition, the first condition is true only for non-empty disks.

The collection of all valid 5-tuples define the vertex set of a graph G. We now put weighted directed edges between vertices in G. Let $v = (D_{prev}, D_{curr}, D_{upper\_prev}, D_{upper\_curr}, D_{extra})$ be a vertex in graph G. Let $b_1 = q_{prev,curr}$. For a disk $D_{new} \in \mathcal{D}_B$, let $b_2 = q_{curr,new}$. If $s_1 \leq b_1 < b_2 \leq s_2$ and disk D_{curr} covers all blue points between $x = b_1$ and $x = b_2$, then there are two possible outgoing edges from vertex $v = (D_{prev}, D_{curr}, D_{upper\_prev}, D_{upper\_curr}, D_{extra})$.

1. *Only one new disk appears in the new vertex.* If $v_1 = (D_{curr}, D_{new}, D_{upper\_prev}, D_{upper\_curr}, D_{extra})$ is a valid tuple, then there is a directed edge from vertex v to vertex v_1. The cost of edge (v, v_1) is the number of red points covered by D_{new} which are not already covered by any disk at vertex v, D_{left}, and D_{right}.
2. *Two new disks appear in the new vertex.* If $v_2 = (D_{curr}, D_{new}, D_{upper\_curr}, D_{upper\_new}, D_{extra})$ is a valid tuple for some $D_{upper\_new} \in \mathcal{D}_B$, then edge (v, v_2) is added to graph G. Further, the cost of edge (v, v_2) is same as the number of red points covered by D_{new} and D_{upper_new} which are already not covered by any disk at vertex v, D_{left}, and D_{right}.

Further, we add one source node S and one sink node T to graph G. We next add the following edges to graph G:

1. For every valid tuple $v_3 = (D_{left}, D_{s_1}, D_{left}, D, D_{extra})$ there is a directed edge from S to v_3 if there are no blue points between $x = s_1$ and $x = b_3$

which are not covered by D_{left} and D_{right} where $b_3 = q_{left,s_1}$. The cost of edge (S, v_3) is the number of red points covered by D, D_{s_1}, and D_{extra} which are not already covered by D_{left} and D_{right}.
2. There is a directed edge of cost zero from every valid tuple $v_4 = (D_{s_2}, D_{right}, D, D_{right}, D_{extra})$ to T if there are no blue points between $x = b_4$ and $x = s_2$ which are not covered by D_{left} and D_{right} where $b_4 = q_{s_2,right}$.
3. Further, add a directed edge of cost zero from S to T if there exist no blue point between $x = s_1$ and $x = s_2$ which are not covered by D_{left} and D_{right}.

The number of vertices in graph G is $O(n^5)$. From each vertex in G, there are $O(n^2)$ possible outgoing edges. Further, to check the validity of each possible edge and finding its cost, we require $O(m)$ time. Hence, the graph has $O(n^7)$ edges and it can be constructed in $O(n^7 m)$ time. From the construction of the edges, we can note that G is a directed acyclic graph. We note that the cost of the shortest path from S to T is equal to the cost of an optimal solution to block red-blue set cover problem (see the details in [13]). Hence, we have the following theorem.

Theorem 2. *We can compute the optimal solution for block red-blue set cover problem in $O(n^7 m)$ time.*

2.2.2 Line-Separable Red-Blue Set Cover Problem

We now use the block red-blue set cover problem as a subroutine to obtain a polynomial time algorithm for the line-separable red-blue set cover problem. As in any sweep-line based algorithm [6,9], we reduce the problem to a shortest path from source to sink on a directed acyclic graph $\mathcal{G}$. The set of vertices of $\mathcal{G}$ is a collection of all possible block red-blue set cover problems over the disks in $\mathcal{D}$. First we add two dummy disks D^1_{dummy} and D^2_{dummy} to the set $\mathcal{D}$ such that $R(D^1_{dummy}) = \min_{D \in \mathcal{D}}\{L(D)\}$ and $L(D^2_{dummy}) = \max_{D \in \mathcal{D}}\{R(D)\}$.

Any vertex in $\mathcal{G}$ is a 4-tuple $(D_{left}, D_{right}, s_1, s_2)$ where

1. D_{left} and D_{right} are disks in $\mathcal{D}$ such that $L(D_{left}) < L(D_{right})$.
2. $s_1 = q_{left,s_1}$ for some $D_{s_1} \in \mathcal{D}$ and $s_2 = q_{s_2,right}$ for some $D_{s_2} \in \mathcal{D}$.

We now define the outgoing edges from each vertex in $\mathcal{G}$. Let $V = (D_{left}, D_{right}, s_1, s_2)$ and $V' = (D_{right}, D_{next}, s'_1, s'_2)$ be two vertices in $\mathcal{G}$. There exists a directed edge from V to V' in $\mathcal{G}$ if and only if $s_2 \leq s'_1$ and all blue points between $x = s_2$ and $x = s'_1$ are covered by disk D_{right}. If block red-blue set cover problem $Block(D_{left}, D_{right}, s_1, s_2, \mathcal{D})$ has no feasible solution, then remove edge (V, V') from graph $\mathcal{G}$. The cost of edge (V, V') is equal to the optimal cost of $Block(D_{left}, D_{right}, s_1, s_2, \mathcal{D})$ plus the number of red points covered by D_{next} which are not covered by D_{right}.

We add a source S' and a sink T' to graph $\mathcal{G}$. We now add the following extra edges to graph $\mathcal{G}$. For every vertex of the form $V_1 = (D^1_{dummy}, D, R(D^1_{dummy}), L(D))$ for some $D \in \mathcal{D}$ such that no blue point exists between $x = R(D^1_{dummy})$ and $x = L(D)$, add a directed edge from S' to

V_1. The cost of edge (S', V_1) is the number of red points covered by D. Further, add a directed edge of cost zero to T' from every vertex of the form $V_2 = (D, D^2_{dummy}, R(D), L(D^2_{dummy}))$ for some $D \in \mathcal{D}$ such that no blue point exists between $x = R(D)$ and $x = L(D^2_{dummy})$.

From the above construction, we note that the size of vertex set of $\mathcal{G}$ is $O(n^4)$ and from any vertex in $\mathcal{G}$ there are at most $O(n^3)$ edges. Further, to check the covering requirement of blue points, we require $O(m)$ time and to compute the cost of the edge we require $O(n^7m)$ time (see Theorem 2). Hence, graph $\mathcal{G}$ has $O(n^7)$ edges and can be constructed in $O(n^{14}m)$ time. Hence we have the following theorem and the proof can be found in [13].

Theorem 3. *The optimal cost of line-separable red-blue set cover with unit disks is the same as the cost of the shortest path from S' to T' in graph $\mathcal{G}$ and further, can be computed in $O(n^{14}m)$ time.*

3 2-Factor Algorithm for Strip-Separable Red-Blue Set Cover

In this section, we assume that all blue points in $\mathcal{B}$ are inside a horizontal strip $\mathcal{T}$ bounded by horizontal lines l_1 and l_2 where the y-coordinate of l_1 is greater than y-coordinate of l_2. Further, no disk in $\mathcal{D}$ has center inside $\mathcal{T}$. Let l_1^+ (resp. l_2^+) and l_1^- (resp. l_2^-) be the regions above and below l_1 (resp. l_2). We note that $\mathcal{T} = l_1^- \cap l_2^+$. Let $\mathcal{D}_1$ and $\mathcal{D}_2$ be the subsets of $\mathcal{D}$ such that $\mathcal{D}_1$ is the set of disks whose centers are in region l_1^+ and $\mathcal{D}_2$ is the set of disks whose centers are in region l_2^-. Every disk in $\mathcal{D}$ intersects each line l_1 and l_2 either zero or two times. For a disk $D \in \mathcal{D}_1$, we define $L_1(D)$ (resp. $R_1(D)$) as the x-coordinate of the left (resp. right) intersection point of D and line l_1. Similarly, we define $L_2(D)$ and $R_2(D)$ for disk D in $\mathcal{D}_2$ with respect to line l_2.

For any given vertical line s, we define a disk $r \in \mathcal{D}_1$ which interests s at the lowest (resp. highest) y-coordinate among all disks in $\mathcal{D}_1$ as an *upper active* (resp. *lower active*) disk with respect to s. We note that for any given vertical line s, there exist at most one upper active disk and at most one lower active disk with respect to s. Let $\mathcal{O} \subseteq \mathcal{D}$ be an optimal solution for strip-separable case. Let $\mathcal{O}_1 \subseteq \mathcal{O}$ (resp. $\mathcal{O}_2 \subseteq \mathcal{O}$) be the set of disks whose centers are above (resp. below) the strip $\mathcal{T}$. One can obtain block decomposition of sets $\mathcal{O}_1$ and $\mathcal{O}_2$ as in Sect. 2.1. As noted above, at any sweep-line position we have at most one upper active disk $r_1 \in \mathcal{O}_1$ as well at most one lower active disk $r_2 \in \mathcal{O}_2$. Further, r_1 and r_2 belong to some blocks in the above block decomposition of sets $\mathcal{O}_1$ and $\mathcal{O}_2$ respectively. We call such blocks as *active blocks* with respect to the sweep-line position.

We now give a sweep-line based algorithm which is similar to the sweep-line based algorithm given for line-separable red-blue set cover problem in Sect. 2. We can interpret this algorithm as running two line-separable algorithms simultaneously with a common sweep-line. At any point of time, we maintain information about current (active) blocks for line-separable problems on both lines l_1 and l_2.

As before, we remember a total of seven disks, two disks which are the boundary disks of the current block, and 5 more disks for the current block (as in block red-blue set cover problem in Sect. 2.2.1) of each line l_1 and l_2.

For $i = 1, 2$, let $(D^i_{left}, D^i_{right}, D^i_{prev}, D^i_{curr}, D^i_{upper\_prev}, D^i_{upper\_curr}, D^i_{extra})$ be a valid 7-tuple of disks in $\mathcal{D}_i$ where the 5-tuple $(D^i_{prev}, D^i_{curr}, D^i_{upper\_prev}, D^i_{upper\_curr}, D^i_{extra})$ satisfies the properties of the 5-tuple in Definition 3 for a block red-blue set cover problem (see Sect. 2.2.1) which is defined by D^i_{left} and D^i_{right} $(L_i(D^i_{left}) < L_i(D^i_{right}))$. Further, let $\mathscr{V}_i$ be the set of all such possible valid 7-tuples. Since we maintain a single sweep-line, the snapshot of the current position of the sweep-line can be captured as a tuple consisting of two 7-tuples, one 7-tuple for each line l_1 and l_2.

As before, for two disks D^1_a and D^1_b in $\mathcal{D}_1$ such that $L_1(D^1_a) < L_1(D^1_b)$, define $q^1_{a,b}$ as the x-coordinate of the unique intersection point of D^1_a and D^1_b in region l_1^- if both intersect in region l_1^-, otherwise $q^1_{a,b} = L_1(D^1_b)$. Similarly, for two disks D^2_a and D^2_b in $\mathcal{D}_2$ such that $L_2(D^2_a) < L_2(D^2_b)$, let $q^2_{a,b}$ be the x-coordinate of the unique intersection point of D^2_a and D^2_b in region l_2^+ if both intersect in region l_2^+, otherwise $q^2_{a,b} = L_2(D^2_b)$.

Definition 4. *Let* $(v_1, v_2) \in \mathscr{V}_1 \times \mathscr{V}_2$[1], *where* $v_i = (D^i_{left}, D^i_{right}, D^i_{prev}, D^i_{curr}, D^i_{upper\_prev}, D^i_{upper\_curr}, D^i_{extra})$ *(for* $i = 1, 2$*) and let* $b = \max\{q^1_{prev,curr}, q^2_{prev,curr}\}$. *We say* (v_1, v_2) *is a valid tuple if and only if it satisfies the following properties:*

1. *If* $b = q^1_{prev,curr}$, *disk* D^2_{curr} *intersects the line* $x = b$ *at the highest* y*-coordinate, in region* l_2^+, *among all the disks in* v_2.
2. *If* $b = q^2_{prev,curr}$ *then among all disks in* v_1 *disk* D^1_{curr} *intersects the line* $x = b$ *at the lowest* y*-coordinate in region* l_1^-.

Further, we assume that the sweep-line is placed at $x = b$ *for the tuple* (v_1, v_2).

As the sweep-line moves to the right, there are three possibilities. (i) There is a change in the current block of line l_1, but the snapshot of the current block of line l_2 remains the same. (ii) There is a change in the current block of line l_2, but the snapshot of the current block of line l_1 remains the same. (iii) Boundary disks of both blocks remain the same, but disks in exactly one of the blocks change (few disks from the last 5 disks in a 7-tuple change).

We now construct a directed acyclic graph $\mathcal{G}_\mathcal{T}$. The vertex set of $\mathcal{G}_\mathcal{T}$ is the collection of all valid tuples from $\mathscr{V}_1 \times \mathscr{V}_2$. For every vertex $v = (v_1, v_2)$ in $\mathcal{G}_\mathcal{T}$ where $v_i = (D^i_{left}, D^i_{right}, D^i_{prev}, D^i_{curr}, D^i_{upper\_prev}, D^i_{upper\_curr}, D^i_{extra}) \in \mathscr{V}_i$ for $i = 1, 2$, we define the outgoing edges from v as follows. Let s_0 be the x-coordinate of the position of the sweep-line at vertex v. As noted above, when the sweep-line moves forward only disks in either v_1 or v_2 change but not from both v_1 and v_2. In the following, we give outgoing edges for change in v_1. The other case when there is a change in v_2 is similar.

[1] $\mathscr{V}_1 \times \mathscr{V}_2$ means the cartesian product of $\mathscr{V}_1$ and $\mathscr{V}_2$.

Let D^1_{new} be a disk in $\mathcal{D}_1$. Let $s_1 = q^1_{curr,new}$. If $s_0 < s_1$ and all blue points between $x = s_0$ and $x = s_1$ are covered by $D^1_{curr} \cup D^2_{curr}$, then three types of outgoing edges are possible from v.

1. *Active block remains same, but exactly one new disk appears.* There exists an outgoing edge from v to a vertex $v' = (v'_1, v'_2)$ (a valid tuple), where $v'_1 \in \mathscr{V}_1$ and $v'_2 = \mathscr{V}_2$, if the following conditions are satisfied.
   (a) In v_1, $D^1_{curr} \neq D^1_{upper\_curr}$ and $D^1_{curr} \neq D^1_{right}$.
   (b) $v'_1 = (D^1_{left}, D^1_{right}, D^1_{curr}, D^1_{new}, D^1_{upper\_prev}, D^1_{upper\_curr}, D^1_{extra})$ and $v'_2 = v_2$.
   The weight of edge (v, v') is equal to the number of red points covered by D^1_{new} which are not already covered by any disk in v_1.
2. *Active block remains the same, but two new disks appear.* A directed edge is added from v to a vertex $v'' = (v''_1, v''_2)$ (a valid tuple), where $v''_1 \in \mathscr{V}_1$ and $v''_2 \in \mathscr{V}_2$, if the following are true.
   (a) In v_1, $D^1_{curr} = D^1_{upper\_curr}$ and $D^1_{curr} \neq D^1_{right}$.
   (b) $v''_1 = (D^1_{left}, D^1_{right}, D^1_{curr}, D^1_{new}, D^1_{upper\_curr}, D^{1,new}_{upper\_curr}, D^1_{extra})$ for some $D^1_{upper\_curr} \in \mathcal{D}_1$ and $v''_2 = v_2$.
   The number of red points covered by D^1_{new} and $D^{1,new}_{upper\_curr}$ which are not already covered by any disk in v_1 is the weight of edge (v, v'').
3. *The active block of line l_1 changes.* We add a directed edge (v, v''') in $\mathcal{G}_T$ on the following conditions.
   (a) $v''' = (v'''_1, v'''_2)$ is a valid tuple where $v'''_1 \in \mathscr{V}_1$ and $v'''_2 \in \mathscr{V}_2$.
   (b) In v_1, $D^1_{curr} = D^1_{upper\_curr}$ and $D^1_{curr} = D^1_{right}$.
   (c) $v'''_1 = (D^1_{right}, D^{1,new}_{right}, D^1_{right}, D^1_{new}, D^{1,new}_{upper\_prev}, D^{1,new}_{upper\_curr}, D^{1,new}_{extra})$ for some $D^{1,new}_{upper\_prev}$, $D^{1,new}_{upper\_curr}$, $D^{1,new}_{extra}$, $D^{1,new}_{right} \in \mathcal{D}_1$ and $v'''_2 = v_2$.
   The edge (v, v''') is assigned weight equal to the number of red points covered by D^1_{new}, $D^{1,new}_{upper\_prev}$, $D^{1,new}_{upper\_curr}$, $D^{1,new}_{extra}$, and $D^{1,new}_{right}$ which are not already covered by D^1_{right}.

We note that the directed acyclic graph $\mathcal{G}_T$ has $O(n^{14})$ vertices and $O(n^{19})$ edges. Further, the graph can be construed in $O(n^{19}m)$ time. By suitably adding dummy disks, we can create the source S and the sink T to graph $\mathcal{G}_T$. We can also connect these vertices to other vertices in $\mathcal{G}_T$ with appropriate weight. We have the following theorem and the proof can be found in [13].

Theorem 4. *The shortest path from S to T in graph $\mathcal{G}_T$ gives a solution for Problem 3 with cost at most 2 times the cost of optimal solution in $O(n^{19}m)$ time.*

4 Red-Blue Set Cover with Unit Disks

In this section, we describe a constant factor approximation algorithm for the red-blue set cover problem (Problem 1). We proceed on the same lines of Ambühl et al. [1]. First, we give a 4-factor approximation algorithm for the case when all points in $\mathcal{B}$ are inside a rectangular box of size $\delta \times \delta$ for $\delta < 1/2$. Using this result, we obtain a 368-factor approximation algorithm for Problem 1.

Let $(\mathcal{P}, \mathcal{D})$ be an instance of geometric set cover problem (unweighted version) where all points in $\mathcal{P}$ are strictly inside a rectangular box of size $\delta \times \delta$ for some $\delta < 1/2$. For this instance Ambühl et al. [1] gave a 2-factor algorithm which consists of the following steps:

1. First guess $k = O(1)$ disks from set $\mathcal{D}$. We note that at most $K = O(n^k)$ different possible values exist for the guessed set of disks. Let $\mathcal{Z}_1, \mathcal{Z}_2, \ldots, \mathcal{Z}_K$ be the collection of all such sets of disks, each of size at most k.
2. For each i $(1 \leq i \leq K)$, do the following:
   (a) Let $\mathcal{D}_i = \mathcal{D} \setminus \mathcal{Z}_i$ and let $\mathcal{P}_i$ be the set of points in $\mathcal{P}$ which are not covered by disks in $\mathcal{Z}_i$.
   (b) Find two strip-separable set cover instances $(\mathcal{P}_{i1}, \mathcal{D}_{i1})$ and $(\mathcal{P}_{i2}, \mathcal{D}_{i2})$ such that $\mathcal{P}_{i1} \cup \mathcal{P}_{i2} = \mathcal{P}_i$, $\mathcal{P}_{i1} \cap \mathcal{P}_{i2} = \emptyset$, and $\mathcal{D}_{i1} \cup \mathcal{D}_{i2} = \mathcal{D}_i$. (The reader can refer Sect. 3 in [1] for this construction.)
   (c) Optimally solve both instances $(\mathcal{P}_{i1}, \mathcal{D}_{i1})$ and $(\mathcal{P}_{i2}, \mathcal{D}_{i2})$ independently using polynomial-time algorithm for strip-separable geometric set cover problem with unit disks [1] . Let OPT_{i1} and OPT_{i2} be the sizes of the optimal solutions of strip-separable geometric set cover instances $(\mathcal{P}_{i1}, \mathcal{D}_{i1})$ and $(\mathcal{P}_{i2}, \mathcal{D}_{i2})$ respectively.
3. Take $\min_{1 \leq i \leq K}\{|\mathcal{Z}_i| + OPT_{i1} + OPT_{i2}\}$ as the solution to the geometric set cover instance $(\mathcal{P}, \mathcal{D})$.

Ambühl et al. [1] show the correctness of the above procedure by showing that an optimal solution $\mathcal{O}$ can be broken into two strip-separable problems $(\mathcal{P}_1, \mathcal{O}_1)$, $(\mathcal{P}_2, \mathcal{O}_2)$, and a set $\mathcal{O}_3$ of maximum size k such that $\mathcal{P}_1 \cup \mathcal{P}_2$ is the subset of points in $\mathcal{P}$ which are not covered by disks in $\mathcal{O}_3$, $\mathcal{P}_1 \cap \mathcal{P}_2 = \emptyset$, $(\mathcal{O}_1 \cup \mathcal{O}_2) \cap \mathcal{O}_3 = \emptyset$, and $\mathcal{O}_1 \cup \mathcal{O}_2 \cup \mathcal{O}_3 = \mathcal{O}$. We note that $\mathcal{O}_3$ must be same as $\mathcal{Z}_{i_0}$ from some i_0 and further $\mathcal{P}_{i_0 1} = \mathcal{P}_1$ and $\mathcal{P}_{i_0 2} = \mathcal{P}_2$. Thus, $OPT_{i_0 1} = |\mathcal{O}_1|$ and $OPT_{i_0 2} = |\mathcal{O}_2|$. Since $\mathcal{O}_1$ and $\mathcal{O}_2$ need not be disjoint, the above algorithm is a 2-factor approximation algorithm. We use a similar procedure to prove the following lemma.

Lemma 1. *There exists a 4-factor algorithm for red-blue set cover with unit disks when all blue points in $\mathcal{B}$ are inside a rectangular box of size $\delta \times \delta$ for $\delta < 1/2$.*

Proof. In the following, we use $\mathscr{R}(\mathcal{X})$ to denote the region covered by disks in set $\mathcal{X}$ in the plane and $\mathscr{R}(\mathcal{X}) \setminus \mathscr{R}(\mathcal{Y})$ denotes the sub-region of $\mathscr{R}(\mathcal{X})$ obtained by removing all points in the plane from $\mathscr{R}(\mathcal{X})$ which are also in $\mathscr{R}(\mathcal{Y})$. Further, let $cost(\mathscr{R}(\mathcal{X}))$ be the number of red points in region $\mathscr{R}(\mathcal{X})$.

Let $(\mathcal{R}, \mathcal{B}, \mathcal{D})$ be an instance of red-blue set cover problem with unit disks where all points in $\mathcal{B}$ are inside a $\delta \times \delta$ rectangular box for some $\delta < 1/2$.

We proceed as follows:

1. Consider the instance $(\mathcal{B}, \mathcal{D})$ of geometric set cover with unit disks problem.

2. Use the method of Ambühl et al. [1] to obtain $O(n^k)$ pairs of instances of strip-separable geometric set cover with unit disks problem. Let the i-th problem be given by $(\mathcal{B}_{i1}, \mathcal{D}_{i1})$, $(\mathcal{B}_{i2}, \mathcal{D}_{i2})$, and $\mathcal{Z}_i$.
3. For each i, solve the two strip-separable red-blue set cover problems $(\mathcal{R}, \mathcal{B}_{i1}, \mathcal{D})$ and $(\mathcal{R}, \mathcal{B}_{i2}, \mathcal{D})$ independently by using the 2-factor algorithm in Sect. 3 without counting the red points covered by disks in $\mathcal{Z}_i$. Let $\mathcal{D}^*_{i1}$ and $\mathcal{D}^*_{i2}$ be the solutions returned by the 2-factor algorithm for both instances $(\mathcal{R}, \mathcal{B}_{i1}, \mathcal{D})$ and $(\mathcal{R}, \mathcal{B}_{i2}, \mathcal{D})$ respectively.
4. Take $value = \min_i\{cost(\mathscr{R}(\mathcal{D}^*_{i1} \cup \mathcal{D}^*_{i2} \cup \mathcal{Z}_i))\}$ and return $value$ as a solution.

We now prove the approximation factor. Let $\mathcal{O} \subseteq \mathcal{D}$ be an optimal solution (with no redundant disk) to red-blue set cover instance $(\mathcal{R}, \mathcal{B}, \mathcal{D})$. We note that the set $\mathcal{O}$ is itself an optimal solution for geometric set cover with unit disks instance $(\mathcal{B}, \mathcal{O})$. Let $(\mathcal{B}_1, \mathcal{O}_1)$, $(\mathcal{B}_2, \mathcal{O}_2)$, and set $\mathcal{O}_3$ be the optimal partition of optimal solution given by Ambühl et al. [1] as discussed above. We note that $(\mathcal{R}, \mathcal{B}_1, \mathcal{D})$, $(\mathcal{R}, \mathcal{B}_2, \mathcal{D})$, and set $\mathcal{O}_3$ form one of the pairs of strip-separable problems constructed and solved by us in the step 2 and step 3 above without counting red points covered by disks in $\mathcal{O}_3$. Let $\mathcal{D}^*_1$ and $\mathcal{D}^*_2$ be the solutions of $(\mathcal{R}, \mathcal{B}_1, \mathcal{D})$ and $(\mathcal{R}, \mathcal{B}_2, \mathcal{D})$, respectively, returned by our algorithm in step 3 above. Hence, $cost(\mathscr{R}(\mathcal{D}^*_1) \setminus \mathscr{R}(\mathcal{O}_3)) \leq 2 \cdot cost(\mathscr{R}(\mathcal{O}_1) \setminus \mathscr{R}(\mathcal{O}_3))$ and $cost(\mathscr{R}(\mathcal{D}^*_2) \setminus \mathscr{R}(\mathcal{O}_3)) \leq 2 \cdot cost(\mathscr{R}(\mathcal{O}_2) \setminus \mathscr{R}(\mathcal{O}_3))$. Further, let $\mathcal{O}'_1$ and $\mathcal{O}'_2$ be optimal solutions of strip-separable red-blue set cover problems $(\mathcal{R}, \mathcal{B}_1, \mathcal{D})$ and $(\mathcal{R}, \mathcal{B}_2, \mathcal{D})$, respectively without counting red points covered by disks in $\mathcal{O}_3$. We note that $cost(\mathscr{R}(\mathcal{O}'_1) \setminus \mathscr{R}(\mathcal{O}_3)) \leq cost(\mathscr{R}(\mathcal{O}_1) \setminus \mathscr{R}(\mathcal{O}_3))$ and $cost(\mathscr{R}(\mathcal{O}'_2) \setminus \mathscr{R}(\mathcal{O}_3)) \leq cost(\mathscr{R}(\mathcal{O}_2) \setminus \mathscr{R}(\mathcal{O}_3))$. Further, $cost(\mathscr{R}(\mathcal{O}'_1) \setminus \mathscr{R}(\mathcal{O}_3)) + cost(\mathscr{R}(\mathcal{O}'_2) \setminus \mathscr{R}(\mathcal{O}_3)) \leq 2 \cdot cost(\mathscr{R}(\mathcal{O}_1 \cup \mathcal{O}_2) \setminus \mathscr{R}(\mathcal{O}_3))$.

Therefore,

$$\begin{aligned} value &\leq cost(\mathscr{R}(\mathcal{D}^*_1 \cup \mathcal{D}^*_2 \cup \mathcal{O}_3)) \\ &\leq cost(\mathscr{R}(\mathcal{D}^*_1) \setminus \mathscr{R}(\mathcal{O}_3)) + cost(\mathscr{R}(\mathcal{D}^*_2) \setminus \mathscr{R}(\mathcal{O}_3)) + cost(\mathscr{R}(\mathcal{O}_3)) \\ &\leq 2 \cdot \{cost(\mathscr{R}(\mathcal{O}'_1) \setminus \mathscr{R}(\mathcal{O}_3)) + cost(\mathscr{R}(\mathcal{O}'_2) \setminus \mathscr{R}(\mathcal{O}_3)) + cost(\mathscr{R}(\mathcal{O}_3))\} \\ &\leq 2 \cdot \{2 \cdot cost(\mathscr{R}(\mathcal{O}_1 \cup \mathcal{O}_2) \setminus \mathscr{R}(\mathcal{O}_3)) + cost(\mathscr{R}(\mathcal{O}_3))\} \\ &\leq 4 \cdot \{cost(\mathscr{R}(\mathcal{O}_1 \cup \mathcal{O}_2 \cup \mathcal{O}_3))\} = 4 \cdot cost(\mathscr{R}(\mathcal{O})) \end{aligned}$$

Hence, the lemma is proved. □

Theorem 5. *There exists a 368-factor algorithm for Problem 1.*

Proof. We proceed as in Ambühl et al. [1]. Choose $\delta = 0.999/2$. We partition the plane into rectangular boxes, each of size $\delta \times \delta$ and solve the red-blue set cover problem independently for blue points in each box of size $\delta \times \delta$ by using Lemma 1. Hence, we conclude that if a red point is covered by solutions of at most τ boxes, then the union of solutions of each box of size $\delta \times \delta$ in the plane will give us a 4τ-factor approximation algorithm. We now show that the maximum value of τ is 92. Note that a red point r can be covered by a solution of a box of size

$\delta \times \delta$ only if the box is completely contained inside a circle of radius $(2 + \sqrt{2}\delta)$ centered at r. Hence, r can participate in solutions of $\lfloor \frac{\pi(2+\sqrt{2}\delta)^2}{\delta^2} \rfloor$ number of small boxes of size $\delta \times \delta$. Since $\delta = 0.999/2$, we conclude that red point r can be covered by solutions of at most 92 small boxes. Hence, the theorem is proved. □

References

1. Ambühl, C., Erlebach, T., Mihalák, M., Nunkesser, M.: Constant-factor approximation for minimum-weight (connected) dominating sets in unit graphs. In: 9th International Workshop on Approximation Algorithms for Combinatorial Optimization Problems, APPROX, pp. 3–14 (2006)
2. Basappa, M., Acharyya, R., Das, G.K.: Cover problem in 2D. J. Disc. Algor. **33**, 193–201 (2015)
3. Călinescu, G., Măndoiu, I.I., Wan, P.J., Zelikovsky, A.Z.: Selecting forwarding neighbors in wireless ad hoc networks. Mob. Netw. Appl. **9**(2), 101–111 (2004)
4. Carmi, P., Katz, M.J., Lev-Tov, N.: Covering points by unit disks of fixed location. In: Tokuyama, T. (ed.) ISAAC 2007. LNCS, vol. 4835, pp. 644–655. Springer, Heidelberg (2007). https://doi.org/10.1007/978-3-540-77120-3_56
5. Carr, R.D., Doddi, S., Konjevod, G., Marathe, M.: On the red-blue set cover problem. In: Proceedings of the Eleventh Annual ACM-SIAM Symposium on Discrete Algorithms, pp. 345–353. SODA '00, Society for Industrial and Applied Mathematics, Philadelphia (2000)
6. Chan, T.M., Hu, N.: Geometric red blue set cover for unit squares and related problems. Comput. Geom. **48**(5), 380–385 (2015)
7. Claude, F., Dorrigiv, R., Durocher, S., Fraser, R., López-Ortiz, A., Salinger, A.: Practical discrete unit disk cover using an exact line-separable algorithm. In: Dong, Y., Du, D.-Z., Ibarra, O. (eds.) ISAAC 2009. LNCS, vol. 5878, pp. 45–54. Springer, Heidelberg (2009). https://doi.org/10.1007/978-3-642-10631-6_7
8. Das, G.K., Fraser, R., Lòpez-Ortiz, A., Nickerson, B.G.: On the discrete unit disk cover problem. In: Katoh, N., Kumar, A. (eds.) WALCOM 2011. LNCS, vol. 6552, pp. 146–157. Springer, Heidelberg (2011). https://doi.org/10.1007/978-3-642-19094-0_16
9. Erlebach, T., van Leeuwen, E.J.: PTAS for weighted set cover on unit squares. In: Serna, M., Shaltiel, R., Jansen, K., Rolim, J. (eds.) APPROX/RANDOM -2010. LNCS, vol. 6302, pp. 166–177. Springer, Heidelberg (2010). https://doi.org/10.1007/978-3-642-15369-3_13
10. Fowler, R.J., Paterson, M.S., Tanimoto, S.L.: Optimal packing and covering in the plane are NP-complete. Inf. Process. Lett. **12**(3), 133–137 (1981)
11. Fraser, R., Lòpez-Ortiz, A.: The within-strip discrete unit disk cover problem. Theor. Comput. Sci. **674**, 99–115 (2017)
12. Li, J., Jin, Y.: A PTAS for the weighted unit disk cover problem. In: Halldórsson, M.M., Iwama, K., Kobayashi, N., Speckmann, B. (eds.) ICALP 2015. LNCS, vol. 9134, pp. 898–909. Springer, Heidelberg (2015). https://doi.org/10.1007/978-3-662-47672-7_73
13. Madireddy, R.R.: Approximability and hardness of geometric covering and hitting problems. PhD thesis, Indian Institute of Technology Ropar, India (2018)
14. Mustafa, N.H., Ray, S.: Improved results on geometric hitting set problems. Disc. Comput. Geom. **44**(4), 883–895 (2010)

2-Node-Connectivity Network Design

Zeev Nutov(✉)

The Open University of Israel, Ra'anana, Israel
nutov@openu.ac.il

Abstract. We consider network design problems in which we are given a graph $G = (V, E)$ with edge costs, and seek a min-cost/size 2-*node*-connected subgraph $G' = (V', E')$ that satisfies a prescribed property.

- In the Block-Tree Augmentation problem the goal is to augment a tree T by a min-size edge set $F \subseteq E$ such that $G' = T \cup F$ is 2-*node*-connected. We break the natural ratio of 2 for this problem and show that it admits approximation ratio 1.91. This result extends to the related Crossing Family Augmentation problem.
- In the 2-Connected Dominating Set problem G' should dominate V. We give the first non-trivial approximation algorithm for this problem, with expected ratio $\tilde{O}(\log^4 |V|)$.
- In the 2-Connected Quota Subgraph problem we are given node profits $p(v)$ and G' should have profit at least a given quota Q. We show expected ratio $\tilde{O}(\log^2 |V|)$, almost matching the best known ratio $O(\log^2 |V|)$.

Our algorithms are very simple, and they combine three main ingredients:

1. A probabilistic *spanning tree* embedding with distortion $\tilde{O}(\log |V|)$ results in a variant of the Block-Tree Augmentation problem.
2. An approximation ratio preserving reduction of Block-Tree Augmentation variants to Node Weighted Steiner Tree problems.
3. Using existing approximation algorithms for variants of the Node Weighted Steiner Tree problem.

Keywords: 2-connectivity · Dominating set · Approximation algorithm · Block-tree augmentation · Symmetric crossing family

1 Introduction

A graph is k-**connected** if it contains k internally disjoint paths between every pair of nodes; if the paths are only required to be edge disjoint then the graph is k-**edge-connected**. A subset S of nodes in a graph is a **dominating set** if every node $v \notin S$ has a neighbor in S; if v has m neighbors in S then S is an m-**dominating set**. We consider problems when we are given a graph G, possibly with edge costs (or with node weights), and the goal is to find a minimum size/cost/weight 2-connected (or 2-edge-connected) subgraph of G that satisfies a given property. Specifically, we consider the following problems.

C. Kaklamanis and A. Levin (Eds.): WAOA 2020, LNCS 12806, pp. 220–235, 2021.
https://doi.org/10.1007/978-3-030-80879-2_15

Block-Tree Augmentation
Input: A tree $T = (V, E_T)$ and an edge set E on V.
Output: A min-size edge set $F \subseteq E$ such that $T \cup F$ is 2-connected.

This problem is equivalent to the problem of augmenting an arbitrary connected graph to be 2-connected, by considering the block-tree of the graph, c.f. [26].

2-Connected Dominating Set
Input: A graph $G = (V, E)$.
Output: A 2-connected subgraph (S, F) of G that dominates V (namely, S dominates V) and has minimum number of edges/nodes.

Note that since $|S| \le |F| \le 2|S|$ holds for any edge-minimal 2-connected graph (S, F), then the approximability of the objectives of minimizing the number of nodes and the number of edges are equivalent, up to a factor of 2.

Two additional problems that we consider are as follows, and note that the former is a particular case of the latter.

2-Connected k-Subgraph
Input: A graph $G = (V, E)$ with edge costs $\{c_e : e \in E\}$ and an integer k.
Output: A min-cost 2-connected subgraph (S, F) of G with $|S| \ge k$.

2-Connected Quota Subgraph
Input: A graph $G = (V, E)$ with edge costs $\{c_e : e \in E\}$, node profits $\{p_v : v \in V\}$, and a quota Q.
Output: A min-cost 2-connected subgraph of G that has profit at least Q.

Using ideas from the papers of Gupta, Krishnaswamy, and Ravi [24], Basavarju et al. [3] (see also [32]), and a recent paper of Byrka, Grandoni, and Ameli [6], we give simple approximation algorithms that are based on a generic reduction of the above problems to variants of the following problem:

Node Weighted Steiner Tree
Input: A graph $H = (U, I)$, a set $R \subseteq U$ of terminals, and node weights $\{w_v : v \in U \setminus R\}$.
Output: A min-weight subtree of H that contains R.

In general, the Node Weighted Steiner Tree problem is Set Cover hard to approximate, and the best known ratio for it is $O(\ln |R|)$ [23,28]. However, often the Node Weighted Steiner Tree instances arising from the reduction have special properties. For example, in the case of the Block-Tree Augmentation problem, non-terminals have unit weights, the neighbors of every terminal induce a clique, and every $v \in V \setminus R$ has at most two neighbors in R. This enables us to use the algorithm of [6] that achieves ratio 1.91 for such instances. For the 2-Connected Dominating Set problem the reduction is to the following more general problem:

SUBSET STEINER CONNECTED DOMINATING SET
Input: A graph $H = (U, I)$ and a set $R \subseteq U$ of (independent) terminals.
Output: A min-size node set $F \subseteq U \setminus R$ such that $H[F]$ is connected and F dominates R.

For other problems we use the results for the plain connectivity quota problem, that achieves ratio $O(\log \min\{n, Q\})$ [4]. We note that often the graph H arising from the reduction is the incidence graph of the paths and edges/nodes of a tree, and thus such instances may admit better ratios than the ones known for general graphs.

Given a distribution $\mathcal{T}$ over spanning trees of G, the **stretch** of $\mathcal{T}$ is $\max_{uv \in E} \mathbb{E}_{T \sim \mathcal{T}} \left[\frac{d_T(u,v)}{d_G(u,v)} \right]$, where $d_H(u, v)$ denotes the distance between u and v in a graph H. Let $\sigma = \sigma(n)$ denote the lowest known upper bound on the stretch that can be achieved by a polynomial time construction of such $\mathcal{T}$ for a graph on n nodes. By the work of Abraham, Bartal, and Neiman [1], that is in turn based on the work of Elkin, Emek, Spielman, and Teng [14], we have:

$$\sigma(n) = O(\log n \cdot \log \log n \cdot (\log \log \log n)^3) = \tilde{O}(\log n) .$$

Table 1. Previous and our ratios for 2-connectivity network design problems. The ratios for plain connectivity are for the corresponding node weighted problems.

	Connected	2-edge-connected		2-connected	
Problem		Previous	This paper	Previous	This paper
AUGMENTATION	–	1.458 [21]	–	2 [18]	1.91
DOMINATING	$O(\log n)$ [22]	$O(\sigma \log n)$ [5]	–	–	$O(\sigma \log^3 n)$
k-SUBGRAPH	$O(\log k)$ [4]	$O(\log k)$ [25]	–	$O(\log n \log k)$ [7]	$O(\sigma \log k)$
QUOTA	$O(\log \min\{n, Q\})$ [4]	$O(\log^2 n)$ [7]	$O(\sigma \log n)$	$O(\log^2 n)$ [7]	$O(\sigma \log n)$

Our ratios and previous best known ratios are summarized in Table 1. The first two rows give the main results in the paper. The purpose of the other two rows is just to demonstrate the method, that enables to obtain easily polylogarithmic approximation ratios that are not too far from the best known ones. We also give a bi-criteria approximation $\left((1+\epsilon)\sigma, \Omega\left(\frac{\epsilon^2}{\log n}\right)\right)$ for the 2-CONNECTED BUDGETED SUBGRAPH problem, where the goal is to maximize the profit under cost budget constraints. But for this problem the algorithm of Chekuri and Korula [7], that achieves bi-criteria approximation $\left(2+\epsilon, \Omega\left(\frac{\epsilon}{\log^2 n}\right)\right)$ for any $\epsilon \in (0, 1]$, has a much smaller budget violation.

The main purpose of this paper is to resolve open questions on the approximability of the two old fundamental problems – BLOCK-TREE AUGMENTATION [26] and 2-CONNECTED DOMINATING SET [11]. We now briefly survey some literature on these problems, and on the edge-connectivity versions of these problems – TREE AUGMENTATION and 2-EDGE-CONNECTED DOMINATING SET.

TREE AUGMENTATION and BLOCK-TREE AUGMENTATION problems admit several simple 2-approximation algorithms even when the edges have costs [18,27]. The problem of obtaining ratio below 2 for the min-size case for these problems was posed by Khuller [26, p. 263]. For TREE AUGMENTATION this was resolved by Nagamochi [33]. Subsequently, the TREE AUGMENTATION problem was studied in many papers [2,3,8–10,15,16,21,30,31,33,35]. It admits a simple efficient 1.5-approximation algorithm [30], and also several LP-based algorithms, with the current best ratio 1.458 [21]. When costs are bounded by a constant the min-cost version admits ratio $1.5+\epsilon$ [16]; this is so also for logarithmic edge costs [35].

A **cactus** is a "tree-of-cycles", namely, a 2-edge-connected graph in which every block is a cycle (equivalently - every edge belongs to exactly one simple cycle). In the CACTUS AUGMENTATION problem we seek a min-size edge set $F \subseteq E$ to make a cactus 3-edge-connected. TREE AUGMENTATION is a particular case, when every cycle has length 2. Basavaraju et al. [3] gave an approximation ratio preserving reduction from CACTUS AUGMENTATION to special instances of the NODE WEIGHTED STEINER TREE problem. Recently, Byrka, Grandoni, and Ameli [6] showed that such instances admit ratio $2\ln 4 - \frac{967}{1120} + \epsilon < 1.91$, thus obtaining the first ratio below 2 also for CACTUS AUGMENTATION. However, no ratio below 2 was known for the BLOCK-TREE AUGMENTATION problem.

As was mentioned, we extend the reduction of [3] to the BLOCK-TREE AUGMENTATION problem. We also prove a related general result in Sect. 4. Two sets A, B **cross** if $A \cap B \neq \emptyset$ and $A \cup B \neq V$. A set family $\mathcal{F}$ on a groundset V is a **crossing family** if $A \cap B, A \cup B \in \mathcal{F}$ whenever $A, B \in \mathcal{F}$ cross; $\mathcal{F}$ is a **symmetric family** if $V \setminus A \in \mathcal{F}$ whenever $A \in \mathcal{F}$. The 2-edge-cuts of a cactus form a symmetric crossing family, with the additional property that whenever $A, B \in \mathcal{F}$ cross and $A \setminus B, B \setminus A$ are both non-empty, the set $(A \setminus B) \cup (B \setminus A)$ is not in $\mathcal{F}$; such a symmetric crossing family is called **proper**. Dinitz, Karzanov, and Lomonosov [12] showed that the family of minimum edge cuts of a graph G can be represented by 2-edge cuts of a cactus. Furthermore, when the edge-connectivity of G is odd, the min-cuts form a laminar family and thus can be represented by a tree. Dinitz and Nutov [13, Theorem 4.2] (see also [34, Theorem 2.7]) extended this by showing that an arbitrary symmetric crossing family $\mathcal{F}$ can be represented by 2-edge cuts and specified 1-node cuts of a cactus; when $\mathcal{F}$ is a proper crossing family this reduces to the cactus representation of [12]. (A representation identical to the one of [13] was announced later by Fleiner and Jordán [17].)

We say that an edge f **covers** a set A if f has exactly one end in A. We show that our result for BLOCK-TREE AUGMENTATION extends to the following problem, that combines the difficulties of the CACTUS AUGMENTATION and the BLOCK-TREE AUGMENTATION problems. All in all, we widely generalize the reduction of [3], by a substantially simpler proof.

CROSSING FAMILY AUGMENTATION
Input: A symmetric crossing family $\mathcal{F}$ on V and an edge set E on V.
Output: A min-size edge set $J \subseteq E$ that covers $\mathcal{F}$.

Problems of finding low size/weight k-connected m-dominating set were vastly studied, but in general graphs non-trivial approximation algorithms are known only for the case $m \geq k$, c.f. [5,19,22,36,38] why the case $m \geq k$ is easier. For $m < k$, constant ratios are known for unit disk graphs and $k \leq 3$ and $m \leq 2$ [37], and ratio $\tilde{O}(\log^2 n)$ for 2-Edge-Connected Dominating Set in general graphs [5]. It was an open problem to obtain a non-trivial ratio for 2-Connected Dominating Set in general graphs. Note that our result is valid for unit node-weights only. For general node weights, the problem is still open.

2 Two Simple Reductions

Our reductions are independent and they can be combined or used separately. In the first reduction, the problem is reduced to a variant of the tree augmentation problem. This reduction is inspired by the one of [24], and it invokes a factor of $\sigma = \tilde{O}(\log n)$ in the ratio. In the second reduction, an augmentation problem is reduced to the corresponding Node Weighted Steiner Tree problem - this reduction is inspired by that of [3].

2.1 Reduction to the Augmentation Problem

Suppose that we have a distribution $\mathcal{T}$ over spanning trees of G with stretch σ. For simplicity of exposition we will assume throughout the paper that we are given a single spanning tree $T = (V, E_T)$ with stretch σ, namely that

$$c(T_f) \leq \sigma c(f) \quad \forall f \in E$$

where T_f denotes the path in the tree between the endnodes of f. We say that $f \in E$ **covers** $e \in E_T$ if $e \in T_f$. For $F \subseteq E$ let $T_F = \cup_{f \in F} T_f$ denote the forest formed by the tree edges of T that are covered by the edges of F. Let us consider the following problem.

Subtree Augmentation Network Design
Input: A tree $T = (V, E_T)$, an edge set E on V with costs, and property Π.
Output: A min-cost edge set $F \subseteq E$ such that T_F is a subtree of T and the graph $T_F \cup F$ is 2-connected and satisfies Π.

Let $\mathcal{I} = (G, c, \Pi)$ be a 2-Connected Network Design instance, where $G = (V, E_G)$ is a graph, c is a cost function on E_G, and Π is a family of subgraphs of G that are feasible solutions (may not be given explicitly), such that $G'' \in \Pi$ implies $G' \in \Pi$ whenever $G'' \subseteq G' \subseteq G$. Given a spanning tree T in G the corresponding Subtree Augmentation Network Design instance $\hat{\mathcal{I}} = (T, E, \hat{c}, \Pi)$ has costs $\hat{c}(e) = 0$ if $e \in E_T$ and $\hat{c}(e) = c(e)$ otherwise, where $E = E_G \setminus E_T$. The reduction is given in the following lemma.

Lemma 1. *A* 2-Connected Network Design *instance* $\mathcal{I}$ *admits ratio* $\rho(\sigma + 1)$ *if the corresponding augmentation instance* $\hat{\mathcal{I}}$ *admits ratio* ρ.

Proof. Clearly, if F is a feasible $\hat{\mathcal{I}}$ solution then $T_F \cup F$ is a feasible $\mathcal{I}$ solution. We show that if (S,J) is a feasible $\mathcal{I}$ solution then $J \setminus E_T$ is a feasible $\hat{\mathcal{I}}$ solution. Hence if F is a ρ-approximate $\hat{\mathcal{I}}$ solution, then $T_F \cup F$ is a feasible $\mathcal{I}$ solution and

$$c(F \cup T_F) \leq c(F) + \sum_{f \in F} c(T_f) \leq c(F) + \sum_{f \in F} \sigma c(f) = (\sigma+1)\hat{c}(F)$$
$$\leq (\sigma+1)\rho\hat{c}(J \setminus E_T) = (\sigma+1)\rho c(J \setminus E_T) \leq (\sigma+1)\rho c(J) \ .$$

It remains to show that $T_{J\setminus E_T}$ is a tree and that $T_{J\setminus E_T} \cup (J \setminus E_T)$ is 2-connected. Suppose to the contrary that $T_{J\setminus E_T}$ has distinct connected components C, C'. Let e be the first edge on the path in T from C to C'. No edge in $J \setminus E_T$ has exactly one end in C, hence $(T \cup J) \setminus \{e\}$ is disconnected, and one of its connected components contains C and the other contains C'. This implies that $(S, J \setminus \{e\})$ is disconnected, contradicting that (S,J) is 2-connected. To see that $T_{J\setminus E_T} \cup (J \setminus E_T)$ is 2-connected, note that the graph $T_{J\setminus E_T} \cup (J \setminus E_T)$ is obtained from the 2-connected graph (S,J) by sequentially adding for each $f \in J \setminus E_T$ the path T_f. Adding a simple path P between two distinct nodes of a 2-connected graph results in a 2-connected graph; this is so also if P contains some edges of the graph. It follows by induction that $(S,J) \cup (\cup_{f \in J\setminus E_T} T_f)$ is 2-connected, concluding the proof of the lemma.

2.2 Reduction to Node Weighted Steiner Tree

We need some definitions before presenting the reduction. Given a set F of paths in a graph $G=(V,E_G)$, the **paths-nodes incidence graph** has node set $F \cup V$ and edge set $\{fv : v \in f \in F\}$, and the **paths-edges incidence graph** has node set $F \cup E_G$ and edge set $\{fe : e \in f \in F\}$. Let $T=(V,E_T)$ be a tree and F an edge set on V (Fig. 1(a)). The (F,V)**-incidence graph** (Fig. 1(b)) is the paths-nodes incidence graph for the set of paths $\{T_f : f \in F\}$, where T_f denotes the unique path in T between the endnodes of f; we let V be the set of terminals of this graph. Similarly the (F,E_T)**-incidence graph** (Fig. 1(c)) is defined as the path-edges incidence graph for $\{T_f : f \in F\}$, and we let E_T to be the set of terminals of this graph. We now define modifications of these graphs.

- The **short-cut** (F,V)**-incidence graph** is obtained from the (F,V)-incidence graph by short-cutting all terminals (nodes in V), where **short-cutting a node** v means adding a clique on the set of neighbors of v.
- The **reduced** (F,V)**-incidence graph** (Fig. 1(d)) is obtained from the short-cut (F,V)-incidence graph by removing $V \setminus L$, where L is the set of leaves of T.

The **short-cut** (F,E_T)**-incidence graph** and the **reduced** (F,E_T)**-incidence graph** (Fig. 1(e)) are obtained from the (F,E_T)-incidence graph in a similar way, by short-cutting terminals (nodes in E_T) and removing $V \setminus L$, respectively, with L being the set of leaf edges of T; in both cases, we let $R=L$ to be the set of terminals of these "reduced" graphs.

Let $J = T \cup F$, and let $\lambda_J(s,t)$ and $\kappa_J(s,t)$ denote the maximum number of edge disjoint and internally disjoint st-paths in J, respectively. By [3], $\lambda_J(s,t) \geq 2$ if and only if the (F,V)-incidence graph H has an st-path. Thus a MIN-COST TREE AUGMENTATION instance $(T=(V,E_T),E,c)$ is equivalent to the NODE-WEIGHTED STEINER TREE instance $(H=(V \cup E_T, I), w, R)$, where H is the (E,V)-incidence graph, $R=V$ is the set of terminals, node weights $w(f)=c(f)$ if $f \in E$ and $w(v)=0$ for $v \in V$. One can modify this NODE-WEIGHTED STEINER TREE instance to obtain a more compact instance, where H is the reduced (E,V)-incidence graph and $R=L$ is the set of terminals.

1. Since the nodes in V have weight zero, they can be short-cut.
2. It is easy to see that $T \cup F$ is 2-edge-connected if and only if $\lambda_J(s,t) \geq 2$ for all $s,t \in L$. Thus we can remove $V \setminus L$ and let $R=L$ be the set of terminals.

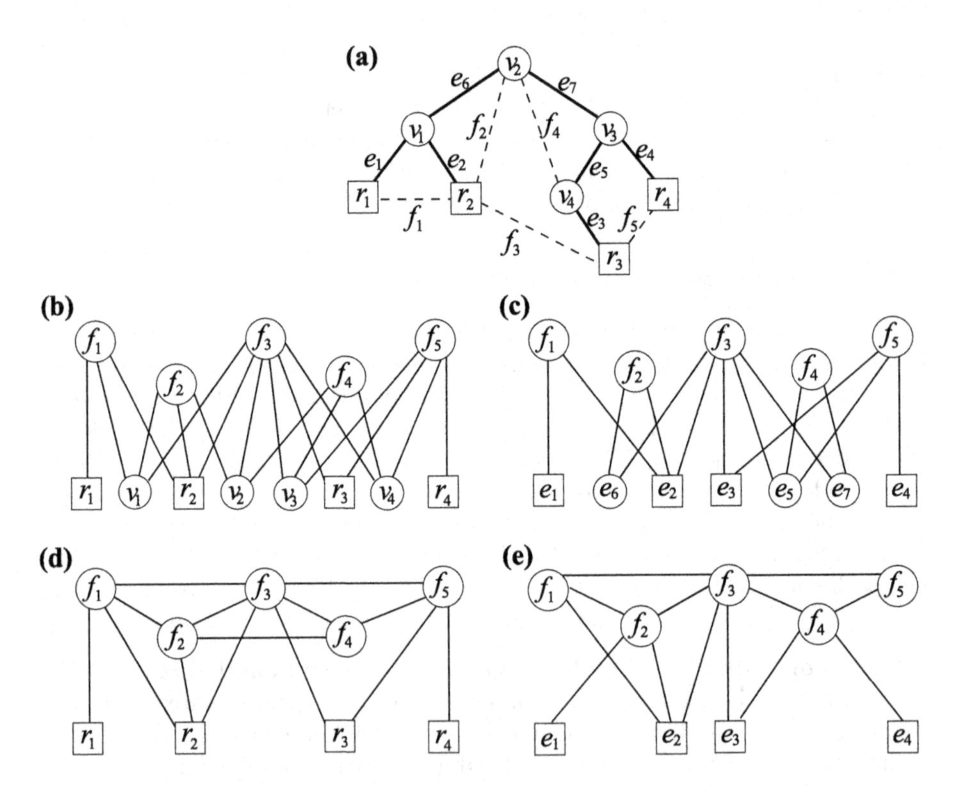

Fig. 1. (a) T and F; (b) the (F,V)-incidence graph ; (c) the (F,E_T)-incidence graph; (d) the reduced (F,V)-incidence graph; (e) the reduced (F,E_T)-incidence graph. Here square nodes correspond to leaves or to leaf edges of T, and they are terminals of the graphs in (d,e).

Summarizing, we have the following:

Lemma 2 (Basavaraju et al. [3]). *Let $T = (V, E_T)$ be a tree, F an edge set on V, and $J = T \cup F$. Then $\lambda_J(s,t) \geq 2$ if and only if the (F,V)-incidence graph has an st-path. J is 2-edge-connected if and only if the reduced (F,V)-incidence graph has a path between every two terminals.*

We note that [3] proved the first part of the lemma for the case when s,t are leaves of T, but the proof easily extends to the case of arbitrary $s,t \in V$. Our main result in this section is the following extension of this lemma to the node-connectivity case.

Lemma 3. *Let $T = (V, E_T)$ be a tree, F an edge set on V, and $J = T \cup F$. Let H be the (F, E_T)-incidence graph, and let us say that distinct $s,t \in V$ are* **H-reachable** *if H has an $e_s e_t$-path (possibly $e_s = e_t$), where e_s and e_t are the corresponding end edges of the st-path in T. Then $\kappa_J(s,t) \geq 2$ if and only if s,t are H reachable. Furthermore, J is 2-connected if and only if the reduced (F, E_T)-incidence graph H_R has a path between every two terminals.*

It is easy to see that $J = T \cup F$ is 2-connected if and only if $\kappa_J(s,t) \geq 2$ for every pair of leaves of T. Thus J is 2-connected if and only if any two terminals are H-reachable. Since H_R is obtained from H by short-cutting the nodes that correspond to non-leaf edges of T, nodes $r, r' \in R$ are H_R-reachable if and only if $H_R[F \cup R]$ has an rr'-path.

Let P be the st-path in T. In the rest of this section we prove the first statement of Lemma 3 by induction on the number of edges $|P|$ in P. The case $|P| = 1$ is obvious, since then $e_s = e_t$ (note that $s \neq t$), while the case $|P| = 2$ is given in the following lemma.

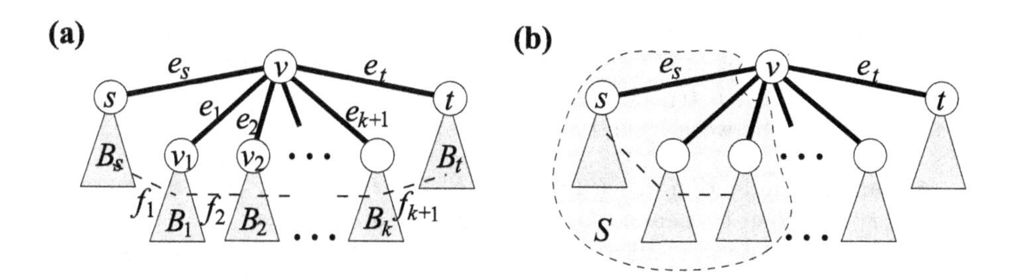

Fig. 2. Illustration to the proof of Lemma 4. Edges in T are shown by bold lines.

Lemma 4. *Lemma 3 holds if $|P| = 2$.*

Proof. Let v be the middle node of P. Consider the graph $J \setminus \{v\}$ and the branches hanging on v in T (see Fig. 2), where we denote by B_s and B_t the branches that contain s and t, respectively. We need to prove that:

- If $\kappa_J(s,t) \geq 2$ then s,t are H-reachable.
- If $\kappa_J(s,t) = 1$ then s,t are not H-reachable.

Suppose that $\kappa_J(s,t) \geq 2$. Then $J \setminus \{v\}$ has an st-path, and there exists such a path that visits every branch at most once; say the visited branches are $B_s, B_1, \ldots, B_k, B_t$, where $s \in B_s$ and $t \in B_t$ (see Fig. 2(a)). Then the sequence $e_s, f_1, e_1, f_2, \ldots, e_{k+1}, f_{k+1}, e_t$ of nodes of H forms an $e_s e_t$-path Q in H.

Suppose that $\kappa_J(s,t) = 1$. Then $J \setminus \{v\}$ has no st-path. Let S be the union of branches reachable from s in $J \setminus \{v\}$ (see Fig. 2(b)). Let F_s and F_t be the set of edges in F with both ends in $S \cup \{v\}$ and in $V \setminus S$, respectively. Then F_s and F_t is a partition of F. In H, there is no path between F_s and F_t, and thus s can reach only nodes in S. Since $t \notin S$, t is not H-reachable from s.

For nodes a, b in P let P_{ab} denote the subpath of P between a and b. To apply the induction we use the following lemma.

Lemma 5. *If $|P| \geq 3$, and uv and $vt = e_t$ are the last two edges of P, then:*

(i) *If $\kappa_J(s,t) \geq 2$ then $\kappa_J(s,v) \geq 2$.*
(ii) *If s,v and u,t are both H-reachable then s,t are H-reachable.*
(iii) *If s,t are H-reachable then s,v are H-reachable.*

Proof. We prove (i). Let C be the cycle in J formed by the 2 internally disjoint st-paths. If $v \in C$ then we are done. Otherwise, let $a \in P_{sv}$ such that P_{av} has only the node a in common with C. One can easily verify that $C \cup P_{av} \cup \{vt\}$ contains 2 internally disjoint sv-paths.

For (ii), note that in H, there is a path between e_s and uv and there is a path between uv and e_t; the union of these paths contains an $e_s e_t$-path.

We prove (iii). In H, let $Q = Q(e_s, e_t)$ be a path between e_s and e_t. We claim that there exists $f \in F$ that belongs to this path and is adjacent to uv. Otherwise, let Q_s and Q_t be the set of edges in F with both ends in the connected component of $T \setminus \{uv\}$ that contains s and t, respectively. Then F_s and F_t partition F. In H, there is no path between F_s and F_t. This contradicts that Q is a path between e_s and e_t.

To carry the induction we assume that $|P| \geq 3$ and let u, v be as in Lemma 5. If $\kappa_J(s,t) \geq 2$ then by Lemma 5(i), $\kappa_J(s,v) \geq 2$ and $\kappa_J(u,t) \geq 2$, thus by the induction hypothesis s,v and u,t are both H-reachable. By Lemma 5(ii) s,t are H-reachable.

If s,t are H-reachable then by Lemma 5(iii), both s,v and u,t are H-reachable. Thus by the induction hypothesis, $\kappa_J(s,v) \geq 2$ and $\kappa_J(u,t) \geq 2$. We claim that then $\kappa_J(s,t) \geq 2$. Otherwise, there is a node $a \in P \setminus \{s,t\}$ such that s,t belong to distinct connected components of $J \setminus \{a\}$. However $a \in P_{su}$ contradicts $\kappa_J(s,v) \geq 2$, while $a = v$ contradicts $\kappa_J(u,t) \geq 2$. Thus no such a exists, and the proof of Lemma 3 is complete.

3 Applications

3.1 Block-Tree Augmentation

Theorem 1. Block-Tree Augmentation *admits ratio* $2\ln 4 - \frac{967}{1120} + \epsilon < 1.91$.

Proof. We construct the reduced (F, E_T)-incidence graph H_R. The corresponding Node Weighted Steiner Tree instance has unit non-terminal weights, and the following two properties:

1. The neighbors of every $r \in R$ induce a clique (by the construction).
2. Every non-terminal has at most two neighbors in R (since every edge connects at most two leaves).

Byrka, Grandoni, and Ameli [6] showed that such Node Weighted Steiner Tree instances admit approximation ratio as in the theorem.

3.2 2-Connected Dominating Set

Theorem 2. 2-Connected Dominating Set *admits (expected) ratio* $O(\sigma \log^3 n)$.

Proof. Since $|S| \leq |F| \leq 2(|S| - 1)$ holds for any edge-minimal 2-connected graph (S, J), we may consider the objective of minimizing the number of edges. Applying the reduction from Sect. 2.1 invokes a loss of σ in the ratio and gives the following problem:

Dominating Block-Tree
Input: A tree $T = (V, E_T)$ and an edge set E on V disjoint to E_T.
Output: A min-size edge set $F \subseteq E$ such that T_F is a dominating tree and $T_F \cup F$ is 2-connected.

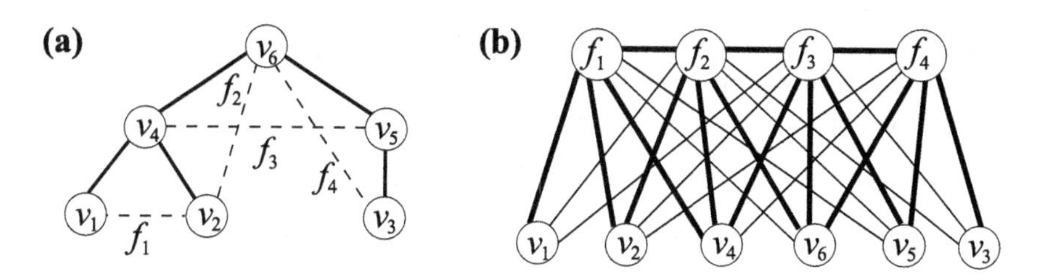

Fig. 3. Illustration to the construction of a Subset Steiner Connected Dominating Set instance. (a) Edges in T are shown by bold lines and edges in E by dashed lines. (b) Edges of H_{con} are shown by bold lines and edges in I_{dom} are shown by thin lines.

Given a Dominating Block-Tree instance $(T = (V, E_T), E)$ construct a Subset Steiner Connected Dominating Set instance (H, R) as follows (see

Fig. 3). The graph H (see Fig. 3(b)) is a union of the short-cut (E, V)-incidence graph H_{con} (that encodes 2-connectivity) and the edge set $I_{dom} = \{fv : f \in E, v \in \Gamma(T_f)\}$ (that encodes domination), where $\Gamma(T_f)$ is the set of neighbors of T_f in $T \cup E$ (e.g., in Fig. 3, $\Gamma(T_{f_1}) = \{v_6, v_5\}$ and $\Gamma(T_{f_3}) = \{v_1, v_2, v_3\}$). The set of terminals of H is $R = V$.

Let $F \subseteq E$. Then $F \cup T_F$ is 2-connected if and only if $H[F]$ is connected; this follows from Lemma 3 and the definition of the short-cut (E, V)-incidence graph. By the construction, T_F dominates V in $T \cup E$ if and only if F dominates V in H. Thus F is a feasible solution to the Dominating Block-Tree instance if and only if F is a feasible solution to the constructed Subset Steiner Connected Dominating Set instance. To get overall (expected) ratio $O(\sigma \log^3 n)$, we show that the obtained instance admits ratio $O(\log^3 n)$. We give a reduction to the Group Steiner Tree problem: given a graph with edge costs and a collection of subsets (groups) of the node set, find a min-cost subtree that contains a node from every group. Obtain an equivalent Group Steiner Tree instance with *unit node weights* as follows: for every $r \in R$, introduce a group S_r that consists of the neighbors of r in H, and then remove r. This is equivalent to Group Steiner Tree instance with *unit edge costs*, since feasible solutions are trees. By [20] Group Steiner Tree (with edge costs) admits ratio $O(\log k \log M \log N) = O(\log^3 n)$, where k is the number of groups, M is the maximum group size, and N is the number of nodes; in our case, $k = |R| = |V|$, $M = \max_{r \in R} \deg_H(r) \leq |E|$, and $N = |E|$.

3.3 2-Connected *k*-Subgraph

We give a simple proof that the 2-Connected k-Subgraph problem admits (expected) ratio $O(\sigma \log k)$. Consider the Subtree Augmentation Network Design instance $(T = (V, E_T), E, c, \Pi)$ obtained by applying the reduction from Sect. 2.1, that incurs a factor of σ in the ratio. Here we need to find a min-cost edge set $F \subseteq E$ such that $T_F \cup F$ is 2-connected and has at least k nodes; note that $T_F \cup F$ has at least k nodes if and only if T_F has at least $k-1$ edges. Next, consider the corresponding node weighted problem defined by the short-cut (E, E_T)-incidence graph H with weights $w(e) = c(e)$ for all $f \in E$, see Sect. 2.2. In this problem we need to find a min-weight subtree with at least $k-1$ terminals, and it admits ratio $O(\log k)$, by [4,29]. The overall (expected) ratio is $O(\sigma \log k)$.

3.4 2-Connected Quota Subgraph

Here we give a simple proof that 2-Connected Quota Subgraph admits (expected) ratio $O(\sigma \log n)$. We may consider the "rooted" version of the problem, when the subgraph should contain a specified node s of zero profit, c.f. [7]. We apply the reduction from Sect. 2.1 with stretch σ on the costs, and root T at s. For every node $v \neq s$, its parent edge in T gets the profit $p(v)$ of v. Next, consider the corresponding node weighted problem defined by the short-cut (E, E_T)-incidence graph H, with weights $w(f) = c(f)$ and profits $p(f) = 0$

for all $f \in E$, while every node in H that corresponds to an edge e of T has cost zero and inherits the profit of e. Here we need to find a minimum weight subtree of profit at least Q, and it admits ratio $O(\log n)$, by [4,29]. The overall ratio is $O(\sigma \log n)$.

A similar method can be used to obtain a bi-criteria approximation for the 2-Connected Budgeted Subgraph problem. The reduction is the same as above, but we need to find a maximum profit subtree of weight at most a given budget B. This problem admits a bi-criteria approximation $(1+\epsilon, \Omega(\epsilon^2/\log n))$ [4], giving the overall ratio $\left((1+\epsilon)\sigma, \Omega\left(\frac{\epsilon^2}{\log n}\right)\right)$.

4 Crossing Family Augmentation

Recall that in the Crossing Family Augmentation problem we need to cover a symmetric crossing family $\mathcal{F}$ by a min-size edge set $J \subseteq E$. Here $\mathcal{F}$ may not be given explicitly, and a polynomial time implementation requires that some queries related to $\mathcal{F}$ can be answered in polynomial time. We make a standard assumption that for any $s, t \in V$ and any edge set J, we can check in polynomial time whether there is $A \in \mathcal{F}$ that is not covered by J such that $|A \cap \{s,t\}| = 1$. Under this assumption we prove the following generalization of Theorem 1.

Theorem 3. Crossing Family Augmentation *admits ratio* $2\ln 4 - \frac{967}{1120} + \epsilon < 1.91$.

Let $\mathcal{F}$ be a symmetric crossing family on V. Note that by the symmetry of $\mathcal{F}$, if $A, B \in \mathcal{F}$ then $A \setminus B, B \setminus A \in \mathcal{F}$ if $A \setminus B, B \setminus A$ are non-empty. Let $\mathcal{C}$ denote the family of inclusion-minimal members of $\mathcal{F}$. For any $C \in \mathcal{C}$ and any $A \in \mathcal{F}$ we have $C \subseteq A$ or $C \cap A = \emptyset$; otherwise A, C or $V \setminus A, V \setminus C$ cross, and one of $C \cap A, C \setminus A$ is a member of $\mathcal{F}$ that is properly contained in C, contradicting the minimality of C. Thus we may contract every set $C \in \mathcal{C}$ into a single element, and assume w.l.o.g. that all members of $\mathcal{C}$ are singleton sets. To simplify the notation we write $v \in \mathcal{C}$ instead of $\{v\} \in \mathcal{C}$, and treat $\mathcal{C}$ as a set of elements.

Definition 1. *Let J be an edge set on V. We say that a set $A \in \mathcal{F}$* **separates** *edges $f, g \in J$ if one of f, g has both ends in A and the other has no end in A; if such A exists then f, g are* **separable**, *and f, g are* **inseparable** *otherwise. The* **separability graph** *$H = (U, I)$ of $\mathcal{F}, J$ has node set $U = \mathcal{C} \cup J$ and edge set*

$$I = \{fv : f \in J \text{ covers } v \in \mathcal{C}\} \cup \{fg : f, g \in J \text{ are inseparable}\} .$$

We prove the following theorem, and then explain how it implies Theorem 3.

Theorem 4. *Let $\mathcal{F}$ be a symmetric crossing family and J an edge set on V. Then J covers $\mathcal{F}$ if and only if the separability graph $H = (U, I)$ of $\mathcal{F}, J$ has an st-path for all $s, t \in \mathcal{C}$.*

Proof. Suppose that J does not cover some $A \in \mathcal{F}$. Since $\mathcal{F}$ is symmetric, $\mathcal{C} \cap A$ and $\mathcal{C} \setminus A$ are both non-empty. Let S be the set of edges in J that have both ends in A. Then $(\mathcal{C} \cap A) \cup S, (\mathcal{C} \setminus A) \cup (J \setminus S)$ is a proper partition of $U = \mathcal{C} \cup J$. In H, there is no edge between these parts, hence H has no st-path for $s \in \mathcal{C} \cap A$ and $t \in \mathcal{C} \setminus A$.

It remains to prove that if H has no st-path for some $s, t \in \mathcal{C}$ then there is a set in $\mathcal{F}$ that is not covered by J. For $A \in \mathcal{F}$ and $f, g \in J$, we say that A **separates f from g** if f has both ends in A and g has no end in A. Similarly, for $S, Q \subseteq J$, we say that A **separates S from Q** if all edges in S have ends in A and all edges in Q have ends in $V \setminus A$. One can easily verify that if $A \in \mathcal{F}$ separates f from g and $A' \in \mathcal{F}$ separates f' from g then:

(i) If $A \cap A' \neq \emptyset$ then $A \cup A' \in \mathcal{F}$ separates $\{f, f'\}$ from g.
(ii) If $A \cap A' = \emptyset$ then $A \setminus A' \in \mathcal{F}$ separates f from $\{f', g\}$ and $A' \setminus A \in \mathcal{F}$ separates f' from $\{f, g\}$.

In particular, note that if f, f' are inseparable then (i) must hold (since (ii) cannot hold).

Let S be the the set of those members of J that belong to the connected component of H that contains s. Assume that $S \neq \emptyset$ and $J \setminus S \neq \emptyset$, as otherwise one of s, t is not covered by J. Let $g \in J \setminus S$. For any $f \in S$ there is a set $A_{fg} \in \mathcal{F}$ that separates f from g. For any inseparable (adjacent in H) $f, f' \in S$, the set $A_{fg} \cup A_{f'g}$ belongs to $\mathcal{F}$ and separates $\{f, f'\}$ from g. Also note that any f, f' that cover the same $v \in \mathcal{C}$ are inseparable, hence $H[S]$ is connected. By induction we get that the set $\cup_{f \in S} A_{fg}$ is in $\mathcal{F}$ and separates S from g.

Let A be an inclusion minimal set in $\mathcal{F}$ that separates S from g. We claim that A is not covered by J. Suppose that there is $g' \in J$ that covers A. Note that $g' \in J \setminus S$, thus there exists $B \in \mathcal{F}$ that separates S from g'. As g' has both ends in $V \setminus B$ and covers A, it has one end in $V \setminus (A \cup B)$ and the other in $A \setminus B$. Thus A, B cross, hence $A \cap B \in \mathcal{F}$. Moreover, $A \cap B$ separates S from g and $A \cap B \subset A$, contradicting the minimality of A.

Let us show that Theorem 4 implies Theorem 3. The algorithm will construct the auxiliary graph H for the edge set E. This reduces the problem to the Node Weighted Steiner Tree problem with unit weights, in which the neighbors of every $r \in R$ induce a clique and every non-terminal has at most two neighbors in R. This allows us to apply the algorithm of [6] in the same way as in Sect. 3.1. It remains to show that H can be constructed in polynomial time, namely, that for any pair $f = uv$ and $f' = u'v'$ we can check in polynomial time whether f, f' are separable. This is equivalent to checking whether there is $A \in \mathcal{F}$ that is not covered by $F = \{f, f'\}$ such that $|A \cap \{s, t\}| = 1$ for some $(s, t) \in \{u, v\} \times \{u'v'\}$.

As was mentioned in the Introduction, Basavaraju et al. [3] gave an approximation ratio preserving reduction from Cactus Augmentation to special instances of the Node Weighted Steiner Tree problem. Now we explain how Theorem 4 generalizes this reduction of [3], and how it is related to Lemma 3. A node x is a **cut-node** of G if $G \setminus \{x\}$ is disconnected, and a union of a proper subset of connected components of $G \setminus \{x\}$ is called an **x-bundle**. Let

$\mathcal{F}(G)$ denote the family of minimum edge cuts of G and $\mathcal{B}(G, X)$ the set of the x-bundles of nodes in X. Dinitz, Karzanov, and Lomonosov [12] showed that the family $\mathcal{F}(G)$ of minimum edge cuts of a graph G can be represented by the family $\mathcal{F}(\hat{G})$ of minimum edge cuts of a cactus $\hat{G} = (\hat{V}, \hat{E})$ and a mapping $\psi : V \longrightarrow \hat{V}$, such that $\mathcal{F}(G) = \{\psi^{-1}(\hat{A}) : \hat{A} \in \mathcal{F}(\hat{G})\}$. As was observed in [12] (see also [13,34]), the above representation is almost bijective (except one case when a min-cut of G can be represented twice). In [13] this representation was extended to arbitrary crossing families, as follows.

Theorem 5 (Dinitz and Nutov [13]). *Let $\mathcal{F}$ be a crossing family on V. Then there exists a cactus $\hat{G} = (\hat{V}, \hat{E})$, a mapping $\psi : V \longrightarrow \hat{V}$, and a set X of cut-nodes of $\hat{G}$ with $\psi^{-1}(X) = \emptyset$, such that $\mathcal{F} = \{\psi^{-1}(\hat{A}) : \hat{A} \in \mathcal{F}(\hat{G}) \cup \mathcal{B}(\hat{G}, X)\}$. Furthermore, if $\mathcal{F}$ is a proper crossing family then $X = \emptyset$.*

Suppose that $X = \emptyset$ in this representation. Then $\hat{G}$ is an ordinary cactus; let J be an edge set on its nodes and $f = uv \in J$. If u, v belong to distinct cycles, then there is a "path of cycles and cut-nodes" $\mathcal{P}(f) = C_1, x_1, C_2, x_2, \ldots x_{p-1}, C_p$ that connects the cycle C_1 that contains u to the cycle C_p that contains v, where each x_i is a cut node that belongs to both C_i and C_{i+1}; we refer the reader for more precise definitions to [3] and [6]. Every consecutive pair x_i, x_{i+1} of cut nodes on this "path" divides the cycle C_i into exactly two paths; say $P_C(f)$ and $\bar{P}_C(f)$. In [3] was defined the following relation: f, g **cross** if $\mathcal{P}(f)$ and $\mathcal{P}(g)$ have a common cycle C such that each of $P_C(f)$ and $\bar{P}_C(f)$ has a node in common with each of $P_C(g)$ and $\bar{P}_C(g)$. In the auxiliary graph H in [3], leaves of the cactus (nodes of degree exactly 2) are connected to the edges in J incident to these leaves, and $f, g \in J$ are connected if they cross. With these definitions and construction, they proved a result analogous to Theorem 4. One can however verify that f, g cross if and only if they are non-separable. To see this, consider the transformation of the cactus after "squeezing" the cycles of $\mathcal{P}(f)$ (namely, identifying the endnodes of f and the cut-nodes along $\mathcal{P}(f)$ into one new node), and then performing a similar operation for g (see [13] for such transformation of a cactus). It is not hard to see that if there is one new node then f, g are both inseparable and cross, while if there are 2 new nodes then f, g are both non-crossing and separable. Thus in the case of a cactus, our result reduces to the one of [3], but our definitions and proof are simpler.

Now consider the case when every cycle of $\hat{G}$ has length 2. Removing one edge from each cycle gives an equivalent representation by a tree T. Consider the case when X is the set of non-leaf nodes of T. One can verify that in this case f, g are inseparable if and only if the paths T_f and T_g have an edge in common. Still, even if there is a way to deduce Lemma 3 from Theorems 4, this is not immediate, since the representation in Theorem 5 does not allow for edges in J to have an endnode in X.

Acknowledgment. I thank an anonymous referee for many useful comments.

References

1. Abraham, I., Bartal, Y., Neiman, O.: Nearly tight low stretch spanning trees. In: FOCS, pp. 781–790 (2008)
2. Adjiashvili, D.: Beating approximation factor two for weighted tree augmentation with bounded costs. In: SODA, pp. 2384–2399 (2017)
3. Basavaraju, M., Fomin, F.V., Golovach, P.A., Misra, P., Ramanujan, M.S., Saurabh, S.: Parameterized algorithms to preserve connectivity. ICALP, Part **I**, 800–811 (2014)
4. Bateni, M.H., Hajiaghayi, M.T., Liaghat, V.: Improved approximation algorithms for (budgeted) node-weighted Steiner problems. SIAM J. Comput. **47**(4), 1275–1293 (2018)
5. Belgi, A., Nutov, Z.: An $\tilde{O}(\log^2 n)$-approximation algorithm for 2-edge-connected dominating set. CoRR abs/1912.09662 (2019). arxiv.org/abs/1912.09662
6. Byrka, J., Grandoni, F., Ameli, A.J.: Breaching the 2-approximation barrier for connectivity augmentation: a reduction to Steiner tree. CoRR abs/1911.02259 (2019), arxiv.org/abs/1911.02259
7. Chekuri, C., Korula, N.: Pruning 2-connected graphs. Algorithmica **62**(1–2), 436–463 (2012)
8. Cheriyan, J., Jordán, T., Ravi, R.: On 2-coverings and 2-packing of laminar families. In: ESA, pp. 510–520 (1999)
9. Cheriyan, J., Karloff, H., Khandekar, R., Koenemann, J.: On the integrality ratio for tree augmentation. Oper. Res. Lett. **36**(4), 399–401 (2008)
10. Cohen, N., Nutov, Z.: A $(1 + \ln 2)$-approximation algorithm for minimum-cost 2-edge-connectivity augmentation of trees with constant radius. Theor. Comput. Sci. **489–490**, 67–74 (2013)
11. Dai, F., Wu, J.: On constructing k-connected k-dominating set in wireless network. In: Parallel and Distributed Processing Symposium, p. 10 (2005)
12. Dinic, E.A., Karzanov, A.V., Lomonosov, M.V.: On the structure of a family of minimal weighted cuts in a graph. Stud. Discrete Optim. 290–306 (1976)
13. Dinitz, Y., Nutov, Z.: A 2-level cactus model for the system of minimum and minimum+1 edge-cuts in a graph and its incremental maintenance. In: STOC, pp. 509–518 (1995)
14. Elkin, M., Emek, Y., Spielman, D., Teng, S.H.: Lower-stretch spanning trees. SIAM J. Comput. **38**(2), 608–628 (2008)
15. Even, G., Feldman, J., Kortsarz, G., Nutov, Z.: A 1.8-approximation algorithm for augmenting edge-connectivity of a graph from 1 to 2. ACM Trans. Algorithms **5**(2) (2009)
16. Fiorini, S., Groß, M., Könemann, J., Sanitá, L.: A $\frac{3}{2}$-approximation algorithm for tree augmentation via chvátal-gomory cuts, 27 February 2017. https://arxiv.org/abs/1702.05567
17. Fleiner, T., Jordán, T.: Coverings and structure of crossing families. Math. Programm. **84**(3), 505–518 (1999). https://doi.org/10.1007/s101070050035
18. Frederickson, G., Jájá, J.: Approximation algorithms for several graph augmentation problems. SIAM J. Comput. **10**, 270–283 (1981)
19. Fukunaga, T.: Approximation algorithms for highly connected multi-dominating sets in unit disk graphs. Algorithmica **80**(11), 3270–3292 (2018)
20. Garg, N., Konjevod, G., Ravi, R.: A polylogarithmic approximation algorithm for the group Steiner tree problem. J. Algorithms **37**, 66–84 (2002)

21. Grandoni, F., Kalaitzis, C., Zenklusen, R.: Improved approximation for tree augmentation: saving by rewiring. In: STOC, pp. 632–645 (2018)
22. Guha, S., Khuller, S.: Approximation algorithms for connected dominating sets. Algorithmica **20**(4), 374–387 (1998)
23. Guha, S., Khuller, S.: Improved methods for approximating node weighted Steiner trees and connected dominating sets. Inf. Comput. **150**(1), 57–74 (1999)
24. Gupta, A., Krishnaswamy, R., Ravi, R.: Tree embeddings for two-edge-connected network design. In: SODA, pp. 1521–1538 (2010)
25. Khani, M., Salavatipour, M.: Improved approximations for buy-at-bulk and shallow-light k-Steiner trees and $(k, 2)$-subgraph. J. Comb. Optim. **31**(2), 669–685 (2016)
26. Khuller, S.: Approximation algorithms for finding highly connected subgraphs. In: Hochbaum, D. (ed.) Approximation Algorithms for NP-hard Problems, chap. 6, pp. 236–265. PWS (1995)
27. Khuller, S., Thurimella, R.: Approximation algorithms for graph augmentation. J. Algorithms **14**(2), 214–225 (1993)
28. Klein, P.N., Ravi, R.: A nearly best-possible approximation algorithm for node-weighted Steiner trees. J. Algorithms **19**(1), 104–115 (1995)
29. Könemann, J., Sadeghabad, S., Sanitá, L.: An LMP $o(logn)$-approximation algorithm for node weighted prize collecting Steiner tree. In: FOCS, pp. 568–577 (2013)
30. Kortsarz, G., Nutov, Z.: A simplified 1.5-approximation algorithm for augmenting edge-connectivity of a graph from 1 to 2. ACM Trans. Algorithms **12**(2), 23 (2016)
31. Maduel, Y., Nutov, Z.: Covering a laminar family by leaf to leaf links. Discrete Appl. Math. **158**(13), 1424–1432 (2010)
32. Misra, P.: Parameterized Algorithms for Network Design. Ph.D. thesis, The Institute of Mathematical Sciences, Chennai (2016)
33. Nagamochi, H.: An approximation for finding a smallest 2-edge connected subgraph containing a specified spanning tree. Discrete Appl. Math. **126**, 83–113 (2003)
34. Nutov, Z.: Structures of Cuts and Cycles in Graphs; Algorithms and Applications. Ph.D. thesis, Technion, Israel Institute of Technology (1997)
35. Nutov, Z.: On the tree augmentation problem. In: ESA, pp. 61:1–61:14 (2017)
36. Nutov, Z.: Approximating k-connected m-dominating sets. CoRR abs/1902.03548 (2019). arxiv.org/abs/1902.03548
37. Wang, W., et al.: On construction of quality fault-tolerant virtual backbone in wireless networks. IEEE/ACM Trans. Netw. **21**, 1499–1510 (2013)
38. Zhang, Z., Zhou, J., Mo, Y., Du, D.Z.: Approximation algorithm for minimum weight fault-tolerant virtual backbone in unit disk graphs. IEEE/ACM Trans. Netw. **25**(2), 925–933 (2017)

Author Index

Albers, Susanne 127
Ashur, Stav 81

Bampis, Evripidis 1
Byrka, Jarosław 113, 174

Christou, Dimitris 143
Cygan, Marek 159

Dinitz, Michael 97
Dudycz, Szymon 113

Eckl, Alexander 127
Escoffier, Bruno 1

Filtser, Omrit 81
Fotakis, Dimitris 143

Halldórsson, Magnús M. 159
Hanguir, Oussama 30
Held, Stephan 189

Katz, Matthew J. 81
Khazraei, Ardalan 189
Kononov, Alexander 1
Kortsarz, Guy 159
Koumoutsos, Grigorios 143

Lee, Joohyun 16
Lewandowski, Mateusz 174
Li, Wenxin 16
Li, Yaqiao 47

Madireddy, Raghunath Reddy 204
Manurangsi, Pasin 113
Marcinkowski, Jan 113
Mudgal, Apurva 204

Narayan, Vishnu V. 47
Nazari, Yasamin 97
Nutov, Zeev 220

Pankratov, Denis 47

Schäfer, Guido 63
Shroff, Ness 16
Stein, Clifford 30

Włodarczyk, Michał 113

Zhang, Zeyu 97
Zweers, Bernard G. 63

Printed in the United States
by Baker & Taylor Publisher Services